Johann Friedrich Blumenbach

The major significance of the German naturalist-physician Johann Friedrich Blumenbach (1752–1840) as a topic of historical study is the fact that he was one of the first anthropologists to investigate humankind as part of natural history. Moreover, Blumenbach was, and continues to be, a central figure in debates about race and racism.

How exactly did Blumenbach define race and races? What were his scientific criteria? And which cultural values did he bring to bear on his scheme? Little historical work has been done on Blumenbach's fundamental, influential race work. From his own time till today, several different pronouncements have been made by either followers or opponents, some accusing Blumenbach of being the fountainhead of scientific racism. By contrast, across early nineteenth-century Europe, not least in France, Blumenbach was lionized as an anti-racist whose work supported the unity of humankind and the abolition of slavery.

This collection of essays considers how, with Blumenbach and those around him, the study of natural history and, by extension, that of science came to dominate the Western discourse of race.

Nicolaas Rupke is Professor Emeritus of the History of Science at the University of Göttingen, Germany, and Rupert H. Johnson Jr. Professor at Washington and Lee University, USA.

Gerhard Lauer is Professor of Digital Humanities at the University of Basel, Switzerland, and the current director of the Blumenbach – Online project.

Routledge Studies in the History of Science, Technology and Medicine

Global Science and National Sovereignty
Studies in Historical Sociology of Science
Edited by Grégoire Mallard, Catherine Paradeise and Ashveen Peerbaye

Vaccinations and Public Concern in History
Legend, Rumor, and Risk Perception
Andrea Kitta

Spatializing the History of Ecology
Sites, Journeys, Mappings
Edited by Raf de Bont and Jens Lachmund

Cancer, Radiation Therapy, and the Market
Barbara Bridgman Perkins

Science and Ideology
A Comparative History
Mark Walker

Closing the Door on Globalization
Internationalism, Nationalism, Culture and Science in the Nineteenth and Twentieth Centuries
Edited by Cláudia Ninhos and Fernando Clara

Johann Friedrich Blumenbach
Race and Natural History, 1750–1850
Edited by Nicolaas Rupke and Gerhard Lauer

For more information about this series, please visit: https://www.routledge.com/Routledge-Studies-in-the-History-of-Science-Technology-and-Medicine/book-series/HISTSCI

Johann Friedrich Blumenbach

Race and Natural History, 1750–1850

Edited by
Nicolaas Rupke and Gerhard Lauer

Routledge
Taylor & Francis Group
LONDON AND NEW YORK

First published 2019 by Routledge

2 Park Square, Milton Park, Abingdon, Oxfordshire OX14 4RN
52 Vanderbilt Avenue, New York, NY 10017

Routledge is an imprint of the Taylor & Francis Group, an informa business

First issued in paperback 2020

British Library Cataloguing-in-Publication Data
A catalogue record for this book is available from the British Library

Library of Congress Cataloging-in-Publication Data
Names: Rupke, Nicolaas A., editor. | Lauer, Gerhard, editor.
Title: Johann Friedrich Blumenbach : race and natural history, 1750–1850 / edited by Nicolaas Rupke and Gerhard Lauer.
Description: Abingdon, Oxon ; New York, NY : Routledge, 2018. | Series: Routledge studies in the history of science, technology and medicine ; 33 | Includes bibliographical references and index.
Identifiers: LCCN 2018005205 | ISBN 9781138738423 (hardback : alk. paper) | ISBN 9781315184777 (e-book)
Subjects: LCSH: Race—Philosophy. | Anthropology—Philosophy. | Natural history—Philosophy. | Blumenbach, Johann Friedrich, 1752–1840.
Classification: LCC GN269 .J64 2018 | DDC 305.801—dc23
LC record available at https://lccn.loc.gov/2018005205

ISBN: 978-1-138-73842-3 (hbk)
ISBN: 978-0-367-58882-3 (pbk)

Typeset in Times New Roman
by Apex CoVantage, LLC

Contents

Figures

Tables

Contributors

Wolfgang Böker is an academic staff member of the "Johann Friedrich Blumenbach – Online" project at the Göttingen Academy of Sciences and Humanities. Previously, he worked on the "Monumenta Germaniae Historica" and for many years was a research assistant at Göttingen University's Institute for the History of Science, when he contributed to a series of projects, among which "Alexander von Humboldt," "Medical Geography," and "Science and Religion." His publications include articles on Christian Gottlob Heyne, Johann Friedrich Blumenbach, and Rudolph Wagner. Currently, he works on the digital linking of Blumenbach's publications with items from the contemporaneous collections of natural history.

Thomas Junker is Adjunct Professor of the History of Biology at Tübingen University. He has worked on *The Correspondence of Charles Darwin* in Cambridge (England) and as a postdoc with Ernst Mayr at Harvard University. Among his many books are *Die zweite Darwinsche Revolution: Geschichte des Synthetischen Darwinismus in Deutschland 1924 bis 1950* (Basilisken 2004), *Die Entdeckung der Evolution* (with Uwe Hossfeld, Wissenschaftliche Buchgesellschaft 2001; 2009), *Die Evolution der Phantasie: Wie der Mensch zum Künstler wurde* (Hirzel 2013), and *Die verborgene Natur der Liebe: Sex und Leidenschaft und wie wir die Richtigen finden* (Beck 2016). His current interest is in the field of evolution and human behavior.

Terence Keel is Associate Professor at the University of California, Santa Barbara, where he serves as Vice Chair to the Department of History and holds an appointment in the Black Studies Department. He earned his PhD from Harvard University. An interdisciplinary scholar with training in religious studies, the history of science, African American history, and science and technology studies, he has written widely about the history of racism and its connections with the modern biological sciences, religious intellectual history, law, medicine, and public health. A recent monograph by him is entitled *Divine Variations: How Christian Thought Became Racial Science* (Stanford University Press 2018). Keel currently works on how science educators and health care professionals working in historically Black institutions spread ideas from evolutionary biology and the eugenics movement into the public imagination.

Gerhard Lauer is Professor of Digital Humanities at the University of Basel, having previously held the Professorship of Modern German Literature at the University of Göttingen. The topics of his books range from literary studies, education, and authorship to the Lisbon earthquake and catastrophe studies. Lauer's monograph titles include *Die verspätete Revolution. Erich von Kahler: Wissenschaftsgeschichte zwischen konservativer Revolution und Exil* (de Gruyter 1995) and *Die Rückseite der Haskala: Geschichte einer kleinen Aufklärung* (Wallstein 2008). He is a fellow of the Göttingen Academy of Sciences and Humanities, and serves as the director of the Blumenbach – Online project.

Renato G. Mazzolini is Professor Emeritus at the Università degli studi di Trento, where he has been active as a historian of the life sciences, covering the period 1600–1850. He combines editing the correspondence of particular scientists with general historical analyses and has published on Charles Bonnet, Albrecht von Haller, Felice Fontana, Joseph Needham, and Rudolf Virchow. Additionally, his approach involves bibliometrics and the comparative study of different national traditions of science. Typical are Mazzolini's early classics *The Iris in Eighteenth-Century Physiology* (Huber 1980) and *Politisch-biologische Analogien im Frühwerk Rudolf Virchows* (Basilisken-Presse 1988). At present his research focuses on the biology of skin color, race, and somatic identity in Western Europe.

Carl Niekerk is Professor of German, French, and Comparative and World Literature at the University of Illinois at Urbana-Champaign. The history of early anthropology is one of his research interests. Among Niekerk's publications are a study of Lichtenberg and Blumenbach, *Zwischen Naturgeschichte und Anthropologie. Lichtenberg im Kontext der Spätaufklärung* (Niemeyer 2005), and the essays "The Romantics and Other Cultures," in *The Cambridge Companion to German Romanticism*, ed. Nicholas Saul, 2009, 147–162, and "The Problem of China: Asia and Enlightenment Anthropology (Buffon, de Pauw, Blumenbach, Herder)," in *China in the German Enlightenment*, ed. Bettina Brandt/Daniel L. Purdy, Toronto UP 2016, 97–117.

Peter Hanns Reill is Distinguished Research Professor Emeritus of European History at the University of California, Los Angeles. He received both Fulbright and Guggenheim Fellowships, was a fellow at the Wissenschaftskolleg, Berlin, and over many years regularly worked as a visiting fellow at the Max-Planck-Institut für Geschichte in Göttingen. His research interests cover the history of historical thought, the history of the life and social sciences in the Enlightenment and Romanticism, and the role of hermetic thought in the Enlightenment and Romanticism. Reill's major monographs range from *The German Enlightenment and the Rise of Historicism* (University of California Press 1975) to *Vitalizing Nature in the Enlightenment* (University of California Press 2005).

Robert J. Richards is the Morris Fishbein Distinguished Service Professor of the History of Science and Medicine at the University of Chicago. He is a professor in the Departments of History, Philosophy, and Psychology. Richards has authored several books dealing with evolutionary theory and German Romanticism, including: *Darwin and the Emergence of Evolutionary Theories of Mind and Behavior* (Chicago 1987), *The Romantic Conception of Life: Science and Philosophy in the Age of Goethe* (Chicago 2002), *The Tragic Sense of Life: Ernst Haeckel and the Struggle over Evolutionary Theory* (Chicago 2008), and *Was Hitler a Darwinian?* (Chicago 2013). His current project concerns the historical and philosophical relationships of science and aesthetics.

Nicolaas Rupke is Professor Emeritus of the History of Science at Göttingen University and former director of the Blumenbach – Online project. Currently, he holds the Rupert H. Johnson Jr. Professorship in the College at Washington & Lee University. Among his publications are *Alexander von Humboldt: A Metabiography* (Chicago 2007) and *Richard Owen: Biology Without Darwin* (Chicago 2009). At present, Rupke is rounding off a project on the non-Darwinian tradition in evolutionary biology (1750–today), Blumenbach being included as an early representative of this tradition.

E. C. Spary holds a readership at the Faculty of History, University of Cambridge. She obtained her PhD from the University of Cambridge, and then worked at the University of Warwick, the Max-Planck-Institut für Wissenschaftsgeschichte, Berlin, and University College London before returning to Cambridge. She is the author of *Utopia's Garden: French Natural History From Old Regime to Revolution* (Chicago 2000), *Eating the Enlightenment: Food and the Sciences in Paris, 1670–1760* (Chicago 2012), *Feeding France: New Sciences of Food, 1760–1815* (Cambridge 2014), and numerous shorter pieces. She has also co-edited collections of essays on the history of natural history, chemistry, food, and Habsburg medicine.

Heiko Weber is Project Leader and Coordinator of the Blumenbach – Online project at the Göttingen Academy of Sciences and Humanities. With a doctorate from the University of Jena, he previously worked on the German Research Society (DFG) project "Empirie versus Spekulation? Begriffene und erfahrene Natur" at Jena's Institute for the History of Medicine, Science and Technology "Ernst-Haeckel-Haus." Among his publications are the introduction to *Johann Wilhelm Ritter: Beyträge zur nähern Kenntnis des Galvanismus und der Resultate seiner Untersuchung. Ersten Bandes erstes und zweytes Stück* (Olms 2010) and *Die Elektrisiermaschinen im 18. Jahrhundert* (Verlag für Wissenschaft und Bildung 2011).

John H. Zammito is John Antony Weir Professor of History at Rice University. He works in history of philosophy and science, concentrating on the German Enlightenment, especially Kant and Herder, and the emergence of life sciences in eighteenth-century Germany. He also works on theoretical issues in science studies and in historical theory. Zammito's publications include: *The Genesis of Kant's Critique of Judgment* (Chicago 1992), *Kant, Herder and the Birth of Anthropology* (Chicago 2002), *A Nice Derangement of Epistemes: Post-Positivism in the Study of Science From Quine to Latour* (Chicago 2004), and *The Gestation of German Biology* (Chicago 2018).

Preface

Speaking to *Newsweek* magazine in January 1988, Stephen Jay Gould commented on the announcement a few months earlier that a group of scientists had found DNA evidence tracing all human beings to a common matrilineal ancestor. Mitochondrial Eve, they christened her, and she roamed the hot savannahs 200,000 years ago. "If it's correct, and I'd put money on it," Gould mused,

> this idea is tremendously important. It makes us realise that all human beings, despite differences in external appearance, are really members of a single entity that's had a very recent origin in one place. There is a kind of biological brotherhood that's much more profound than we ever realized.

The politics of human equality that Gould discerned in the science of mitochondrial DNA and that gave new voice to his own credo that "human equality is a contingent fact of history" were echoed a year or two later when researchers encountered Y-Chromosome Adam – Eve's male equivalent. Peter Underhill, the geneticist who pioneered the laboratory technique involved, declared, "We are all Africans at the Y chromosome level and we are really all brothers." In one way or the other, anthropological genetics found itself mobilized in the cultural politics of contemporary race relations.

This is no new development. To the contrary, anthropological science has routinely been freighted with cultural significance. The moral ecology of human rights, the ethics of racial equality, and the grounding of moral values have all surfaced in debates – often between monogenists and polygenists – about human origins and human variation. Coming to terms with the complex history of this turn of events is of crucial importance. And *Johann Friedrich Blumenbach: Race and Natural History, 1750–1850* is an ideal place to begin, for it not only brings to prominence one crucial moment in the birth of what has come to be called 'scientific racism' but also illustrates just how contested that narrative has proven to be.

For one reason or another Johann Friedrich Blumenbach has often been implicated in the emergence of scientific racism, even though he explicitly affirmed the common origin and unity of the human species, insisted that all human varieties were equally worthy of respect, championed the spiritual and moral equivalence of all human beings, and sought to preserve non-Europeans from abuse at the hands of colonizers. At the same time, by placing the Caucasian at the center of a racial cartography that expressed a presumptive anthropological normativism, he resorted to a mode of classification that later served to reinforce a far more tyrannical set of judgments on human alterity.

Reflecting on the thought of Blumenbach two and a half centuries ago, and on his fate in the subsequent annals of cultural history, has salutary lessons for our own day. First, the different stagings of Blumenbach as champion of egalitarianism or unwitting accomplice in the fixing of racial hierarchy serve to highlight the fragility of intellectual legacy. Was he a champion of racial parity or a precursor of Nazi eugenics? In different settings Blumenbach and his work have been judged in markedly different ways, delivering a variegated reputational geography. Blumenbach has meant and has been made to mean different things in different settings. As the essays which follow make clear, later readers have imposed anachronistic meanings on past terminology; they have sought to mobilize or repudiate the intellectual prestige of an eminent figure for their own purposes; they have failed to historicize past judgments by placing them in the cultural context of the times in which they were issued. For these, and no doubt many other reasons, Blumenbach's standing has been shaped and reshaped by the vicissitudes of historical circumstance.

Second, these essays reveal just how contested the origins of racism in general, and scientific racism in particular, really are. In what follows we see different genealogies at work. Is the genesis of racial hierarchy to be located in Enlightenment projects to measure human difference? Or is it rooted in Christian thought and a supersessionist reading of Judaism? Is it a product of secular reason or theological dogma? Is it more profoundly anchored in Darwinian biology and the evolutionary anthropology of Thomas Henry Huxley than in the Enlightenment cultivation of the natural history of 'man'? Was the term 'race' itself stabilized by the accidents of politics, the ruminations of philosophers, or the experiments of naturalists? Even if we cannot definitively adjudicate on these competing narratives – and whether we *should* remains an open question – getting to grips with them ought to widen the scope of our understanding and inform the ongoing task of working for universal human flourishing.

Reexamining Blumenbach and his influence delivers a third salutary lesson: language has the power to spiral well beyond the control of any author and to acquire associations operating to subvert or undermine writers' declared intentions. Blumenbach's use of the very term 'Caucasian' displays something of the hazards involved in the naming of entities. Regardless of his motives, his deployment of the expression meant that later readers, sensitive to the un-innocence of language, fastened on it as disclosing a blameworthy Eurocentrism shaped by a pernicious geographical imaginary. In a comparable vein, his efforts to bring the human species within the sphere of objective inquiry by mathematizing the study of the human form could not prevent successors from harnessing his findings for purposes alien to his own aspirations.

Finally, these revisionist essays reveal just how critical aesthetic judgment was to Blumenbach's cultivation of a science of the human subject (as indeed it was to other contemporaries), and bear witness to the power of material objects and artistic illustrations as rhetorical devices of persuasion. In the end, it was Blumenbach's sense of beauty that led him to place the skull of a Georgian girl, perhaps reflecting ancient ideals of female beauty, at the apex of human excellence. The painstaking geometrical scrutiny of cranial shape, proportions, morphology, and the like was used to aestheticize, as much as to scientize, the anthropological analysis of the human physical form. Subjective judgment and aesthetic appeal, pleasure and taste, it seems, all had a crucial role to play in the cultivation of the science of anthropology and the measurement of the body of humanity.

One contributor to the essays ahead observes, "To a large extent Blumenbach's impact was a consequence of the fact that he took up a political debate that was passionately discussed in the eighteenth century, and presented a scientific solution." In our own day, governments, funding agencies, and the like press upon scholars and scientists the need to demonstrate the impact of research outside academia. The concern is to foster research that has an effect on, and brings benefit to, the economy, society, culture, public policy, health, environment, and quality of life. A careful reading of this collection of essays on Blumenbach will reveal something of the complexities and consequences of seeking to fulfill that mandate.

David N. Livingstone
Queen's University Belfast
November 6, 2017

Acknowledgments

The majority of the chapters in this volume is based on papers presented at the international symposium "Johann Friedrich Blumenbach and the Culture of Science in Europe Around 1800," organized by the Blumenbach – Online project and its steering committee at the Göttingen Academy of Sciences, April 23–24, 2015. Funding was generously provided by the Fritz Thyssen Foundation, Cologne. A few of this volume's chapters were added at a later stage. The editors wish to express their gratitude to the symposium participants for their lively and substantive input and in particular to Nell Irvin Painter for her memorable keynote address "Five Skulls That Made Human Taxonomy." Access was kindly provided by Michael Schultz to the Blumenbach collection of human skulls and by Joachim Reitner and Alexander Gehler to Blumenbach's many original mineralogical and paleontological specimens in Göttingen University's Geological Museum. Last but not least, we warmly thank the staff of the Blumenbach – Online project (Wolfgang Böker, Markus Hellmann, Claudia Kroke, Nadine Schäfer, Heiko Weber); their collaboration in organizing the symposium and solidifying the scholarly quality of this volume has been invaluable.

The Editors

Part I

Blumenbach studies

1 Introduction

A brief history of Blumenbach representation

Nicolaas Rupke and Gerhard Lauer

Introduction

Johann Friedrich Blumenbach was one of the most celebrated practitioners of natural history of his generation, who from 1776 till the year of his death taught in the medical faculty of the University of Göttingen. His fame has been shaped on the several occasions of commemorative celebrations at the university. During his life time, Blumenbach enjoyed two such events, which happened to fall on dates of consecutive years: 1825 marked the fiftieth anniversary of Blumenbach's doctorate and the appearance of the first edition of his highly acclaimed doctoral dissertation, *De generis humani varietate nativa* (*On the natural variety of mankind*), whereas 1826 represented the fiftieth anniversary of his career as an internationally eulogized university professor. Additionally, Blumenbach was honored during the university's centennial celebrations of 1837, and commemorated in the context of the 1887, 1937, and 1987 events, as well as during the 150th anniversary of the Göttingen Academy in 1901.

Exactly why has Blumenbach been commemorated? And how has he been remembered? What biographical portrait has been depicted of him? Intriguingly, no single answer has been or can be given to these questions, but several different ones exist, each reflecting the place and time of the particular publications about him. Even just at his own university, Blumenbach's representation has undergone noticeable changes; and if we add to the Göttingen location the ones of London, Paris, Cambridge, Massachusetts, and others – all places of authorship about him – the collective body of literature on Blumenbach shows an even broader range of interpretations of his scientific identity. Taking a bird's eye view of the entire body of Blumenbach scholarship, from his own time until today, we discern some four surprisingly different portraits. Let us take a brief look at each, and compare and contrast them with what we are doing in this volume, which, perhaps, is adding up to a fifth portrait.

The revered teacher of natural history who upheld the unity of humankind

Today, internationally, Blumenbach's significance is commonly narrowed down to his work on human skulls. *The Cambridge Biographical Encyclopedia* offers no more than the following single and incorrect (Böker, this volume) sentence: "By his study of comparative skull measurements, he established a quantitative basis for racial classification" (Crystal 1994, 119). Yet in 1825/1826, at Göttingen, Blumenbach was honored for a much wider range of contributions. He was seen as a founder of what today we call the biomedical and paleontological sciences, and highly regarded for his research, teaching, writing of textbooks, and indefatigable collecting of natural history objects. His *Handbuch der Naturgeschichte* – to cite just one title – went through twelve editions (from 1779 till 1830) and was translated into Danish, Dutch, English, French, Italian, and Russian (Kroke 2010, 25–31).

Colleagues at the Georgia Augusta and other universities, some of them former students, celebrated him as much for his physiological insights as for his comparative anatomical work. Karl Friedrich Burdach, founder of a major school of physiology at Königsberg, in his published tribute to Blumenbach focused on the latter's ideas about a *Bildungstrieb* (1825) (*nisus formativus*). Jan Evangelista Purkinje, famous for his physiological work at the University of Breslau, offered a gratulatory essay that dealt with the development of a bird's egg before incubation (1825); the Swedish-born Karl Asmund Rudolphi at the University of Berlin contributed a booklet on numismatics (1825); and the following year, broadening the scope of relevant topics yet further, Karl Ernst Adolf von Hoff, gentleman-geologist in the civil service of the duchy of Gotha, emphasized the importance of his former teacher for the geological sciences (1826). At Göttingen, upon Blumenbach's death, a similarly broad appreciation was shown, among others by Karl Friedrich Heinrich Marx, a medical colleague and protégé of Blumenbach, in an obituary address to the Göttingen Academy (1840). This inclusive portrayal of him as the consummate professor of natural history persisted into the early part of the twentieth century (Ehlers 1901).

Also Blumenbach's highly original dissertation and his contributions to physical anthropology were repeatedly addressed.[1] Influential was his classification of humankind into five (initially four) varieties: the Caucasian, or white; the Asiatic, or yellow; the African, or black; the American, or red; and the Malay, or brown (for the original names used by Blumenbach, see Böker, this volume). Yet more important than this fivefold division was the fact that Blumenbach stressed their essential sameness – the unity of the human species. Such unity served the humanitarian politics of anti-slavery, the emancipation of black slaves in the New World, and in particular the acknowledgment of the equality of "the Negro." At the time that Blumenbach produced the four editions of his dissertation (1775; 1776; 1781; 1795 – the first two of 1775/1776 were essentially the same) ideas of polygenism were gaining scientific ground (Livingstone 2011, passim). Blumenbach's monogenist stance, the view of unity amid variety, was highlighted by Marx:

> At the time when Negroes and savages were still thought of as half animals and the idea of the emancipation of slaves had not even been raised, Blumenbach spoke out to make known, how their mental aptitude was not inferior to that of Europeans, how even between those tribes major differences exist, and how their higher faculties merely lack the opportunity for development.[2]
> (Marx 1840, 10; author's translation)

Others later echoed a similar view: "It was not just the study of humans as part of natural history that drew attention, rather more his advocacy of the unity of mankind"[3] (Ehlers 1901, 400; author's translation). Not only in Göttingen, however, but throughout the Western world, during the first half of the nineteenth century, Blumenbach's name became synonymous with the liberal notions of "unity of mankind" and "Negro emancipation." Among the great names of biomedical science who saw him that way and cited him accordingly were – to select merely two examples – the English physician, anthropologist, and ethnologist James Cowles Prichard, and one of Georges Cuvier's protégés, the French physiologist Marie Jean Pierre Flourens.[4] This portrait of Blumenbach was given its definitive brushstroke by his most famous student, Alexander von Humboldt, who since the publication of the *Essai politique sur l'île de Cuba* (1826) was publicly known for his abolitionist stance; later, in the first volume of *Kosmos* (1845), having referred to his Göttingen teacher as "my master," Humboldt canonically stated, "While we maintain the unity of the human species, we at the same time repel the depressing assumption of superior and inferior races" (Humboldt 1997, 356, 358).

When Blumenbach's dissertation was translated into French, it was given a new, interpretative title, *De l'unité du genre humain et de ses variétés* (1804; 1806; 1808). In his *Éloge historique* of Blumenbach (1847), Flourens opined that Blumenbach's fivefold division was not perfect; yet, referencing the French translation, he continued:

> But the idea, the grand idea, which reigns and rules and predominates throughout in the admirable studies of Blumenbach is the idea of the unity of the human species, or, as it has also been expressed, of the human genus. Blumenbach was the first who wrote a book under the express title of the *Unity of the Human Genus*. The *Unity of Mankind* is the great result of the science of Blumenbach, and the great result of all natural history. [. . .] The human race had forgotten its original unity, and Blumenbach restored it. [. . .] According indeed to his doctrine, all men are born, or might have been born, from the same man. He calls the negroes *our black brothers*. It is an admirable thing that science seems to add to Christian charity, or, at all events, to extend it, and invent what may be called *human charity*. The word Humanity has its whole effect in Blumenbach alone.
>
> (Flourens 1865, 56, 58, 60)

The "odious traffic" in black slaves would come to an end "in our age," Flourens predicted (1865, 57), and humanitarian science would join politics in a crusade for abolition, led by Blumenbach,

> a man of high intellect, an almost universal scholar, philosopher and sage; a naturalist, who had the glory, or rather the good fortune, of making natural history the means of proclaiming the noblest and, without doubt, the highest truth that natural history ever had proclaimed, *The Physical Unity*, and through the *physical unity* the *moral unity*, of the human race.
>
> (Flourens 1865, 63)

Europe's leading physical anthropologist who validated racist politics

Yet precisely this conclusion of "moral unity" was contentious and objected to by many. At about the time that the English translation of Flourens's *Éloge historique* of Blumenbach appeared, his identity was radically changed by being made part of a racist discourse. The new interpretation originated in London, spread to the United States and, finally, in our own day, also found an audience in Germany. Blumenbach's image was reconstructed and turned from the great scientific advocate of human equality into the leading authority of anthropology – in fact the founding father of physical anthropology – whose work on human varieties legitimized racism and, according to some, the politics of segregation and slavery. This reframing of Blumenbach took place within the Anthropological Society of London. The group was founded in 1863 in a breakaway move from the older Ethnological Society of London (Rainger 1978). The differences between the two societies were of a profound political-scientific kind. The fellows of the Ethnological Society, influenced by Prichard, were by and large monogenists and tended to be politically liberal in matters related to race. The Anthropological Society, by contrast, advocated polygenism and supported the Confederacy in the American Civil War (Desmond and Moore 2009, 332–333, 413). The issue that most sharply divided the two societies was the so-called Negro question. The president of the new society was James Hunt, author of *On the Negro's Place in Nature* (1863b), a book written in opposition to "that gigantic imposture known by the name of 'Negro Emancipation'" (Hunt 1863b, viii). He believed that Africans belong to a different species from Caucasians, that they are irredeemably inferior, and that slavery is the role for which they are best suited. In his programmatic "On the Study of Anthropology," Hunt bracketed Blumenbach's foundational work with the mission of the Anthropological Society:

> Whatever may be the conclusion to which our scientific inquiries may lead us, we should always remember, that by whatever means the Negro, for instance, acquired his present physical, mental, and moral character, whether he has risen from an ape or descended from a perfect man, we still know that the Races of Europe have now much in their mental and moral nature which the races of Africa have not got. We have hitherto devoted our attention almost exclusively to physical Anthropology, which Blumenbach first founded. We now require to investigate the mental and moral characteristics of mankind generally. The difference between the European and the African is not so great physically as it is mentally and morally.
>
> (Hunt 1863a, 3)

In Hunt's opinion, "Blumenbach saw, in his five varieties of man, nothing but degeneracy from some ideal perfect type" (Hunt 1863b, 4). When in 1865 the Anthropological Society had grown to no fewer than 500 fellows/members, the vice president, Thomas Bendyshe, continued the appropriation of Blumenbach's legacy by producing an English translation entitled *The Anthropological Treatises of Johann Friedrich Blumenbach*, which included the obituary essays by Flourens and Marx. In the preface, Bendyshe criticized Flourens for having interpreted/translated Blumenbach's doctoral thesis in the sense of "The unity of the human genus," which he characterized as a "singular mistake" (Bendyshe 1865, xiii). The volume made some of Blumenbach's work available to an English readership, but the selection was limited and the translation, in crucial respects, tendentious (Michael 2017). This partial source of information about Blumenbach, produced in the context of the racist polygenism of the Anthropological Society of London, has continued to be the primary source of information about Blumenbach in the English language. Till the present day, the portrayal of him as a founder of scientific racism has remained in vogue in different circles and for different purposes.

The German skull collector bearing historical blame for twentieth-century eugenicist atrocities

One of these circles took shape in the wake of World War II, in the context of which Blumenbach was included in a third, wholly different discourse. Two points of contrast stand out. First, the issue of race was now discussed less in relation to black slavery and more in relation to eugenics and the Holocaust. Second, although the main source of information about Blumenbach continued to be the inadequate Bendyshe volume, the purpose of painting a racist image of Blumenbach was no longer to adulate and appropriate him but to use his picture as a dart board. In a search for the historical causes of the eugenicist atrocities as promoted by National Socialist anthropology, the finger was pointed, in the international arena, at Blumenbach's dissertation and his collection of human skulls – his "Golgotha."

An early accusatory finger was that of the British-Polish mathematician and science popularizer Jacob Bronowski. In the book that accompanied his highly acclaimed BBC documentary *The Ascent of Man* (1973), he took special notice of Göttingen and paid tribute to the mathematician Carl Friedrich Gauß and the mathematical physicists Max Born and Werner Heisenberg, even telling his audience/readership about the "Gänseliesel" (a fountain statue in the town square) and the "Rathausspruch" (a saying painted on the wall of an entrance to the town hall): "Extra Gottingam non est vita" (outside Göttingen there's no life); but then he went on to hint at a link between Blumenbach and Nazi racism:

> The sky was darkening all over Europe. But there was one particular cloud which had been hanging over Göttingen for a hundred years. Early in the 1800s Johann Friedrich Blumenbach had put together a collection of skulls that he got from distinguished gentlemen with whom he corresponded throughout Europe. There was no suggestion in Blumenbach's work that the skulls were to support a racist division of humanity, although he did use anatomical measurements to classify the families of man. All the same, from the time of Blumenbach's death in 1840 the collection was added to and added

> to and became a core of racist, pan-Germanic theory, which was officially sanctioned by the National Socialist Party when it came into power.
>
> (Bronowski 1973, 367)

In the course of the 1960s, 1970s, and 1980s, as part of a wider discourse of post-colonialism in Europe and of civil rights debates in the United States, the very notion of race became a contested and discredited concept. Among many historians, sociologists, and some scientists, "race" was a social construct that needed deconstruction. Specifically, social anthropologists from Franz Boas onward argued for cultural relativism in a move to end racism. Blumenbach became a whipping boy for those engaged in that activity; unambiguously, although hesitantly, it was intimated that *On the Natural Variety of Mankind* was materially implicated in the Holocaust. The Harvard paleontologist and science popularizer Stephen J. Gould, well known for his left-wing Marxist leanings, addressed these issues in his bestselling *The Mismeasure of Man* (1981). Especially in the second edition of 1996, Gould lay the accusation of scientific racism at Blumenbach's feet, even though, like Bronowski, he called him a genial person who had acted without ill intent. Blumenbach meant well, but ended up affirming racial hierarchy by way of geometry and aesthetics, not by any overt viciousness" (Gould 1996, 49). His classification was

> the foundation of so much that continued to influence and disturb us today, [. . .] a scheme that has promoted conventional racism ever since. [. . .] Blumenbach lived as a cloistered professor all his life, but his ideas reverberate through our wars, our conquests, our sufferings, and our hopes. [. . .] Where would Hitler have been without racism?
>
> (Gould 1996, 402, 405, 412)

In an instance of scholarly fraud, Gould tried to back up his contention with fabricated evidence by reprinting a doctored version of Blumenbach's famous illustration of a row of five skulls (Gould 1996, 409, 1998; Junker 1998, this volume).

In a further twist to the story, the Princeton historian of the US South, Nell Irvin Painter, reconnected Blumenbach with slavery – in part sex slavery – by focusing on the actual individuals whose skulls ended up as part of the Göttingen collection and were depicted in *On the Natural Variety of Mankind.* Papers that culminated in Painter's bestselling *The History of White People* (2011) outlined the case against Blumenbach as follows: Blumenbach's five skulls came from real people who entered science through sexual exploitation and imperialist wars. His choice of the name "Caucasian" for the European, white race "came from traditions about sexual vulnerability and beauty rooted in an ancient form of human subjugation: slavery" (Painter 2015, 36). Why should Blumenbach have adopted the term "Caucasian" for the white race, Painter asked? The reason was that the Europeans regarded their own, white race as the most beautiful and that the epitome of that beauty traditionally was embodied by girls and young women in the Caucasus. They were the highly prized "odalisques" of the Black Sea white slave trade, famously portrayed by Jean-Léon Gérôme, Jean-Auguste Dominique Ingres, Henri Matisse, and others who provided respectable Europeans with soft porn paintings in their art collections.

Blumenbach enthused about the beauty of his Caucasian skull, which had been sent to him from Russia by the Göttingen alumnus Georg Thomas von Asch. That skull, Painter explained, belonged to "a sex slave who had been raped to death" (Painter 2015, 20). "[H]is use of 'Caucasian,' with its connotations of slavery, endlessly elaborated and depicted during the nineteenth century, tightens the relationship between concepts of race, on the one hand, and slavery on the other." Blumenbach's Russian-derived skull collection and the artistic representations that followed reinforced the tie between concepts of race and white slavery – "enslaved femininity" (Painter 2001, 29).

Göttingen's Enlightenment humanist and internationalist

In the meantime, a fourth refashioning of Blumenbach began taking shape, which, like the first, was centered on Göttingen. It was carried out as part of a new discourse about the university, intended to face up to, and deal with, its Third Reich past. During the 1937 centennial celebrations, which doubled as the bicentenary, the university in no uncertain terms affirmed its commitment to National Socialism. The Georgia Augusta putatively was a national, Germanic institution (Petke 1987; Wegeler 1996) that had cleansed itself from foreign racial impurities by firing its Jewish professors and burning many of its "un-German" books. Half a century later, as part of the 1987 observances, the university, aided by the Göttingen Academy and the Max Planck Institute for History, confronted its Nazi past, documenting parts of its involvement, and expropriating the university's academic identity from the 1937 Nazi claims (Becker, Dahms, and Wegeler 1987; Kamp 1987). The Georgia Augusta should be defined in terms of eighteenth-century Enlightenment ideals that had been the fertile soil of its early blossoming. Göttingen should be seen less as a German national institution and more of a European international university (Boockmann 1997, passim). The historical importance of the university's cultivation of the natural sciences, the international networks of correspondence of professors, the multinational student body, the encouragement of expeditions to parts unknown – these and similar signs and symbols of Enlightenment openness to the world should be accentuated. Among the eighteenth-century linchpin professors at Göttingen used to give substance to this revisionist discourse were the Swiss physiologist and polymath Albrecht von Haller, the physicist, satirist, and Anglophile Georg Christoph Lichtenberg and Blumenbach (Rupke 2002, 24–26 and passim).

Blumenbach proved an ideal figure for adding content to the discourse of Göttingen's Enlightenment identity (Rupke and Böker 2010). Luckily, the Nazis had not made him one of their heroes nor loudly appropriated his heritage on behalf of racialist anthropology. One indication of this was that the centenary of Blumenbach's death in 1940 was not marked by any commemorative events that led to publications. Admittedly, as part of the university's 1937 bicentenary, the soon-to-be rector, Hans Plischke, an avowed National Socialist, produced *Johann Friedrich Blumenbachs Einfluß auf die Entdeckungsreisenden seiner Zeit* (1937) (Johann Friedrich Blumenbach's influence on the explorers of his time). Yet, as in the case of the Nazi appropriation of Humboldt's journey of exploration of the Americas, and the all-encompassing scope of his *Kosmos*, the interest in global breadth was construed in terms of German world supremacy (Rupke 2008, 87–92), not of Enlightenment cultivation of a brotherhood of men. The latter exegesis was now put forward. Blumenbach's network of foreign contacts, his many students from abroad, and his efforts in collecting objects of natural history from around the world, including skulls, reflected enlightened and liberal ideals. Blumenbach's place in the so-called German Enlightenment, his friendships with London's long-time president of the Royal Society Joseph Banks, who took part in James Cook's first scientific expedition to the South Pacific, and with Georg Forster, who along with his father, Johann, took part in Cook's second voyage, became "hot topics" of Blumenbach scholarship (Gascoigne 1994, 2002), as did "the Göttingen moment" of the university's participation in "the science of man" (Bödeker, Büttgen, and Espagne 2008; Vermeulen 2015).

Around the time of the 1987 university commemoration, the Göttingen historian of medicine Ulrich Tröhler solicited the services of the Canadian historian of science Frank Dougherty in starting an editing project of Blumenbach's correspondence, which at the time amounted to some 4,500 letters. As Tröhler put it:

> The correspondence of JFB, professor of medicine and natural history at the University of Göttingen, is a unique testimony to the European dimension of the republic of letters of the Late Enlightenment. In succession to Albrecht von Haller he materially contributes in this context to the blossoming of German science. Blumenbach cultivated contacts with scientists throughout Europe – with those of scientific centres such as Paris and London, but also for instance with those in Amsterdam, Copenhagen, Padua and St. Petersburg. His correspondence shows exemplarily the endeavor of the scientists to work together, beyond political, social and religious boundaries, on behalf of science and the public good.
>
> (Tröhler 1994, unpublished progress report)[5]

This project came to a premature end by the tragically early death of Dougherty, and by Tröhler's departure from Göttingen. Yet indirectly, it led to a major digitization project, Blumenbach – Online, which brings us to our present volume.

The digital Blumenbach

The starting point of our volume's joint effort is the Blumenbach – Online project, described in some detail in Chapter 2. This comprehensive and large-scale documentation of Blumenbachiana is guided by the scholarly interests and needs of a collective of experts to help set priorities. The project's early focus has been the issue of race, and this collection in part is a spring harvest of these efforts. We look at a broad range of issues about race and racism in order to address how Blumenbach has been implicated in their historical development. It recognizes that the conflicting differences between the various Blumenbach interpretations of the past two hundred years or so in part are attributable to the inaccessibility and inadequacy of information – a situation Blumenbach – Online is meant to rectify.

The Blumenbach identity to which we are contributing is a contested, fractured one and, for the first time, the product of professional historians of science and medicine. A broader set of interests, questions, and fields of expertise than before is brought to bear on his physical anthropology. The race issue is explored in the wider context of Blumenbach's own complete works, of contemporaneous figures who may have influenced him – for example, Georges-Louis Leclerc de Buffon, Immanuel Kant, and Samuel Thomas Soemmerring – and of the then novel academic subjects of biology and geology. Blumenbach's anthropology increasingly has become seen in the light of Franco-German life sciences and as a product of the institutional and intellectual dynamics that marked the transition from Enlightenment to Romantic thought. Several of our contributors have been leading participants in this contextualizing and deepening of Blumenbach studies.

A milestone in these developments was the enunciation of the so-called Lenoir thesis (Lenoir 1980, 1982; see also Zammito 2012), which argues that the rise of German biology was engendered by the "Göttingen School," and that Blumenbach was instrumental in bringing Kantian thought – especially the *Kritik der Urteilskraft* (1790) (Critique of Judgment) – to bear on a "teleomechanical" research program – naturalistic yet non-reductionist. Several experts on Enlightenment science as well as the Romantic period have critically expanded upon the thesis (Richards 2000, 2002; Reill 2005; Zammito 2018; see also Marino 1975, and contributions in Mann and Dumont 1985, 1990, Mazzolini 2007, and Mensch 2018). Among the canonical texts of Blumenbach scholarship, his *Über den Bildungstrieb* (1791, Tröhler 1993) has now joined the *De generis humani varietate*.

The Blumenbach – Online initiative synergizes with this comprehensive approach and so does our volume. We address largely unanswered questions, such as: how exactly did Blumenbach define race and races? What were his scientific criteria? And which cultural values did he bring to bear on his scheme? Various thematic threads run through the collection. A series of essays considers the ways in which, with Blumenbach and others, the study of natural history – that is, of biology – came to dominate the Western discourse of race. Another thread follows the way in which Eurocentric exceptionalism – scientific and sociopolitical – affected racial classification. Collectively, this volume develops a new narrative of race and its historiography that highlights anti-racism, not just racist science and its representatives. Belatedly, we now are following up on William F. Bynum's programmatic exhortation:

> We need a full-scale study of Blumenbach, for many of his anthropological and biological concepts in the 1780s and '90s foreshadowed much that was to come. He explicitly discarded the chain of being, consolidated the human

species, championed the equality of Negroes, stressed the fundamental unity of man, and definitively discredited the philosophical importance attached to Peter the Wild Boy and other 'wild children.'

(Bynum 1975, 22)

Notes

1 The lectures on anthropology at Göttingen, as announced in the *Göttingische Gelehrte Anzeigen*, were not given by Blumenbach, but by an adjunct professor, Wilhelm Liebsch, known for his *Grundriß der Anthropologie*, 2 vols. (Göttingen: Vandenhoeck & Ruprecht, 1806, 1808) (although from the winter semester of 1776/77 to that of 1788/89 Blumenbach six times offered a lecture course on the topic of "Naturgeschichte des Menschengeschlechts" [natural history of humankind]).

2 "Zur Zeit als die Neger und Wilden noch für halbe Thiere galten und der Gedanke der Emancipation der Sclaven noch nicht einmal angeregt war, erhob Blumenbach seine Stimme, um bemerklich zu machen, wie ihre psychischen Anlagen denen der Europäer nicht nachständen, wie zwischen jenen Stämmen selbst die grössten Verschiedenheiten obwalteten, und wie ihren höheren Vermögen einzig die Gelegenheit zur Entwicklung mangle" (Marx 1840, 10).

3 "Die naturgeschichtliche Behandlung des Menschen war es nicht allein, die Aufsehen machte, mehr noch das Eintreten für die Einheit des Menschengeschlechtes" (Ehlers 1901, 400).

4 Prichard dedicated the second edition of his *Researches Into the Physical History of Man* (1826) to Blumenbach, just as William Lawrence had done with his *Lectures on Physiology, Zoology and the Natural History of Man* (1822). Other scientific proponents of monogenism included Petrus Camper, Georges Cuvier, Marcel de Serres, and Jean Louis Armand de Quatrefages. Among the clergy, the French bishop and fervent abolitionist Henri Jean-Baptiste Grégoire, known as Abbé Grégoire, in his *De la littérature des nègres* (1808) made frequent mention of Blumenbach, and elsewhere commented, "L'unité de type dans l'espèce humaine, proclamée par la révélation, est en général avouée des naturalistes, sourtout par le célèbre Blumenbach" (Grégoire 1826, 27).

5 "Der Briefwechsel von JFB, Professor der Medizin und Naturgeschichte an der Universität Göttingen, ist ein einmaliges Zeugnis der Europäischen Dimension der Gelehrtenrepublik der Spätaufklärung. In der Nachfolge von Albrecht von Haller ist er in diesem Kontext wesentlich am Aufblühen der deutschen Wissenschaft beteiligt. Blumenbach pflegte Kontakt mit Wissenschaftlern in ganz Europa – mit denen der wissenschaftlichen Zentren wie Paris und London, aber etwa auch mit denen in Amsterdam, Kopenhagen, Padua und St. Petersburg. Sein Briefwechsel zeigt exemplarisch das Bestreben der Wissenschaftler, über die politischen, ständischen und konfessionellen Grenzen hinweg zur Beförderung der Wissenschaft und des Wohls der Menschheit zusammenzuarbeiten" (Tröhler 1994, unpublished progress report).

Bibliography

Becker, Heinrich, Hans-Joachim Dahms, and Cornelia Wegeler, eds. 1987. *Die Universität Göttingen unter dem Nationalsozialismus: Das verdrängte Kapitel ihrer 250jährigen Geschichte*. München: Saur.

Bendyshe, Thomas, trans. 1865. *The Anthropological Treatises of Johann Friedrich Blumenbach, with Memoirs of Him by Marx and Flourens, and an Account of His Anthropological Museum by Professor R. Wagner, and the Inaugural Dissertation of John Hunter, M.D., on the Varieties of Man*. London: Longman, Green, Longman, Roberts & Green.

Bödeker, Hans Erich, Philippe Büttgen, and Michel Espagne. 2008. *Die Wissenschaft vom Menschen in Göttingen um 1800: Wissenschaftliche Praktiken, institutionelle Geographie, europäische Netzwerke*. Göttingen: Vandenhoeck & Ruprecht.

Boockmann, Hartmut. 1997. *Göttingen: Vergangenheit und Gegenwart einer europäischen Universität*. Göttingen: Vandenhoeck & Ruprecht.

Bronowski, Jacob. 1973. *The Ascent of Man*. Boston: Little, Brown.

Burdach, Karl Friedrich. 1825. *Viro perillustri Johanni Friderico Blumenbach solemnium quibus ante hos quinquaginta annos summos in medicina honores adeptus est instaurationem gratulatur Academia Regiomontana*. Regiomonti: Hartung.

Bynum, William F. 1975. "The Great Chain of Being After Forty Years." *History of Science* 13: 1–28.

Crystal, David, ed. 1994. *The Cambridge Biographical Encyclopedia*. Cambridge: Cambridge University Press.

Desmond, Adrian, and James Moore. 2009. *Darwin's Sacred Cause: How a Hatred of Slavery Shaped Darwin's Views on Human Evolution*. Boston: Houghton, Mifflin and Harcourt.

Ehlers, Ernst. 1901. "Göttinger Zoologen." In *Festschrift zur Feier des hundertfünfzigsten Bestehens der Königlichen Gesellschaft der Wissenschaften zu Göttingen*. [Vol. 3] Beiträge zur Gelehrtengeschichte Göttingens, 391–494. Berlin: Weidmann.

Flourens, Pierre Marie Jean. 1847. *Élogue historique de Jean-Frédéric Blumenbach*. Paris: Didot.

———. 1865. "Memoir of Blumenbach." In *The Anthropological Treatises of Johann Friedrich Blumenbach*, edited by Thomas Bendyshe, 49–63. London: Longman.

Gascoigne, John. 1994. *Joseph Banks and the English Enlightenment: Useful Knowledge and Polite Culture*. Cambridge: Cambridge University Press.

———. 2002. "Blumenbach, Banks, and the Beginnings of Anthropology at Göttingen." In *Göttingen and the Development of the Natural Sciences*, edited by Nicolaas Rupke, 86–98. Göttingen: Wallstein.

Gould, Stephen Jay. 1996. *The Mismeasure of Man*, 2nd edn. New York: Norton.

———. 1998. "On Mental and Visual Geometry." *Isis* 89: 502–504.

Grégoire, Henri Jean-Baptiste. 1826. *La noblesse de la peau, ou du préjugé des blancs contre la couleur des Africains et celle de leurs descendans noirs et sang-mêlés*. Paris: Baudouin.

Hoff, Karl Ernst Adolf von. 1826. *Erinnerung an Blumenbach's Verdienste um die Geologie*. Gotha: Engelhard-Reyersche Schriften.

Humboldt, Alexander von. 1997. *Cosmos: A Sketch of the Physical Description of the Universe*. Vol. 1. Baltimore: Johns Hopkins University Press.

Hunt, James. 1863a. "On the Study of Anthropology." *The Anthropological Review* 1: 1–20.

———. 1863b. *On the Negro's Place in Nature*. London: Trübner.

Junker, Thomas. 1998. "Blumenbach's Racial Geometry." *Isis* 89: 498–501.

Kamp, Norbert. 1987. *Das Göttinger Jubiläum von 1937: Glanz und Elend einer Universität*. Göttingen: Vandenhoeck & Ruprecht.

Kroke, Claudia. 2010. *Johann Friedrich Blumenbach: Bibliographie seiner Schriften*. Göttingen: Universitätsverlag.

Lenoir, Timothy. 1980. "Kant, Blumenbach, and Vital Materialism in German Biology." *Isis* 71: 77–108.

Liebsch, Wilhelm. 1806, 1808. *Grundriß der Anthropologie*, 2 vols. Göttingen: Vandenhoeck & Ruprecht.

———. 1982. *The Strategy of Life: Teleology and Mechanics in Nineteenth-Century German Biology*. Dordrecht: Riedel.
Livingstone, David N. 2011. *Adam's Ancestors: Race, Religion, and the Politics of Human Origins*. Baltimore: Johns Hopkins University Press.
Mann, Gunter, and Franz Dumont, eds. 1985. *Samuel Thomas Soemmerring und die Gelehrten der Goethezeit*. Stuttgart: Fischer.
———. 1990. *Die Natur des Menschen. Probleme der Physischen Anthropologie und Rassenkunde (1750–1850)*. Stuttgart: Fischer.
Marino, Luigi. 1975. *I Maestri della Germania*. Turin: G. Einaudi.
Marx, Karl Friedrich Heinrich. 1840. *Zum Andenken an Johann Friedrich Blumenbach: Eine Gedächtniss-Rede gehalten in der Sitzung der Königlichen Societät der Wissenschaften den 8. Februar 1840*. Göttingen, Dieterichschen Buchhandlung.
Mazzolini, Renato. 2007. "Diritti dell'umanità e antropologia fisica: il caso Blumenbach." In *Chiesa cattolica e mondo moderno. Scritti in onore di Paolo Prodi*, edited by Adriano Prosperi, Pierangelo Schiera, und Gabriella Zarri, 487–495. Bologna: Il Mulino.
Mensch, Jennifer. 2018. "Kant and the Skull Collectors: German Anthropology from Blumenbach to Kant." In *Kant and His German Contemporaries*, edited by Corey Dyck and Falk Wunderlich, 192–210. Cambridge: Cambridge University Press.
Michael, John S. 2017. "Nuance Lost in Translation: Interpretations of J.F. Blumenbach's Anthropology in the English Speaking World." *NTM Zeitschrift für Geschichte der Naturwissenschaften, Technik und Medizin* 25: 1–29.
Painter, Nell Irvin. 2001. "Q. Why Are White People Called 'Caucasians'/ A. Beauty + the White Slave Trade + German Science." Princeton University, unpublished paper, 4 December.
———. 2011. *The History of White People*. New York: Norton.
———. 2015. "Five Skulls That Made Human Taxonomy." Göttingen University, unpublished paper, 23 April.
Petke, Wolfgang. 1987. "Karl Brandi und die Geschichtswissenschaft." In *Geschichtswissenschaft in Göttingen*, edited by Hartmut Boockmann and Hermann Wellenreuther, 287–320. Göttingen: Vandenhoeck & Ruprecht.
Plischke, Hans. 1937. *Johann Friedrich Blumenbachs Einfluß auf die Entdeckungsreisenden seiner Zeit*. Göttingen: Vandenhoeck & Ruprecht.
Purkinje, Jan Evangelista. 1825. *Joan. Fried. Blumenbachio eq. Guelph. viro de omni scientia naturali uni omnium maxime merito universitatis Georgiae Augustae decori eximio die XIX. Septembri MDCCCXXV summorum in medicina honorum semisaecularia faustis omnibus celebranti gratulatur Ordo Medicorum Vratislaviensium*. Vratislaviae: Typis Universitatis.
Quatrefages de Bréau, Jean Louis Armand de. 1861. *Unité de l'espèce humaine*. Paris: Hachette.
Rainger, Ronald. 1978. "Race, Politics, and Science: The Anthropological Society of London in the 1860s." *Victorian Studies* 22: 51–70.
Reill, Peter Hanns. 2005. *Vitalizing Nature in the Enlightenment*. Berkeley: University of California Press.
Richards, Robert J. 2000. "Kant and Blumenbach on the Bildungstrieb: A Historical Misunderstanding." *Studies in History and Philosophy of Biological and Biomedical Sciences* 31 (1): 11–32.
———. 2002. *The Romantic Conception of Life*. Chicago, IL, London: University of Chicago Press.

Rudolphi, Karl Asmund. 1825. *Ioanni Fr. Blumenbach . . . Viro Illustri Germaniae Decori diem semisecularem Physiophili Germanici laete gratulantur: Inest Index Numismatum in virorum de rebus medicis vel physicis meritorum memoriam percussorum*. Berolini: s.n.

Rupke, Nicolaas, ed. 2002. *Göttingen and the Development of the Natural Sciences*. Göttingen: Wallstein.

———. 2008. *Alexander von Humboldt: A Metabiography*. Chicago, IL: University of Chicago Press.

———, and Wolfgang Böker. 2010. "' . . . dieses kleine Fleckchen freier Erde': das frühe Göttingen als Freiraum für neues Denken." *Georgia Augusta* 7: 4–11.

Tröhler, Ulrich. 1993. *Blumenbach – Bildungstrieb/On generation*. Göttingen: Institut für Geschichte der Medizin.

———. 1994. Unpublished progress report on the editing project of Blumenbach's correspondence by Frank W.P. Dougherty. Göttingen: Institut für Geschichte der Medizin.

Vermeulen, Han F. 2015. *Before Boas: The Genesis of Ethnography and Ethnology in the German Enlightenment*. Lincoln, NE: University of Nebraska Press.

Wegeler, Claudia. 1996. ". . . wir sagen ab der internationalen Gelehrtenrepublik." In *Altertumswissenschaft und Nationalsozialismus: Das Göttinger Institut für Altertumskunde 1921–1962*. Vienna: Boehlau.

Zammito, John. 2012. "The Lenoir Thesis Revisited: Blumenbach and Kant." *Studies in History and Philosophy of Science, Part C: Biological and Biomedical Sciences* 43(1): 120–132.

———. 2018. *The Gestation of German Biology: Philosophy and Physiology from Stahl to Schelling*. Chicago, IL: University of Chicago Press.

2 Johann Friedrich Blumenbach – Online

Gerhard Lauer and Heiko Weber

Johann Friedrich Blumenbach – Online

Blumenbach's scientific oeuvre was a crucial part of the rise of modern biology, yet no complete edition of his books exists, and his numerous papers in a range of periodicals are difficult to find. Little work has been done on the history of Blumenbach's natural history collections, although in his publications he frequently referred to specimens from these collections. A number of nineteenth-century biographical accounts exist (Marx 1840; Bendyshe 1865; Ehlers 1901), but no comprehensive biography of Blumenbach has so far been produced.

The State and University Library of Lower Saxony, colloquially known as Göttingen University Library, owns copies of almost every text published by Blumenbach and of nearly all the books he used for his research. Its archives contain the majority of Blumenbach's unpublished papers; and the objects in the Academic Museum he curated as well as his private collection of human skulls, with a few exceptions, have been preserved and are present in today's university collections (cf. Böker, this volume). In 2009, a committee of the Göttingen Academy of Sciences, the State and University Library of Lower Saxony, and the Georg-August-Universität Göttingen, chaired by Nicolaas Rupke, put forward to the Union of German Academies of Sciences and Humanities a proposal for a so-called Langzeitvorhaben (long-term project) to produce a digital edition of Blumenbach's work and of his collections of natural history. The application was successful, and one year later "Johann Friedrich Blumenbach – Online" took off on what is anticipated to be a fifteen-year journey of application and innovation of the possibilities provided by the Internet in making Blumenbach and his oeuvre available to an international readership (Kerzel, Reich and Weber 2013).

A digital edition differs in two important respects from traditional edition projects. First, it can provide more than just text and include images of, and information about, physical objects which can readily be hyperlinked to a considerable extent of both breadth and depth. Especially in the history of the earth and life sciences it is of importance to have access not only to a text that describes and interprets a specific object but also to images of, and information about, the object. Blumenbach's writings and his collections serve as an excellent example of this. Second, an online edition makes Blumenbach's work accessible to digital tools that are starting to supplement – not wholly supplant – traditional hermeneutic approaches. Needless to say, it is of key importance that digital editions maintain long-established standards of quality. When necessary, new rules have to be defined – for example, regarding citability and long-term preservation of electronic data.

While the basic tasks of producing electronic editions are the same as in pre-digital times, something has changed since the humanities have gone digital (Rosselli Del Turco 2016). A fast and steadily growing number of techniques and tools connect texts and objects in digital space. Key features of scholarly digital editing are the establishment of relations between data, and processing data, on a scale and with a complexity and precision previously unattainable. The same holds true for interoperability, the ability to share data in different computational environments, which enhances interactions between and within different scientific communities (Jannidis 2010).

Instead of presenting texts only as strings of letters and objects as plain images, "Blumenbach – Online" enriches them with metadata and standardized personal names and geographical names. This makes it possible to connect data with other editions, library catalogues, search engines, and even Wikipedia articles. Traditional editions and even present-day search options for digital texts can help us only little to understand the definition of a mammal at a certain point in history, the relevance of comets for natural history around 1800 or how the term "race" was used in French, English, and German sources of that time. By relating terms, names, and objects in semantic ways scholarly digital editions become nodes in machine-based networks of knowledge, the so-called semantic web, which is already part of the current Internet, even if only relatively few people are aware of it.

Digital editions in the history of science

The story of digital editions began in 1946, when Roberto Busa, an Italian Jesuit and an expert on St. Thomas Aquinas, planned the lemmatization of Thomas's works, intending to produce a concordance of his monumental oeuvre and writings by a few related authors (Winter 1999). Busa's guiding question was, What is the metaphysics of *presence* in Thomas Aquinas? Given the voluminousness of the writings, a complete lemmatization was possible only by using machines; and in 1949, Busa convinced Thomas J. Watson, the founder of IBM, to support the development of a computer-based tool for performing text searches. Much later, in 1967, the card-punching of Thomas Aquinas's texts was completed, and in 1974 the *Corpus Thomisticum* was printed in fifty-six volumes. More recently, in 2005, a web-based version was made available online, while the following year the "Index Thomisticus Treebank," a syntactic annotation of the entire edition, was begun.

What Busa did in the 1940s was pioneering work, considering that the first computers came on the market not before 1953, and FORTRAN was presented to the computing world as late as 1957, when programming languages became a tool for data processing. Today, digital editing is a common practice not only for texts but for cultural heritage in a variety of forms; and digital editions – for example, of the complete musical works of Wolfgang Amadeus Mozart (http://dme.mozarteum.at) and of the paintings, sketches, and writings by Leonardo da Vinci (www.universalleonardo.org) – demonstrate the point.

In the history of science, "The Complete Work of Charles Darwin Online" started in 2002, led by John van Wyhe (http://darwin-online.org.uk). It is the first and probably the most extensive digital edition devoted to a scientist (Wyhe 2002, 2009) and includes Darwin's complete record of writings – that is, his books, articles, published letters, manuscripts, private papers, and a rich array of supplementary material, such as reviews and critiques. Technical and editorial principles work hand in hand to offer easy access to "the whole Darwin." The site's very large number of hits indicates the worldwide interest in Darwin's work and highlights the potential of digital editions for the history of science.

An edition such as "Darwin Online" addresses both experts and laypeople, and so does another digital edition by van Wyhe, which focuses on Darwin's contemporary Alfred Russel Wallace (http://wallace-online.org/). Begun in 2012, it includes the published illustrations of Wallace's animal specimens. Carl Linnaeus has become the subject of two digitization projects: "Linné On Line" by Roland Moberg and his co-editors makes selected research areas of Linnaeus available to a wider audience and informs about his contributions to pharmaceutics, botany and zoology, physics, mathematics, and ecology (http://www2.linnaeus.uu.se/online/index.html). The site "The Linnean Collections," run by the Linnean Society of London, provides access to Linnaeus's books, collections, correspondence, manuscripts, and his annotated private library, supplemented by a variety of non-Linnean manuscripts – for example, by Johann Reinhold Forster (http://linnean-online.org/). In a similar manner "The Newton Project," edited by the historians of science Rob Iliffe and Scott Mandelbrote, publishes an online edition of all of Isaac Newton's writings, including not only his scientific and mathematical texts but also the alchemical and religious ones, featuring the amendments Newton made to his own writings (www.newtonproject.ox.ac.uk/). The edition uses the platform-independent markup language XML, following the recommendations of the Text Encoding Initiative (TEI), to enable a faceted search for, among other things, references to the Bible, changes of hands, and other editorial details.[1]

These various editions have a common understanding of what an edition in the history of science should be like. Just as non-digital editions, digital ones are critical editions, but on a scale traditional editions couldn't possibly attain (Sahle 2013). Digitization can and, in several instances, does present the entire oeuvre of a scientist, which means all published texts in their multiple historical editions. Unpublished manuscripts may be included. In some instances, the editions also provide a wide variety of supplementary material, such as contemporaneous reviews and other responses, diaries, library catalogues, museum records, and descriptions of specimens, not to mention their iconography. Furthermore, digital editions are able to offer side-by-side scans of the original pages and the transcribed and searchable text for critical philological comparisons. An advantage is that, unlike old-time "critical editions," they allow the reader to collate the transcribed text with the original. In the digital age, more than ever before, a "critical" perspective on texts is possible in the humanities and in the history of science in particular. Complex and stable bibliographical databases back up the edited texts.

For the purpose of global accessibility, each edition makes use of international encoding standards. The encoded documents must remain valid XML to allow the texts to be indexed automatically, to be displayed in any browser and to guarantee a precise correspondence between the image of original page and digital transcription. All documents are marked up with a tag set to provide an unambiguous matching between the texts and the bibliographical database. Only a few of the editions also provide the markup of semantic entities such as personal names, place names, specific periodicals or document types to make possible, for example, a combined search for illustrations and descriptions.

None of the current editions includes images of material objects from the working contexts of scientists, be it specimens or instruments, let alone experiments. These objects, described or discussed in the historical texts, have not been part of digital history of science editions so far. Thus these digitizations to a large extent follow in the footsteps of philological editions and make only partial use of the potential of digital editing, something that "Blumenbach – Online" is trying to remedy by integrating physical objects and semantics as part of the edition.

Editorial principles and standards of "Blumenbach – Online"[2]

"Blumenbach – Online" provides digital material of high quality: the digital images of book pages and objects will meet or surpass the standards of the German Research Council (Deutsche Forschungsgemeinschaft) for the quality of images (300 dpi color images) and for the accuracy of character recognition of the electronic texts (Liebetruth 2015). Quality is also an issue with regard to the objects of Blumenbach's collections. Until now, more than 4,000 of these have been identified in various present-day collections at Göttingen University. The project offers high-resolution photos of all objects. If the shape of an object requires this, these are supplemented by rotating and scalable images, as for example in the case of the items in Blumenbach's collection of skulls (Reich, Böhme, and Numberger-Thuy 2012). For each object there will also be a data set with detailed measurements and metadata about its "biography," documenting when, where, and by whom a specimen was collected, owned, sent, or brought to Göttingen. The interconnections between texts and collection items are made visible by hyperlinks: whenever Blumenbach in his texts mentions or provides an engraving of an object in his collections, a link is made available to the digital images and the data set of the present-day collection item, and vice versa. It will be possible to see which features of an object Blumenbach highlighted, neglected or – for whatever reason – did not consider worth mentioning.

The Blumenbach project is not limited to the more or less marketable part of his oeuvre – for example, his most famous works in their most popular edition.

The project is to pass on his *complete* work to the new digital media. In this respect, it will not only guarantee that in the digital cosmos of tomorrow there is as much "Blumenbach" available as in today's analog world; as a digital repository, it additionally will outperform any analog library by holding copies of *all* his publications. Also the accessibility of Blumenbach's scientific collections will increase as the site's virtual museum of visual Blumenbachiana can be accessed everywhere and by everyone.

Our editing project encodes information in a machine-readable form that is not – or not directly – intelligible to humans. These instructions facilitate the application of advanced digital tools that automatically carry out analytical procedures on, and extract information from, texts and data sets of collection items. Crucial for this are, first, standardized "tags" to encode information and, second, the availability of digital authority databases for referencing information – for example, about people and places and especially bibliographical data. In general, the details of standards are pivotal to the sustainability of an edition such as "Blumenbach – Online." The more widely adopted the standards are, the better the edition can interact with a digital environment of databases, catalogues, and browsers.

For tagging information, the Blumenbach project follows the guidelines of the Text Encoding Initiative (TEI), here in the version TEI-XML, Best Practice, Level 5. The recommendations are based on the markup language XML and have become a widely used quasi-standard for projects in the digital humanities. Our project uses TEI-tags to encode information about the structure, form, and content of Blumenbach's texts – for example, headlines, chapters, illustrations, original footnotes, change of language (with identification of the languages), abbreviations (full term added), printing errors (corrected version added), book decoration, uncertain readings, names of people, institutions and places, dates, bibliographical references, quotations, and collection items (see Figure 2.1). More particularly, people, institutions, and places are identified by references to authority databases, and dates are alternatively given in standardized form (ISO 8601: yyyy-mm-dd). All bibliographical references are identified, and if Blumenbach uses a quotation, the original text, enclosed in a tag, is added to allow human readers or digital devices to compare it to Blumenbach's version.

Authority databases pose a problem. Machines may, sooner or later, master normal vocabulary – perhaps even in historical orthography; but they might have difficulty identifying and interpreting proper names, especially historical place names, or identical personal names that refer to different individuals. Whereas in most cases human readers are able to make sense of such ambiguous or opaque information, digital devices still need unambiguous identifications in a machine-readable, standardized form. Unfortunately, the requisite universal authority databases do not yet exist. The "Getty Thesaurus of Geographical Names" serves as an informal standard for place names all over the world, but although strong for places in the United States, it is less so for – say – Eastern Europe; and not very useful at all for historical place names, or bygone politico-geographical entities, such as the German principalities of Blumenbach's day and age. Moreover, the

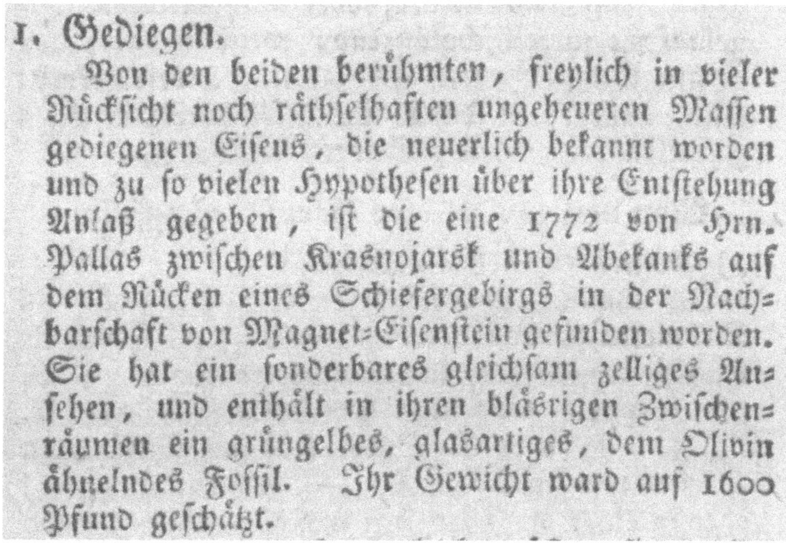
1. Gediegen.
Von den beiden berühmten, freylich in vieler
Rücksicht noch räthselhaften ungeheueren Massen
gediegenen Eisens, die neuerlich bekannt worden
und zu so vielen Hypothesen über ihre Entstehung
Anlaß gegeben, ist die eine 1772 von Hrn.
Pallas zwischen Krasnojarsk und Abekanks auf
dem Rücken eines Schiefergebirgs in der Nach=
barschaft von Magnet=Eisenstein gefunden worden.
Sie hat ein sonderbares gleichsam zelliges An=
sehen, und enthält in ihren bläsrigen Zwischen=
räumen ein grüngelbes, glasartiges, dem Olivin
ähnelndes Fossil. Ihr Gewicht ward auf 1600
Pfund geschätzt.

```
<p rendition="#indent-1">1. Gediegen.</p><p
rendition="#l1em">Von den beiden berühmten, freylich in
vieler<lb/>Rücksicht noch räthselhaften ungeheueren Massen
<lb/><rs type="object" subtype="GeMMe" ref="http://
www.blumenbach-online.de/../..">gediegenen Eisens
</rs>, die neuerlich bekannt worden<lb/>und zu so vielen
Hypothesen über ihre Entstehung<lb/>Anlaß gegeben, ist die
eine <DATE WHEN="1772">1772</DATE> von <PERSNAME
REF="../../PERS/JFB-O_TEI_PERSONEN.XML
#PALLASP">Hrn.<lb/>Pallas</PERSNAME> zwischen
<NAME TYPE="PLACE" REF="GETTYID:7030364">Krasnojarsk
</NAME> und <NAME TYPE="PLACE" REF="GETTYID:1051902">
Abekanks</NAME> auf<lb/>dem Rücken eines Schiefergebirgs
in der Nach-<lb type="inWord"/>barschaft von Magnet-
Eisenstein gefunden worden<lb/>Sie hat ein sonderbares
gleichsam zelliges An-<lb type="inWord"/> sehen, und
enthält in ihren bläsrigen Zwischen-<lb type="inWord"/>
räumen ein grüngelbes, glasartiges, dem Olivin<lb/>ähnelndes
Fossil. Ihr Gewicht ward auf 1600<lb/>Pfund geschätzt.</p>
```

Figure 2.1 (a) (b) Example of TEI-encoded text (from Blumenbach's *Handbuch der Naturgeschichte*. 6th edition. Göttingen: Dieterich, 1799, 655). Synopsis of original and digital text, with additional information encoded in TEI: information about typography and original layout (typeface Courier, not highlighted); identification of dates, names of persons and places (typeface Courier, highlighted light gray); link to the images of a collection item (typeface Courier, highlighted dark gray).

Getty Thesaurus is a commercial project that until recently charged fees. Presently, it can be used for free, but this may change again in years to come, causing sustainability problems for references to the Getty Thesaurus and similar products. For personal names one can use national databases that commonly exist within a library context, such as the German "Gemeinsame Normdatei" (GND) of the "Deutsche Nationalbibliothek." They tend to be most useful for identifying people who were involved in writing, producing, and selling books, yet none of them has achieved the status of a worldwide standard. However, Blumenbach also exchanged information and specimens with collectors and amateur scientists, and refers to them in his publications. For the identification of their names, the existing databases are not always adequate.

Currently, this problem remains unsolved. In the case of place names, the Blumenbach project uses the Getty Thesaurus, if only because the datasets contain geo-coordinates, which allow for automated visualizations – for example, of the geographical extent of the provenance of Blumenbach's collections or of the literature he cites. In order to identify people and publications, the Blumenbach project has created two databases of its own. The one for people, mentioned by Blumenbach or occurring in bibliographical Blumenbachiana (e.g., publishers, translators), offers basic biographical information and if possible the ID number in an authority database. The other, for publications, contains a full bibliographical record of every book or paper referenced by Blumenbach, and offers a link to a digitized version (if available).

Future developments

In a future Blumenbach portal, texts and collection items will be available in various ways and in different digital formats.[3] Users can choose the original printed pages or, if they find the historical black-letter typeface awkward, a transcription ("full text") in modern typography, and they have the additional option of retrieving extra information. A variety of synoptical arrangements will be possible, of original and transcribed versions, from one and the same or different publications/editions/translations, and of text with relevant collection items and data sets. All material will be searchable in different ways. Because it is difficult if not impossible to anticipate future developments of digital tools, the Blumenbach portal itself will not provide such devices. Instead, users will be able to download texts, images, and datasets in different digital standard formats and analyze the data with devices of their own choosing.

Digital editions such as "Blumenbach – Online" are also capable of responding to another innovative development: a change in the understanding of editions. Currently, editions are commonly understood as discrete entities, centered on single authors. However, in the history of science as well as in other fields of scholarly inquiry, practitioners are increasingly interested in specific layers of history, extended across many historical places – in the geography of scientific knowledge. Not just single editions are needed, but also editions as part of a wider corpus, customized to meet specific and thematic research interests. No edition can fully anticipate future research questions. What can be done, however, is to support the integration of data into a corpus individually compiled by a researcher for analysis. Therefore it is increasingly imperative for projects like "Blumenbach – Online" to develop interoperable editions that can be integrated into such research corpora easily. The Blumenbach project has already made its texts available in larger corpora like the German Text Archive (Geyken and Gloning 2015). The Blumenbach portal will support corpus-based research in the history of sciences by offering a complete as well as a selective download of all data. There is no such thing as "the definitive edition" and this holds true also for "Johann Friedrich Blumenbach – Online."

Notes

1 Other digitization projects of the works of remarkable figures in the history of science include "The Humboldt Digital Library" (www.avhumboldt.net/), the "Portal Alexander von Humboldt" (http://humboldt.hs-offenburg.de) and "Journals of the Lewis & Clark Expedition" (https://lewisandclarkjournals.unl.edu/).

2 This section is based on work by Claudia Kroke of the Blumenbach – Online project, whom we would like to thank for her contribution.

3 The Blumenbach project is making some parts of its editorial work – as soon as they become available – accessible on its homepage (www.blumenbach-online.de). An online version of the Blumenbach bibliography by Claudia Kroke (2010) includes links to the digitized versions (PDF) of all texts, and for most of them also a preliminary full text (HTML), searchable in browsers. Like the Darwin portal, "Blumenbach – Online" augments the edition of Blumenbach's publications and specimens with supplementary material.

Bibliography

Bendyshe, Thomas, trans. 1865. *The Anthropological Treatises of Johann Friedrich Blumenbach [. . .]. Unnumbered vol. of Publications of the Anthropological Society of London.* London: Longman, Green, Longman, Roberts, & Green.

Ehlers, Ernst Heinrich. 1901. "Göttinger Zoologen." In *Festschrift zur Feier des hundertfünfzigjährigen Bestehens der Königlichen Gesellschaft der Wissenschaften zu Göttingen.* [Vol. 3:] Beiträge zur Gelehrtengeschichte Göttingens, 391–494. Berlin: Weidmannsche Buchhandlung.

Geyken, Alexander, and Thomas Gloning. 2015. "A Living Text Archive of 15th–19th-Century German: Corpus Strategies, Technology, Organization." In *Historical Corpora: Challenges and Perspectives*, edited by Jost Gippert and Ralf Gehrke, 165–180. Tübingen: Narr.

Jannidis, Fotis. 2010. "Digital Editions on the Net: Perspectives for Scholarly Editing in a Digital World." In *Beyond the Screen: Transformations of Literary Structures, Interfaces and Genres*, edited by Jörgen Schäfer and Peter Gendolla, 543–560. Bielefeld: Transcript.

Kerzel, Martina, Mike Reich, and Heiko Weber. 2013. "Die Edition 'Johann Friedrich Blumenbach – Online' der Akademie der Wissenschaften zu Göttingen." In *Evolution der Informationsinfrastruktur: Kooperation zwischen Bibliothek und Wissenschaft*, edited by Heike Neuroth, Norbert Lossau, and Andrea Rapp, 107–136. Glückstadt: Werner Hülsbusch.

Kroke, Claudia. 2010. *Johann Friedrich Blumenbach. Bibliographie seiner Schriften: Unter Mitarbeit von Wolfgang Böker und Reimer Eck. Vol. 2 of Schriften zur Göttinger Universitätsgeschichte*. Göttingen: Universitätsverlag. URL: http://resolver.sub.uni-goettingen.de/purl?isbn-978-3-941875-64-7.

Liebetruth, Martin. 2015. "Blumenbach digitalisieren: Einblicke in die Digitalisierungswerkstatt." In *TextGrid: Von der Community – für die Community*, edited by Heike Neuroth, Andrea Rapp, and Sibylle Söring, 57–62. Glückstadt: Werner Hülsbusch.

Marx, Karl Friedrich Heinrich. 1840. *Zum Andenken an Johann Friedrich Blumenbach: eine Gedächtniss-Rede gehalten in der Sitzung der Königlichen Societät der Wissenschaften den 8. Februar 1840*. Göttingen: Dieterichsche Buchhandlung. English version cf. Bendyshe 1865, 1–46.

Reich, Mike, Annina Böhme, and Lea Numberger-Thuy. 2012. " 'Preziosen jeglicher Couleur' – Objektdigitalisierung der naturhistorischen Sammlungen von Johann Friedrich Blumenbach (1752–1840)." *Philippia* 15 (2): 153–168.

Rosselli Del Turco, Roberto. 2016. "The Battle We Forgot to Fight. Should We Make a Case for Digital Editions." In *Digital Scholarly Editing. Theories and Practices*, edited by Matthew J. Driscoll and Elena Pierazzo, 219–238. Cambridge: Open Book. URL: http//dx.doi.org/10.11647/OBP.0095.

Sahle, Patrick. 2013. *Digitale Editionsformen: Zum Umgang mit der Überlieferung unter den Bedingungen des Medienwandels*. 3 vols. Norderstedt: BoD. URL: http://kups.ub.uni-koeln.de/5351/, http://kups.ub.uni-koeln.de/5352/, and http://kups.ub.uni-koeln.de/5353/.

Winter, Thomas N. 1999. "Roberto Busa, S.J., and the Invention of the Machine-Generated Concordance." *The Classical Bulletin* 75 (1): 3–20.

Wyhe, John van, ed. 2002. "The Complete Work of Charles Darwin." URL: http://darwin-online.org.uk.

———. 2009. "Darwin Online and the Evolution of the Darwin Industry." *History of Science* 47 (4): 459–476.

Part II

Defining human races

3 Buffon, Blumenbach, Herder, Lichtenberg, and the origins of modern anthropology

Carl Niekerk

"Anthropology" has become a popular concept in contemporary cultural studies. Parallel to this, "culture" has emerged as a key notion not only for describing some of the most visible conflicts in the world today but also when we focus on intergroup dynamics in our increasingly global view of human activity. Anthropology, at least the variety called "cultural anthropology," with some right can claim that culture is central to what it intends to study. As a discipline, anthropology appears to be perfectly suited to examine and do justice to the diversity of the world facing us. Since its "linguistic turn," anthropology also appears to be well equipped to reflect on the question of to what extent discursive structures shape a culturally specific construction of reality (see Clifford and Marcus 1986; Marcus and Fischer 1986). As a result, the field has become much more aware of its own (conscious or unconscious) biases.

In spite of the popularity and ubiquity of the term "anthropology," and the loss of semantic contours inevitably accompanying such an omnipresence, also in eighteenth-century studies, there has been surprisingly little interest in the history of the discipline called "anthropology." Even among anthropologists themselves today there is not only a lack of consensus about important schools and figures in the history of anthropology but also a dearth of factual knowledge about that history. In part this may be explained by the fact that anthropology as a field of inquiry is so diverse. A typical American department of anthropology houses not only cultural anthropologists – usually constituting the majority of the department – but also physical anthropologists, archaeologists, and quite possibly scholars focusing on the intersections between anthropology and linguistics. Many anthropologists, even those primarily studying cultural issues, prefer to think of themselves as doing contemporary and empirical research, and therefore are less interested in the discipline's past (similar maybe to the ways in which empirical psychologists prefer to ignore the legacy of Sigmund Freud, who nevertheless in the popular imagination for many may represent the prototypical psychologist). Another reason is probably that most anthropologists may have a suspicion that the history of anthropology was maybe not the glorious trajectory – a long list of important persons, dates, and achievements – one would imagine one's disciplinary history ideally to be. This has certainly something to do with the normative claims that have been made over the years in the name of "anthropology." Finally, the development of anthropology as a discipline in particular in the nineteenth and twentieth centuries widely varied and depended strongly on national contexts. The formation of the discipline of anthropology in the United States, for instance, was radically different from that in Germany (see Barth et al. 2005). And yet, it would be wrong to say that these national traditions did not intersect with each other. In fact, the German and German-trained anthropologist Franz Boas played a major role in the establishment of American anthropology; in particular, the ideas of Alexander and Wilhelm von Humboldt were important for Boas (Bunzl 1996, 60, 62–63, 66–68; Barth et al. 2005, 88–89, 91, 170).

In the eighteenth century, the field of inquiry that emerged as "anthropology" was a highly interdisciplinary and also still an international enterprise. The semantic development of what "anthropology" meant at the time was not (yet) fundamentally different in France, Germany, England or the Netherlands, as I will discuss ahead. Throughout the eighteenth century there was an eager transnational and transcultural exchange of travel reports, in the original and in translation, that also included smaller cultural communities, like the Netherlands and the Scandinavian countries. The same was true for the work of natural historians and anthropologists that sought to interpret the new "empirical" information produced by these reports. In the following I am especially interested in reconstructing the early history of anthropology and in analyzing the normative contexts of the developments that will lead to the establishment of the discipline that by the late eighteenth century will be named "anthropology." Göttingen plays a key role in this, and for that reason also deserves to be considered one of the leading centers of the European Enlightenment.

What is Enlightenment anthropology?

How can we make sense of the seemingly infinitely different ways in which the term "anthropology" is used in eighteenth-century studies? One could look at all the different ways in which the term "anthropology" is used in relation to eighteenth-century thinking, in order to subsequently identify some common denominators and come to a pragmatic clarification of the term. In the following I will not pursue this road, because it is precisely this approach that has led to a proliferation of the meanings of the term "anthropology" and as a consequence any awareness of the original contexts in which the term was used was lost.

To navigate the debate about the question of what the term "anthropology" means, it is important to distinguish between "anthropology" as part of an object language to be studied and a use of the term as part of the metalanguage with the help of which we study the eighteenth century. Especially this second approach has led to a semantic proliferation of the concept "anthropology."[1] One possible answer to the question of what "anthropology" means in the context of eighteenth-century studies is a precise pragmatic reconstruction of the use of the term during the eighteenth century – who used the term and with which precise meaning? I am not arguing here that we should limit the use of the concept "anthropology" in eighteenth-century studies only to those cases where the term is clearly documented, but I do believe that a sense of how the term was employed at the time may be helpful to understand its (hermeneutically understood) meaning and impact, and maybe also its provocative aspects.

When looking at the conceptual history of the term "anthropology" in the eighteenth century, it is important to be aware that the concept is not a semantically stable term, but rather a word that is part of a discourse that is in the process of developing its own semantics throughout Enlightenment Europe. Such an etymological study has been done, for instance, for France. According to Michèle Duchet the term "anthropology" is rooted in theology and originally meant the attribution of human traits to God (in the current sense of "anthropomorphism"), but by the middle of the eighteenth century had come to be understood anatomically as the "study of the human body" (*étude du corps humain*) (Duchet 1995, 12). In the second half of the eighteenth century the term's meaning broadens and by the end of that century Alexandre-César Chavannes from Lausanne in a text entitled *Anthropologie ou science générale de l'homme* from 1788 lists "physical anthropology" (*anthropologie physique*) as the term's primary meaning, but adds "ethnology or the 'science of man considered as belonging to a species spread out across the globe and divided in multiple bodies of societies'" (*Ethnologie ou 'science de l'homme considéré comme apparent à une espèce répandue sur le globe et divisée en plusieurs corps de société'*) as its second meaning (Duchet 1995, 12). The clearest example of this type of new anthropology was, according to Duchet, Georges-Louis Leclerc de Buffon's *Histoire naturelle de l'homme* (Duchet 1995, 13). The emergence of the discipline of anthropology is simultaneous with the availability and accessibility of a new type of (empirical) information, in particular in the form of (supposedly empirically reliable) travel reports (Duchet 1995, 13). Not only were many older reports factually not reliable but also collecting scientifically accurate information for a long time was simply not a priority for those traveling the world (Duchet 1995, 13–14).

In German-speaking parts of Europe the word "anthropologium," according to Han Vermeulen, first is mentioned in Magnus Hundt's *Ant[h]ropologium de hominis dignitate, natura, et proprietatibus*, a popular introduction to medicine published in Leipzig in 1501 that includes a discussion of anatomy and physiology and also considers philosophical and theological aspects of these fields (Vermeulen 2015, 360). In 1594 and 1596 Otto Casmann published a "psychologia anthropologica" defined as a doctrine of human nature, and covering both physical and spiritual aspects (Vermeulen 2015, 361). According to Vermeulen, "anthropology" did not obtain its modern meaning in German-speaking areas until the 1790s, when it was "used to label a study defined either as the 'natural history of man' (Blumenbach 1790, 1795, 1798) or as the 'pragmatic philosophy of mankind' (Kant 1798)," thus indicating, in both cases, an emphasis on the empirical study of humans in a broad scientific context and leaving its older disciplinary contexts (in theology and medicine) behind (Vermeulen 2015, 358–359). The term "anthropology" had been used with some frequency in the seventeenth and eighteenth centuries, but did not achieve true popularity until the last three decades of the eighteenth century. Between 1770 and 1800 in Europe forty-three books were printed with some version of "anthropology" in its title, not counting those covering anthropological topics but not using the term as part of their title, like Johann Friedrich Blumenbach's *De generis humani varietate nativa* (1776; 1781a; 1795) (Vermeulen 2015, 362).

In addition to exhibiting a lack of awareness of what "anthropology" in practice meant during the eighteenth century, scholars often use the term in almost complete isolation from Enlightenment thinking in general and unconnected to the development of disciplines other than anthropology. In order to counter this, in addition to reconstructing its conceptual history it may be useful to locate anthropology more clearly as an Enlightenment discipline. In the eighteenth century, the developing discipline of anthropology is part of a broader Enlightenment discourse that was interested in rationalizing Europe's view of other, non-European peoples and cultures. In line with Enlightenment philosophy and historiography, and their universalist aspirations, humans were seen as both similar and different across the world. Eighteenth-century anthropology sought to understand alterity as the product of a spatial organization of nature: instead of assuming that the non-European world was populated by bizarre creatures (freaks and monsters) with only a remote resemblance to (Western) humans, Enlightenment anthropology worked under the assumption that in other parts of the world people were different because they lived under different geographical and climatological circumstances. Anatomy and natural history collections documenting other parts of the world during the eighteenth century became less about monstrosity and nature's abnormalities and more about gaining insight into nature's developmental patterns (Hagner 1999, 213). Alterity is also the product of a temporal organization of the order of nature: Enlightenment anthropology is interested in studying the developmental patterns based in part in an innate drive characteristic for all living beings – Blumenbach names this the *Bildungstrieb* or "formative drive" (Blumenbach 1781b) that, however, as we will see ahead, cooperates with and is influenced by environmental factors.

When eighteenth-century anthropology is discussed in connection with other disciplines and specifically in a German context frequently "medicine" is mentioned prominently. At least in part the physician Ernst Platner is responsible for this. Platner, in his text *Anthropologie fuer Aerzte und Weltweise*, first published in 1772, predominantly focused on the relations of body and mind and had claimed the term "anthropology" for his project.[2] This alliance picks up on one of the older meanings of "anthropology," a term that indeed, as Duchet and Vermeulen both demonstrate, had also been used in a medical sense. Nevertheless, Platner was an exception. By the 1770s many of Platner's medical theories were outdated – the assumption, for instance, that the nerve system was inhabited by liquid "Lebensgeister," or his failure to distinguish between the nerve system and blood circulation – something that would make his theories, and the concept of "anthropology" that they promoted, suspect in the eyes of men like Blumenbach and Georg Christoph Lichtenberg, who saw themselves as rigorous empirical scientists (Niekerk 2005, 101, 190, 199). There certainly existed a productive interaction between medicine and anthropology in the late eighteenth century, but a more logical nexus, as my brief summary of the term's semantic history earlier indicates, is the one between anthropology and natural history. This link is, for instance, convincingly made by the Göttingen historian Frank William Peter Dougherty (1996), building in turn on the work of Wolf Lepenies (1976, 1980), who argues that natural history in the eighteenth century is characterized by an epistemological battle between a static, hierarchical, and spatial model, exemplified by the work of Carl von Linné on the one hand, and a newer temporal and developmental model, exemplarily represented by Buffon on the other, who rejected such strict hierarchies. Vermeulen too emphasizes the roots of anthropology in natural history (2015, 6, 367–374).

For the study of the eighteenth-century sciences of man, in particular Michel Foucault's thinking has been influential. His theories have been helpful in positing the emergence in the late eighteenth century of a new type of biological knowledge that uses a temporal model of organization and breaks with an older static paradigm of natural knowledge – the classical paradigm – that is spatially organized (and visualized by the tableau) (see Foucault 1989, 51–85, 136–179, 235ff.). Ernst Cassirer locates this epistemological break a bit earlier, and claims that the tendency to break with the seventeenth century's systematic spirit in the name of a more complex sense of reality that emphasizes temporality and developmental patterns can be found throughout the eighteenth century (1998, 100; see also 8ff.). The popularity of the concept "anthropology" in the last three decades of the eighteenth century following the model offered in Buffon's *Histoire naturelle* indeed backs up Foucault's (and also Cassirer's) ideas. To some extent Foucault's work, however, also stands in the way of understanding the dynamics surrounding the origins of modern anthropology. It is Foucault's interest in discursive framing that leads him to "bracket" truth claims, to use a formulation by Hubert L. Dreyfus and Paul Rabinow (1983, 49, 59), meaning not that Foucault believes in the inability of language to speak the truth but rather that for his project – the identification of patterns of thinking, speaking, and writing – the truth value of what is thought, spoken or written is not relevant. But in the case of anthropology one risks missing an important dimension of what constitutes the discourse of scientific validity if one ignores the truth value of what are meant to be scientific observations (the same goes for the discipline of medicine in an eighteenth-century context).

The 1770s are the time of what historian Jonathan Israel has called a "radical breakthrough" (2011, 648–683). Materialist thinking that had long lingered in the margins of public discourse now moved to its center. This breakthrough of radical thinking interested in a strictly empiricist approach to science coincides with a proliferation of publications in what one could call "early anthropology," in particular in German-speaking areas. In 1768 and 1769 Cornelis de Pauw published his *Recherches philosophiques sur les Américains*, the first anthropological treatise after Buffon and a book that greatly contributed to the popularization of Buffon's thinking (see Duchet 1995, 114; Zantop 1997, 13). Almost immediately, it was translated into German by Karl Lessing. In 1770 the Abbé Raynal, representing a collective of authors, published his *Histoire philosophique et politique des établissements et du commerce des Européens dans les deux Indes* in six volumes;

new and substantially revised editions followed in 1774 and 1780; it was translated into English in 1776, 1783, and 1784, and into German in 1780 (in the form of an excerpt) and in ten volumes from 1783 on.[3] Between 1771 and 1774 a highly influential and popular seven-volume German translation appeared of Buffon's main work, the *Histoire naturelle*. I already mentioned Platner's *Anthropologie fuer Aerzte und Weltweise* from 1772, a text with a dubious reputation among scientists. In 1774 Johann Gottfried Herder published his rather speculative essay *Auch eine Philosophie der Geschichte zur Bildung der Menschheit*. Between 1775 und 1778 Johann Caspar Lavater's four-volume *Physiognomische Fragmente, zur Befoerderung der Menschenkenntniß und Menschenliebe* appeared in print – not in the technical sense an example of scientific anthropology, but nevertheless contributing to the budding new discipline's popularity. In 1776 Blumenbach had his dissertation printed, which he had defended at the University of Göttingen the year before under the title *De generis humani varietate nativa* (republished in expanded editions in 1781 and 1795, and brought out in a German translation by Johann Gottfried Gruber in 1798 as *Über die natürlichen Verschiedenheiten im Menschengeschlechte*). In this text Blumenbach seeks to convert Buffon's methodological principles into an autonomous anthropology, to be clearly separate from his *Handbuch der Naturgeschichte*, the first edition of which appeared in two volumes in 1779 and 1780. In 1785 one of Blumenbach's colleagues in Göttingen, Christoph Meiners, published his *Grundriß der Geschichte der Menschheit*. Between 1784 and 1791 Herder continued and to some extent also provided a summary of the debate about anthropology in his four-volume *Ideen zur Philosophie der Geschichte der Menschheit*.

While at first sight, all of these texts may seem to participate in a trend toward a temporalization of natural history, there are, seen from the perspective of a fact-based model of anthropology, substantial differences. Buffon and Blumenbach sought, and had the institutional help and financial means, to back up their ideas through their own empirical research. De Pauw, Raynal, and also Herder were "Schreibtischgelehrte" (desktop scientists) (Krauss 1987, 109) and "Kompendienschreiber" who did not have the financial means at their disposal to do their own empirical research, but nevertheless put in an honest effort to distinguish between fact and fantasy among the written anthropological and ethnographic sources accessible to them. Lavater explicitly adopted Buffon's criticism of Linné's model of natural history, but nevertheless in practice aimed for a systematic and static form of organization that far more resembles Linné's than Buffon's thinking (Niekerk 2005, 145–148). Meiners shares with Buffon a temporal conception of natural history, but resembles Lavater in his lack of concern about the factual accuracy of the material he discusses and in his tendency to aestheticize natural history by distinguishing between beautiful and ugly populations, equating beauty with a white skin color and a dark skin with ugliness.[4] Blumenbach dedicated an entire essay in his *Beyträge zur Naturgeschichte* to a rebuke of Meiners, emphasizing that his own observations are based in either empirical research or a careful weighing of existing scientific observations by others (Blumenbach 1790, 62–78). Lichtenberg in a letter to Georg Forster refers to the author, discussing his anthropological interests, politically not very correctly as the "Mongole Meiners" and his "unbändiger Eigendünkel" (unlimited arrogance) (1983–2004, III, 919 [letter 1893]). Lichtenberg's reaction may have something to do with the strong normative implications of Meiners's ideas (this may also explain why Blumenbach keeps the discussion of his colleague's work rather short).

Although many turned toward anthropology as an epistemological paradigm in the 1770s, Immanuel Kant, in contrast, of whom a 1768 painting by J. G. Becker exists in which the philosopher holds a book entitled *Anthropologie oder Naturkenntnis des Menschen* (Zammito 2002, 104, 292–293) and who in 1775 published an essay entitled "Über die verschiedenen Rassen der Menschen," which further develops but also critiques Buffon's theories (Zammito 2006, 41–42), turned away from anthropology to focus instead on the development of his transcendental philosophy. As John Zammito shows, Kant wanted to rescue philosophy from psychological and empiricist reductionism (2002, 206, 241). In the late eighteenth century physical anthropology was mainly a scientific and therefore empirical enterprise. Was Kant indeed moved by epistemological discussions to abandon his anthropological ambitions, or did other, unmentioned factors play a role? Major anthropologists such as Buffon, Petrus Camper, and Blumenbach had access to resources not available to Kant. Buffon was overseer of the French king's gardens from 1739 on. Camper had his former students, now working in the colonies, ship monkeys to him for dissection. Blumenbach managed to build up a substantial anatomical collection at the University of Göttingen and was even able to examine some mummies. None of this was economically viable for Kant; therefore, he did not possess the necessary empirical materials for serious research in natural history or anthropology and risked missing the rapid developments around 1770 that led to the academic establishment of anthropology as a distinct discipline. It must have been hard for someone as ambitious as Kant to accept that, but in establishing his transcendental philosophy, he found a different route to intellectual fame.

While allowing for the term "anthropology" to mean many different things when applying it to eighteenth-century thinking, it is important, I believe, to include the emerging global view of mankind in that century and to pay attention to the ways in which this global perspective changed the view of humankind among those working in the fields of natural history and anthropology. It would be helpful, I believe, if we were to look at the history of anthropology simply as the history of a discipline, comparable, for instance, to the history of medicine. This is therefore my approach in the remainder of this essay. I do by no means intend to downplay the importance of Foucault in identifying discursive and therefore cultural patterns underlying scientific thinking around 1800. I do think in fact that Foucault's work in many respects has been eminently helpful for detecting the presence of broader structures in scientific discourse. It is, however, the truth value or empirical validity of many of these scholars' statements that is decisive for their own standing in the eyes of the scientific community surrounding them and therewith the survival of their thinking. Buffon and Blumenbach are key figures in the establishment of modern anthropology.

Buffon and the origins of modern anthropology

The overarching question that anthropology, understood as a scientific discipline and as a modern phenomenon, asks is how science can explain human diversity. But it asks this question in ways different from ethnology, which has roots in the eighteenth century as well and has research interests that overlap with those of anthropology (see Vermeulen 2015, 6–10). Anthropologists are interested in the structure behind humanity's variety, while ethnology is more interested in an empirically accurate description of populations and their cultures. Nevertheless, anthropologists do rely on ethnographic descriptions of other cultures in order to develop their theses. This is in particular clear in the case of Buffon and his *Histoire naturelle*, which, in addition to his own observations, borrows frequently from the literature of exploration and travel that was available at the time, in particular concerning the non-European world.

While one should be cautious crediting one scientist with the creation of a new field of study, there is no doubt that Georges-Louis Leclerc de Buffon played a foundational role in establishing modern anthropology, followed soon after by Camper and Blumenbach. Buffon's main work, the *Histoire naturelle générale et particulière*, appeared between 1749 and 1788 in thirty-six volumes, with eight more volumes being published after Buffon's death. Buffon's text articulates a transition from a generation of scientists who saw it as natural history's primary task to collect all documented information about nature and to classify this information systematically to a generation of natural historians who understood themselves as empirical scientists who based their observations on experience and sought to proceed unprejudiced and without the burden of tradition (Lepenies 1988, 7). As mentioned before, to these scientists Linné represented an old static classificatory model that was visually oriented; Buffon, in contrast, advocated a new dynamic and temporal model that sought to avoid visual elements and instead relied on language to establish causal and temporal connections. Buffon himself in the introduction to his *opus magnum* emphasizes the importance of direct observation and speaks of the different errors one can make in studying natural history – for instance, the mistake

> de s'assujétir à des méthodes trop particulières, de vouloir juger du tout par une seule partie, de réduire la Nature à de petits systèmes qui lui sont étrangers, & de ses ouvrages immenses en former arbitrairement autant d'assemblages détachez; enfin de rendre, en multipliant les noms & les représentations, la langue de la science plus difficile que la Science elle-même.
>
> [to subject oneself to methods that are too specific, to want to judge everything from one sole perspective, to reduce nature to little systems which are alien to it, and to create arbitrarily out of its immense works many compilations that have nothing to do with it; and finally to make, by multiplying the names and representation, the language of science more difficult than Science itself.]
>
> (Buffon 1749a, 9)

Later in his introduction, Buffon follows up on this passage with a critical discussion of Linné's attempts to organize nature, calling his attempts "très-arbitraire & fort incomplète" (very arbitrary and extremely incomplete) (1749a, 37). Buffon describes his own approach – which he bases on ancient natural historians like Aristotle and Pliny, but also contemporary experimental physics and mathematics – as one of a precise observation of facts in combination with a subsequent search for reasons and causes underlying those facts, with the purpose of finding more general laws of nature (1749a, 57). The impact of these methodological innovations, proposed by Buffon, among eighteenth-century intellectuals has been compared to that of René Descartes's *Discours de la méthode* (Sloan 1996, 304; Cassirer 1998, 102–107).

One of the main truths to which an empirical study of nature leads us according to Buffon is "maybe humiliating for man" – namely, the insight that he himself belongs to the "class of animals which he resembles in every material respect" (peut-être humiliante pour l'homme; c'est qu'il doit se ranger lui-même dans la classe des animaux, auxquels il ressemble par tout ce qu'il a de matériel) (1749a, 12). Buffon develops his natural history of man in the third volume of the *Histoire naturelle*, published also in 1749, simultaneously with the influential introduction. Buffon summarizes his theory of human diversity concisely at the end of this volume in the context of his argument that humanity consists of only one species:

> il n'y a eu originairement qu'une seule espèce d'hommes, qui s'étant multipliée et répandue sur toute la surface de la terre, a subi différens changemens par l'influence du climat, par la différence de la nourriture, par celle de la manière de vivre, par les maladies épidémiques, et aussi par le mélange varié à l'infini des individus plus ou moins ressemblans.
>
> [Originally there was no more than one species of humans that multiplied itself and spread out over the entire surface of the earth, and underwent a variety of changes through the influence of climate, by differences in nutrition, different ways of living, epidemic diseases, and also through the various ways of mixing with each other of individuals who more or less resembled each other.]
>
> (1749c, 530)

That humanity consists of only one species is an epistemological point with important political consequences that are implied in Buffon's text, even though not explicitly articulated – doing so would have gone against Buffon's firm intention to limit his studies to empirical knowledge alone. It is because all human beings are part of one species that they deserve to be treated equally, a principle for which Herder later will invent the following formulation:

> Weder der Pongo, noch der Longimanus ist dein Bruder; aber wohl der Amerikaner, der Neger. Ihn also soll[s]t du nicht unterdrücken, nicht morden, nicht stehlen: denn er ist ein Mensch, wie du bist; mit dem Affen darfst du keine Brüderschaft eingehn.

> [Neither the pongo nor the longimanus is your brother; but the American, the Negro is. Him you should not repress, murder or steal [from]; because he is a human being like you; but don't fraternize with monkeys.]
>
> (1989, VI, 255)

However, in spite of his underwriting the principle of a unified mankind, Buffon also allows for differences: it is because of climate, nutrition, the concomitant different ways of living, epidemic diseases, and the general tendency of humans to mix with other humans that local varieties develop. This territorial theory of human difference, which has roots in antiquity, specifically in Hippocrates's treatise "Περί αέρων, υδάτων, τόπων" (*On Airs, Waters, Places*) (Dougherty 1996, 341 fn 40), is fundamental not only for Buffon but also for modern anthropology in general. It is a clear alternative to polygenetic accounts of the origins of humanity that inevitably lead to arguments about the inherent differences among humans and the inferiority of some humans in comparison to others. It is also an alternative to attempts to explain humans' biological variety through assumed intermingling of humans and animals, especially apes (Niekerk 2004). The attraction of climate theory for Enlightenment thinking is that it allows for universal values – a common humanity – while also acknowledging biological and cultural variety, explaining how different populations develop their own mode of rationality (in the sense of a rational response to a specific geographical and climatological environment).

Scholars today often associate eighteenth-century climate theory with Charles-Louis de Montesquieu, who in Book 14 of Part 2 of *De l'esprit des lois* (1748) discusses the effects of climate on society's legal system (see, e.g., Carhart 2008, 230). This preference for Montesquieu over Buffon may be explained by the fact that among scholars and the general public today Montesquieu's text is more frequently read than Buffon's methodological introduction to his *Histoire naturelle*. However, among eighteenth-century natural historians and anthropologists like Blumenbach and Lichtenberg, Buffon was far better known and also more frequently quoted and referred to, while Montesquieu is relatively rarely mentioned (without a doubt at least partly because he was primarily a legal scholar and political scientist, and not a scientist claiming to do empirical work). Also the intentions behind Montesquieu's climate theory are quite different from those of Buffon. For Montesquieu climate is supposed to explain the backwardness and irrationality of certain peoples, their stasis of body and mind, and thus he highlights the negative impact of climate that needs to be overcome (Montesquieu 1979, 378–379; see also Todorov 1998, 95–96). Buffon, in contrast, seeks to prove how different peoples rationally adapt to the climate surrounding them and thereby emphasizes the rational potential of all humans.

Within its eighteenth-century contexts, Buffon's natural history represented clear progress, very much in line with the goals of the radical Paris circles which Buffon frequented, in particular Paul-Henri Thiry d'Holbach's salon, of which Denis Diderot was part as well (Blom 2010, 59–61). Nevertheless, there are also reasons to be cautious with Buffon's theories and the normative potential of some of the descriptively meant vocabulary that Buffon employs. Buffon, for instance, uses a notion of "dégéneration" (degeneration) to describe the effects of an extreme climate on a population over several generations (1766, 312–316; see also Duchet 1995, 246–247, 263–267, and 274–276). Furthermore, while Buffon generally speaks of "variétés" (varieties) of humans that develop under the influence of climate, he also uses the term "race." It is not until his fourth supplement (published in 1777) that he clarifies some of his thoughts on human diversity and also counters some criticism his work has received. The concept "race," as Buffon makes clear, functions within his natural history as a subcategory of "espèce" (species) and in the case of humankind comprises groups of humans who have developed similarly because of living in comparable climates, not because they are closely related biologically (1777, 462–463, 479–480; see also Duchet 1995, 270–271). Buffon uses "race" to indicate differences caused by climatological factors that have become hereditary. While Buffon stays within the emancipatory agenda of his climate theory, arguing that humans are the same species everywhere, his notions of race and degeneration, especially when used together, do enable a view of human difference that is hierarchical and therefore has potentially strong normative implications (see, e.g., Buffon's comments on Greenlanders and Eskimos 1777, 463).

Blumenbach and Buffon

In German-speaking countries Johann Friedrich Blumenbach was without doubt Buffon's most influential supporter. Blumenbach not only helped propagate Buffon's thinking (and was a major influence on Herder) but also developed Buffon's thinking further in his own way. While "anthropology" in the case of Buffon was still an integrated part of his natural history project (Volumes 2 and 3), Blumenbach decided to publish his natural history (*Handbuch der Naturgeschichte*, first edition in two volumes published 1779–1780; twelfth edition 1830) and his anthropology as separate volumes.

In his anthropological dissertation, *De generis humani varietate native* (*On the Natural Varieties of Mankind*), translated as *Über die natürlichen Verschiedenheiten im Menschengeschlechte*, Blumenbach only occasionally engages with Buffon (e.g., 1798, 26, 183, 209–210), whereas Linné is mentioned prominently, albeit critically, in the introduction. In the letter to Joseph Banks that serves as an introduction to the third edition of Blumenbach's main anthropological text and to whom the text is dedicated, Blumenbach is respectful toward Linné, but also indicates that artificial attempts to come to an order of nature like his by necessity will have "sehr große Lücken" ("insignes hiatus"; major gaps) (1798, XIX; cf. 1795, XII). Here and elsewhere Buffon's influence is evident. In Paragraph 34 of his study Blumenbach deals with climate (1795, 88–93; 1798, 73–77). The importance of this section for Blumenbach is clear because of its strategic location immediately after Paragraph 33, in which Blumenbach explains his concept of "Bildungstrieb" ("nisus formativus"; formative drive), his most generic and influential contribution to eighteenth-century anthropology and philosophy of nature.

He first introduced this concept in an essay in the *Göttingisches Magazin der Wissenschaften und Litteratur* (1780b) that he subsequently reworked into a separate publication (1781b) and expanded multiple times (1789a, 1791). With this term Blumenbach creates a conceptual framework for the temporal dimension, the element of development that according to Cassirer and Lepenies constitutes the truly innovative element of Buffon's theory. Climate is one of the factors responsible for the fact that the "formative drive" inherent to every being can manifest itself in a number of varieties; in this context Gruber, the translator of Blumenbach's text, speaks of "Ausartung" to translate the Latin term "degeneratio" (1798, 75; see 1795, 91). Further factors mentioned by Blumenbach are nourishment ("Nahrungsmittel" or "victus") (Paragraph 35), way of living ("Lebensart" or "vitae genus") (Paragraph 36), the production of hybrids ("Bastard=Erzeugung" or "generatio hybrida") (Paragraph 37), and features or characteristics that animals inherit through disease ("Thieren durch Kränklichkeit angeerbte Eigenschaften" or "heriditariae proprietatis animalium ex morbosa intemperie") (Paragraph 38) (1798, 77–85; 1795, 93–105). Interestingly, Blumenbach here follows Buffon's model extremely closely, without, however, mentioning him by name. Buffon had mentioned "l'influence du climat," "différence de la nourriture," "manière de vivre," "maladies épidémiques," and "le mélange varié à l'infini des individus plus ou moins ressemblans" (1749c, 530) – the only thing Blumenbach changed was that he reversed the order of the last two terms.

Like his mentors Buffon and Camper, Blumenbach articulates a shift within natural history and anthropology toward an exclusively developmental explanation of the order of nature that aims to identify material conditions as the cause of change; human beings are the product of their environment, and their dependence on the climate and their surroundings is one of the clearest material indicators of this developmental determination of humanity. Furthermore, Blumenbach introduces two new concepts meant to clarify developmental patterns, both based on Buffon while simultaneously expanding on the Frenchman's ideas. Blumenbach introduces the "Bildungstrieb" concept, adapted from Buffon's idea of a "moule intérieure" (inner form) (1749b, 41–43, 45–46) that joins forces with a "puissance active" (active power) (1749b, 45) guiding a being's development. And the translator of Blumenbach's text uses, as we already saw, the term "Ausartung," sometimes also called "Verartung" (1795, 80; cf. 1795, 96; "degenerandi"), as an equivalent for Buffon's "degeneration" in order to explain human variety.

Does it make sense to talk about a difference between physical and cultural anthropology in the context of eighteenth-century anthropological thinking? There is some – even though a rather limited amount of – space in eighteenth-century anthropological theory for cultural differences. Both Buffon and Blumenbach seek to conceptualize ways of living ("manière de vivre," "Lebensart") within their anthropology as one factor with the help of which they seek to explain varieties among populations and their cultures. Blumenbach uses the term "cultura" in the section on "vitae genus" of his anthropology; Gruber translates the term as "verfeinernde Ausbildung" – a term one could maybe translate as "educational refinement" – and compares it to the power of habit ("vis consuetudinis" or "Macht der Gewohnheit") regarding its effect on human development (1798, 79; 1795, 96). The examples that Blumenbach gives are all taken from the animal world and not the human realm, however, and do not really point to culture as a positive and autonomous force. Humans can try to train horses and other mammals, but often this leads to a bad temper, an increase in deformed births, and new diseases among these animals (1798, 79–80; 1795, 96–98). Humans and their culture, Blumenbach claims, clearly have only limited power when it comes to manipulating nature. Neither in Blumenbach nor in Herder does it make sense to speak of a dichotomy of biology and culture; instead culture is very much understood as a function of material circumstances (climate) or innate drive (the "Bildungstrieb"). While Blumenbach creates space for a cultural component in explaining human diversity within the borders of eighteenth-century anthropology, in practice the potential offered by this cultural component is rarely realized. Culture is a function of biology, geography, and climate.

And yet, in spite of the material and biological determinism that characterizes Enlightenment anthropology, there is a great hesitance to start using the term "race" within anthropological texts. When speaking about humanity's diversity, Blumenbach emphasizes his monogenetic views on the subject: the title of the fourth section of *De generis humani varietate nativa/Über die natürlichen Verschiedenheiten im Menschengeschlechte* states his position very clearly that within humanity there are five varieties, but only one species: "Generis humani varietates quinae principes, species vero unica"/"Es gibt fünf Hauptvarietäten des Menschengeschlechts, jedoch nur Eine Gattung desselben" (1795, 284, 1798, 203).[5] The fact that this text was originally published in Latin meant that Blumenbach could sidestep the issue of race – a term that Kant had used, but that has no direct equivalent in Latin – to use "varietas" or "Varietät" as alternatives instead. Blumenbach long resisted the introduction of the term "race" and avoided using it, although he made some exceptions. In the first three editions of his *Handbuch der Naturgeschichte* he uses the term "Raçe" only once in each edition to refer to humans (1779, 63, 1782, 60, 1788, 60); this is certainly not to be interpreted as an endorsement of the term. Similarly, in the title of a 1789 essay, "Ueber Menschen=Racen und Schweine=Racen, vom Herrn Hofrath Blumenbach," Blumenbach uses the term "race," but his essay from its very beginning is very explicitly a discussion of the race concept of other scientists, and in it Blumenbach defends the position that what others call "races" are "Spielarten" (varieties) of one species that find their origin in climatological differences (see 1789b, 1). Similarly, in the first edition of the first volume of his *Beyträge zur Naturgeschichte* from 1790 we find a chapter on the "Eintheilung des Menschengeschlechts in *fünf* Spielarten" in which "Spielarten" clearly translates "varietas" (1790, 79; italics by Blumenbach); on the other hand, at the end of a chapter in the same volume that engages Meiners's criticism of Blumenbach the latter uses the term "Menschenrassen," but one could argue that here he adopts Meiners's own terminology (1790, 78). In the second edition of the first volume of the *Beyträge*, published in 1806, the chapter that originally had the title "Eintheilung des Menschengeschlechts in *fünf* Spielarten" is now entitled "Eintheilung des Menschengeschlechts in *fünf* Hauptrassen" (1806, 67), clearly indicating a shift in favor of the term "Rasse" over the term "Spielart." Why did Blumenbach eventually embrace the term "race"?

Blumenbach's translator Gruber in *Über die natürlichen Verschiedenheiten im Menschengeschlechte* initially seems to pursue a strategy similar to the first edition of the *Beyträge* when he uses consistently the terms "Hauptvarietäten" and "Varietäten" when referring to Blumenbach's own theories (see 1798, 203–208), while he uses the term "race" when discussing the theories of others – for instance, in the title "Die Eintheilung des Menschengeschlechts in Racen, nach andern Schriftstellern" (The division of the human species in races according to other authors) (1798, 208). However, after these sections Gruber turns inconsistent – for instance, when he starts a description of the "kaukasische Hauptvarietät" (Caucasian main variety) with the words "Diese Race erhielt ihren Namen von dem Berge Kaukasus" (This race received its name from the mountain Caucasus), suggesting that "Race" and "Hauptvarietät" can be used interchangeably (1798, 213; a similar inconsistency can be found in 1798, 214–215). In the explanatory notes added to his translation, "Erläuternde Anmerkungen zu vorstehendem Werke nebst Zusätzen aus den früheren Ausgaben desselben" (Explanatory notes to the preceding work including additions from previous editions of the same), Gruber discusses the terminological difficulties of translating Blumenbach's work, and proposes to follow the terminology of Blumenbach's contemporary Immanuel Kant:

> Uebrigens weiß ich nicht, warum sich die neuern Naturgeschichtschreiber des Menschen nicht der von unserm großen Kant gesetzten Bestimmungen bedienen. Ich zweifle, ob man eine bestimmtere finden würde. [. . .]
>
> Den Abartungen subordinirt er:
>
> 1) Racen
>
> d.h. diejenigen Abartungen, welche sich sowohl bey allen Verpflanzungen in langen Zeugungen unter sich beständig erhalten, als auch in der Vermischung mit anderen Abartungen desselbigen Stammes jederzeit halbschlächtige Junge zeugen.
>
> [By the way I don't know why the newer natural historians do not use the terminology developed by our great Kant. I doubt that one will find a more precise terminology. [. . .]
>
> To the varieties of a species he subordinates
>
> 1) Races
>
> that is: those varieties that remain stable among themselves both when being replanted and reproduced over long periods, and also when mixed with other varieties from the same root consistently produce blended offspring.]
>
> (1798, 259–260)

Gruber indicates little concern about the significant epistemological differences between Kant and Blumenbach, especially when it comes to the notion of "race"; Kant defended the existence of germs and predispositions that determined skin color, while Blumenbach, with the single exception of his published dissertation of 1776, denied the role of germs in generation to instead develop his theories of a "formative drive" and climate as shaping human variety (Zammito 2006, 47–49; Bernasconi 2006, 73–74, 76–78). In essence what Gruber proposes is an attempted reconciliation of the theories of Buffon, Kant, and Blumenbach[6]: the term "race" is used when variations over several generations have become stable and reproduced themselves. In his actual translation, however, Gruber is inconsistent and appears to assume that "Race" and "Hauptvarietät" mean more or less the same thing. Blumenbach's later adaptation of the term "race" may be an indication that he acknowledges that Kantian terminology has won out over the conceptual framework promoted by Buffon and himself, or that the differences between their theories were not that great after all. As Robert Bernasconi has pointed out, the fifth edition of the *Handbuch der Naturgeschichte* from 1797 is the first version to use the term "race" (now spelled in a Germanized form as "Rasse") frequently, instead of incidentally, to refer to humans (Bernasconi 2006, 75, 85). Interestingly, after the 1795/1798 Latin-German third edition of his anthropological treatise Blumenbach published no further editions of the text, while many more editions of his *Handbuch der Naturgeschichte* were published.

That Blumenbach resisted the introduction of the concept "race" in his writings for a long time without doubt had something to do with the term's potentially normative dimensions. By 1797, however, the term had become common enough and Blumenbach clearly felt it could no longer be avoided. The inevitability of "race" made something clear about the political turn anthropological discourse had taken since the 1770s. The element of temporality introduced into natural history and anthropology by Buffon and Blumenbach was originally meant to be emancipatory, but had problematic aspects as well. The anthropologist Johannes Fabian has criticized Western ethnographic practice as creating a temporal gap between the West and the cultures it is studying, with the "other" representing a more primitive stage in comparison to the West's advanced cultural development. In Fabian's view, it is this temporal distancing that keeps the Western ethnographer from seeing the native other as a contemporary and communicating with him or her as such (Fabian 1983, 30–32). To conceive of societies as being part of a developmental continuum means on the one hand in the Enlightenment anthropologist's mind that other societies have the potential to achieve, some day, the same level of civilization as one's own (Western) society. On the other hand, this is not going to happen necessarily; it may very well be that other societies do not achieve the desired level of civilization. I do believe that anthropologists like Buffon and Blumenbach meant "degeneration" and its German equivalents "Ausartung" and "Verartung" primarily as descriptive categories, and not as prescriptive or normative concepts. But the normative potential of these concepts was recognized by at least some of their contemporaries.

Herder and Lichtenberg

How does the story of early anthropology continue after Buffon and Blumenbach? Few scholars have noted that Buffon and Blumenbach are important for Johann

Gottfried Herder's anthropology (which, because of this, is firmly grounded in Enlightenment thinking). In a manner very similar to Blumenbach, Herder in his *Ideen zur Philosophie der Geschichte der Menschheit* (1784–1791) starts out with the assertion that humans in spite of all their differences everywhere on earth belong to one and the same species.[7] Humans are, however, continually subjected to change. Herder speaks out against the assumption that humans and (some) animals are related (see earlier), and also disapproves of the term "race": he sees no reason to speak of "races" for dissimilar groups of humans because this suggests different roots ("eine Verschiedenheit der Abstammung") (1989, VI, 255). Humans have, so the central thesis of the second section of Book 7 of Volume 1 of the *Ideen*, adapted to climate everywhere (1989, VI, 256). Climate for Herder incorporates a variety of things. It includes heat and coldness, and many other forces at work in the air: not only electricity ("der elektrische Feuerstrom") but also the vapors of all bodies ("Ausdünstungen aller Körper") in a specific area have an impact on human life (1989, VI, 265–266). Climate further includes geography and through this affects almost all areas of human activity:

> Endlich die Höhe oder Tiefe eines Erdstrichs, die Beschaffenheit desselben und seiner Produkte, die Speisen und Getränke, die der Mensch genießt, die Lebensweise, der er folgt, die Arbeit, die er verrichtet, Kleidung, gewohnte Stellungen sogar, Vergnügen und Künste, nebst einem Heer andrer Umstände, die in ihrer lebendigen Verbindung viel wirken; alle sie gehören zum Gemälde des vielverändernden Klima.
>
> [Finally the altitude of an area, its condition and products, food and beverages which a human being enjoys, the way of living he pursues, the work he does, the clothes, normal positions even, entertainment and arts, in addition to a mass of other circumstances, that in their lively combination have a major effect; they all belong to image of the always changing climate.]
>
> (1989, VI, 266)

"Climate" is understood here by Herder in the most comprehensive way possible. He does not hesitate to bring up phenomena that the sciences of his time do not yet understand adequately (electricity; contagious diseases) in support of his theory of climate. Furthermore, culture – conceptualized by Herder as "Lebensweise," reminiscent of Buffon's "manière de vivre" and Blumenbach's "Lebensart" – including human work, clothes, habits, entertainment, and the arts, is linked to the effects of climate and thus still very much a function of geography and biology.[8]

In addition, "climate" has a political function for Herder. Herder is the first one to articulate positions that had remained mostly implicit in the work of earlier anthropologists like Buffon, Camper, and Blumenbach, who tended to avoid normative judgments in their anthropological texts. Herder translates climate theory into what could be called an ethics of territoriality: not only are natives who have been taken away from the lands where they were born naturally inclined, as if by necessity, to go back to their home country even if conditions there are very primitive, as is the case with the Greenlander (1989, VI, 259–260), but also Herder disapproves of Europeans taking possession of lands belonging to natives: "warum kamen sie [die Europäer] zu ihrem Lande? warum führten sie sich in demselben als fodernde, gewalttätige, übermächtige Despoten auf? Jahrtausende waren sich die Einwohner desselben das Universum: von ihren Vätern hatten sie es geerbt" (1989, VI, 262) ("why did the Europeans come to their country? why did they act in these lands like demanding, violent, and overly powerful despots? For thousands of years the inhabitants were their own universe; from their fathers they had inherited it"]. In fact, Herder attributes cannibalism to a desire for revenge against outsiders taking possession of what rightfully belongs to the natives.

However, Herder's tolerant views do not extend to every ethnic group. Herder sees Jews and Gypsies ("Zigeuner") rather negatively in the context of his discussion of what he calls "Fremde Völker in Europa" (foreign peoples in Europe) (1989, VI, 699–703) in the *Ideen.* Herder calls the European Jews a "parasitische Pflanze, die sich beinah allen Europäischen Nationen angehängt und mehr oder minder von ihrem Saft an sich gezogen" (1989, VI, 702) (a parasitical plant that attached itself to almost all European nations and drew in more or less of their sap). Similarly, Gypsies are characterized as "[e]ine verworfne Indische Kaste, die von allem, was sich göttlich, anständig und bürgerlich nennet, ihrer Geburt nach entfernt ist, und dieser erniedrigenden Bestimmung noch nach Jahrhunderten treu bleibt" (1989, VI, 703) (a rejected Indian cast, by birth distanced from everything that considers itself divine, decent, and civil, and after centuries still faithful to this humiliating destination). This brings a presupposition to light that is not only inherent to Herder's thinking but also characteristic of Enlightenment anthropology more generally. When thinking through the position of populations and their societies and cultures, it is clear that Herder tends to think of them as sedentary in the sense that they have developed organically as a product of their specific climate and geography, although he also cautions, in a paradoxical gesture that is typical for Herder's thinking as a whole, against privileging one type of society over others, as one critic has pointed out (Muthu 2003, 238–239, 243–244). Mobile and nomadic populations like the Jews and the Gypsies, however, resist the pattern of organic development that Herder has in mind for populations and their cultures. They are at odds with a principle of alterity based on territoriality.

Herder summarizes and to some extent also politicizes eighteenth-century anthropological discourse following Raynal and de Pauw (see Niekerk 2016, 101–103). Herder spells out the normative aspects of the new branch of knowledge by arguing for a notion of tolerance that respects biological and cultural difference as the product of and (rational) response to varying geographical and climatological conditions under which populations live. It is a notion of alterity that in itself, however, is also exclusionary: because its understanding of alterity is tied to territoriality and presumes a sedentary culture, Enlightenment anthropology has trouble accommodating nomadic or mobile societies. To phrase it another way: eighteenth-century anthropological thought is in possession of a notion of alterity, but not one that allows for mobility and interaction among different cultures. Any overview of Herder's anthropology and theory of culture that does not acknowledge his problematic ideas about Jews and Gypsies is incomplete.

Among Enlightenment intellectuals, Georg Christoph Lichtenberg belongs to the most outspoken not only in identifying the epistemological principles underlying his contemporaries' texts on natural history and anthropology, but also in problematizing what I earlier have called the normative potential present in these writings. Traces of Buffon's theory of climate can be found in Lichtenberg's essay "Über Physiognomik; wider die Physiognomen. Zu Beförderung der Menschenliebe und Menschenkenntnis" ("On Physiognomy; Against the Physiognomists. On the Advancement of Philanthropy and the Knowledge of Human Nature") from 1777.[9] Against Lavater's static view of the human body in his physiognomic fragments, Lichtenberg argues for a temporal and thereby flexible view of humanity that is materially conditioned. The human body connects the soul ("Seele") and the rest of the world ("der übrigen Welt") and mirrors the effects of both: one's body tells us not only of "unsere Neigungen und Fähigkeiten, sondern auch die Peitschenschläge des Schicksals, Klima, Krankheit, Nahrung und tausend Ungemach, dem uns nicht immer unser eigner bōser Entschluß, sondern oft Zufall und oft Pflicht aussetzen" (Lichtenberg 1968–1992, III, 266) (dispositions and abilities, but also of the whip lashes of fate, climate, disease, nutrition and a thousand other kinds of adversity, to which we are exposed not only through our own decision, but also through coincidence and duty). Lichtenberg identifies in Lavater's enterprise the dangers of deterministic thinking, stereotyping, and essentializing – highly problematic, especially since Lavater claims scientific credentials for his speculations and his writings thus may have the effect of discrediting science itself. To some extent, Lichtenberg too believes along the lines of Buffon's materialistic theories that man's fate is determined. But while Lichtenberg is a skeptic and maybe even a pessimist, he is convinced that there is space for human liberty. If Lavater's theses were really true, one should hang the children before they commit the crimes for which they would deserve to be hanged (1968–1992, I, 532 F521).

What Lichtenberg offers, in particular in his so-called *Sudelbücher* (*The Waste Books*), in which most of his aphoristic fragments are to be found, often amounts to a kind of counter-reading of contemporaneous natural history and anthropology (not unlike the way one can read Diderot's deliberations as a counter-reading of his contemporaries Buffon and Bougainville). "Stedman Narrative T. II. p. 190 glaubt nicht mit Unrecht (?) daß die Olivenfarbe die eigentliche Farbe des Menschen wäre, und daß Schwärze und Weiße nur die Produkte von Hitze und Kälte wären" (1968–1992, I, 913 L428; Stedman Narrative T. II. p. 190 believes not unjustifiably [?] that the olive color is humans' original color, and that black and white are the products of heat and coldness alone). Lichtenberg, in reference to the Dutch-Scottish soldier and travel writer John Gabriel Stedman, shows that subscribing to a temporal paradigm for understanding the history of humanity does not necessarily have to mean that one also supports the Europe-centered normative framework established by natural historians and anthropologists at the time. In this case Lichtenberg is willing to consider that "olive" instead of "white" is the original color of human skin. To use Buffon and Blumenbach's terminology, "white" and "black" are degenerative forms of the original "olive" color, and it is a mistake to believe that "olive" and "black" derive from an original "white" color, as Blumenbach had claimed (1795, 304, 1798, 214). Stedman, however, in the text to which Lichtenberg refers, goes a step further:

> In her features was displayed that beautiful simplicity, that native unsuspecting innocence, which cannot be put on where there is the slightest consciousness of guilt. Nor is the olive-colour incompatible with beauty, it is certainly the standard complexion of the human race, while the black and white are supposed to be only gradations, produced probably by the extremes of heat and cold.
>
> (1796, II, 190)

In Stedman's passage, the argument for the beauty of an "olive" skin color is combined with an argument for the native other's moral superiority: the girl's beauty is a sign of her sense of innocence, her complete lack of an awareness of guilt. Both Lichtenberg and Stedman point to the normative potential present in aesthetic strategies of organizing natural history. But both also entertain the possibility of subverting the hierarchies inherent in the aesthetics of natural history of their time. Lichtenberg's and Stedman's deliberations are not incidental. They illustrate the radical nature of eighteenth-century anthropology, and such a questioning of normative claims can be found elsewhere as well. The Dutch anatomist and anthropologist Camper, for instance, considers the possibility that Adam was not white, but black, brown, or black and brown, to claim subsequently that the issue matters very little, since the skin color of his descendants will adapt to geography, nutrition, and diseases (Camper 1784, 48). All humans, according to Herder's interpretation of this passage in Camper, have the potential to be black (1989, VI, 233). Here we see Enlightenment anthropology at its most daring and radical.

As much, though, as we may cherish the image of Lichtenberg as a recalcitrant, witty, undogmatic, and highly critical thinker relentlessly exposing the hidden normative dimensions in his contemporaries' thinking, there is another side to Lichtenberg's fragmentary anthropology that is far less likeable, as was the case with Herder as well. The language of natural history returns, for instance, in some of Lichtenberg's thoughts in the *Sudelbücher* on the expulsion of some Jewish families from Göttingen:

> Warum wollen wir unsern Boden anders bearbeiten um eine sehr unnütze Frucht zu nähren, die unter unserm Klima nicht gedeiht, und sich auch nicht nach ihm bequemen will? [. . .] Ich sehe nicht warum wir mit vielem Aufwand eine Pflanze bauen sollen, die sich nicht für unser Klima schickt und die uns wahrlich nichts einträgt, bloß aus dem empfindsamen Prinzip, daß das Pflänzchen nicht verloren gehe.
>
> [Why should we work our grounds differently to feed a rather unusable fruit that does not want to thrive under our climate, and also does not want to adapt to it? (. . .) I do not see why we should cultivate with much energy

> a plant that is not suited for our climate and does not bring us anything, only out of the sentimental principle that the little plant is not lost.]
>
> (1968–1992, I, 934 L593)

Lichtenberg's use of the plant metaphor to refer to Jews is reminiscent of Herder's comparison of Jews with a parasitical plant (1989, VI, 702), mentioned earlier. It is part of the paradox underlying Lichtenberg's thinking that someone who like Blumenbach and many of their eighteenth-century contemporaries was an avid reader of travel reports, in particular about the non-European world, and exhibited a genuine curiosity about other societies, cultures, and the ways in which they were very different from the Western European culture surrounding him could simultaneously strongly reject ethnic and cultural difference when faced with it in his immediate environment. Lichtenberg's thoughts mirror Herder's reservations toward mobile and nomadic populations, supposedly attaching themselves and profiteering from native populations occupying the territory that belongs to them and purportedly is their natural home.

Here too, we find that the Enlightenment, on the one hand, has a concept of alterity – one that connects otherness to territoriality and respects the insight that cultures in different parts of the world under the influence of differing conditions of climate and geography develop their own (rational) patterns of behavior in response to these conditions. On the other hand Enlightenment anthropology does not know how to accommodate mobility and the inter- and transcultural dynamics resulting from it.

Conclusions

Göttingen is central to the early history of the discipline that nowadays is called "anthropology." Anthropology is an Enlightenment invention, and at the time was part of a broader European discourse. Enlightenment anthropology resists a division into a physical and cultural branch because environment, biology, and culture are still understood to be operating as parts of one structure, and in fact one of the principal aims of eighteenth-century anthropology is to show that "culture" – a term that indeed starts to be used parallel to the coming into vogue of the term "anthropology" (see Carhart 2008) – is a rational response to environment, climate, and conditions of living more generally. Anthropology saw itself as an empirical science that aimed to base its theories about human development and diversity on empirically verified information, even if at times it was still hard to decide what information could be proven to be true, empirically. Enlightenment anthropology is part of a radical discourse to the extent that it allows for basic assumptions about humanity to be questioned, among them the biological relationship between man and animal, the biblical account of human development or the centrality of Europe and its inhabitants to natural history. To some extent the questions these early anthropologists ask may be more interesting than the answers they try to formulate. Men like Blumenbach and Lichtenberg had no qualms about questioning the biblical account of the creation when fossils appeared to point to the previous existence of species that had become extinct.[10] Not everyone, however, agreed. In Göttingen, Blumenbach's and Lichtenberg's radicalism was countered by Johann Georg Heinrich Feder, Christoph Meiners, and Johann David Michaelis and their more moderate versions of Enlightenment.

The legacy of Enlightenment anthropology, as I hope to have shown, is highly ambiguous. The men I have discussed here (Buffon, Camper, Blumenbach, Herder, and Lichtenberg) did understand their efforts as cosmopolitan, meaning that they wanted to contribute to a better factual knowledge and understanding of the world. They in general were also motivated by an emancipatory agenda: they wanted to help improve the lot of those in the non-European world who suffered from abuse. However, the temporal and spatial paradigms they used in their efforts to understand the non-Western world are without doubt one source for later theories of alterity that had no emancipatory ambitions (one can think of a concept like "degeneration" in this context, but also the principle of tying the understanding of alterity to territoriality). Eighteenth-century thinkers, among them very unconventional ones, like Herder and Lichtenberg (and Diderot in France), developed an awareness of the normative potential hidden in concepts like the ones discussed here and cautioned their contemporaries against the abuse that could be made of them. At the same time, these thinkers to some extent also fell victim to the normative dimension inherent to Enlightenment anthropology – something that is particularly clear in Herder's and Lichtenberg's statements about their Jewish contemporaries. The history of Enlightenment anthropology is incomplete if we do not take into consideration its views of cultures it sees as non-sedentary, like the Jews and Gypsies.[11] This is, incidentally, why we need to study the history of anthropology and history in general. Only by studying the past in all its complexity can we understand our world today.

Notes

1 For an intelligent example of this approach, see Košenina (2008).

2 For an example of a body/mind approach to anthropology in contemporary scholarship, see Schings (1994).

3 Bibliographical information taken from www.univie.ac.at/Geschichte/China-Bibliographie/blog/2010/12/25/raynal-histoire-philosophique/.

4 See Blumenbach's summary of Meiners's theories in *Über die natürlichen Verschiedenheiten im Menschengeschlechte* (1798, 212). Blumenbach refers to the author's introduction of the second edition of his *Grundriß der Geschichte der Menschheit* (1793, 5–8). Here Meiners states that instead of a model emphasizing descent, as in the first edition of his book, he will make aesthetic criteria into the overarching organizational principle of his anthropology.

5 This is a defining feature of Blumenbach's anthropology from its very beginning; already the 1776 edition of *De generis humani varietate* (still based on Linné's model assuming four human varieties) argues, "Humani generis species saltem unica" and "Varietates autem quator constituuntur" (1776, 97).

6 Such a reconciliation had been proposed by Christoph G. Girtanner, who in 1796 had published *Über das Kantische Prinzip für die Naturgeschichte*, dedicated to Blumenbach, in which he seeks to demonstrate that Kant's and Blumenbach's theories of generation and diversity are compatible (see Bernasconi 2006, 74–75). In his

"Verbesserungen und Zusätze" to the 1797 edition of the *Handbuch der Naturgeschichte*, Blumenbach adds a note regarding Kant's theory of race: "s. hiervon ausführlich Hrn. Geh. Hofr. Girtanner über das Kantische Princip für die Naturgeschichte. Göttingen, 1796. 8" (Blumenbach 1797, xv); from the sixth edition on, this addition is part of the footnote referring to Kant (1799, 24). Bernasconi noticed it in the 1807 edition, but overlooked it in the 1797 edition (which is easily explained by the fact that it is an addendum) (see Bernasconi 2006, 75).

7 See the first subheading of Volume 1, Book 7: "In so verschiedenen Formen das Menschengeschlecht auf der Erde erscheint: so ist's doch überall ein und dieselbe Menschengattung" (Herder 1989, VI, 251).

8 This very much speaks against Michael Carhart's attempt to identify an independent science of culture (as distinguished from anthropology and natural history) in eighteenth-century German thinking (2008). These deliberations also go against Isaiah Berlin's attempts to read a belief not only in the "multiplicity" but also in the "incommensurability of the values of different cultures and societies, and in addition, in the incompatibility of equally valid ideals" in Herder's work that later would be of great importance to the Romantics (see Berlin 2000, 176). It is certainly possible to find a belief in "multiplicity" in Herder's work, but it is firmly linked with theories of climate, geography, and biology, and as such, in contrast to Berlin's identification of Herder as a precursor of Romanticism, firmly rooted in Enlightenment anthropology (see Niekerk 2009, especially147–148).

9 A detailed discussion of Lavater's epistemological assumptions and Lichtenberg's response to them in his essays on physiognomy can be found in Niekerk (2005, 143–164).

10 See Blumenbach, "Ein Blick in die Vorwelt": "Fast jeder Pflasterstein in *Göttingen* zeugt davon, daß Gattungen – ja sogar ganze Geschlechter von Thieren untergegangen seyn müssen. Unser Kalkboden wimmelt gleichsam von den mannigfaltigsten Arten versteinter Seegeschöpfe, unter welchen meines wissens [*sic*] nur eine einzige Gattung ist, wozu wir gegenwärtig ein *wahres* ganz damit übereinkommendes Original kennen" (1790, 6). Regarding this issue, see also Lichtenberg's 1795 essay "Geologische Phantasien (Franklins Geogenie)" (Lichtenberg 1968–1992, III, 112–124).

11 In particular the Romantics picked up on this aspect of Enlightenment anthropology (see Niekerk 2009, 153–156).

Bibliography

Barth, Fredrik, Andre Gingrich, Robert Parkin, and Sydel Silverman. 2005. *One Discipline, Four Ways: British, German, French, and American Anthropology*. Chicago, IL: University of Chicago Press.

Berlin, Isaiah. 2000. *Three Critics of the Enlightenment: Vico, Hamann, Herder*, edited by Henry Hardy. Princeton: Princeton University Press.

Bernasconi, Robert. 2006. "Kant and Blumenbach's Polyps: A Neglected Chapter in the History of the Concept of Race." In *The German Invention of Race*, edited by Sara Eigen and Mark Larrimore, 73–90. Albany: SUNY University Press.

Blom, Philipp. 2010. *A Wicked Company: The Forgotten Radicalism of the European Enlightenment*. New York: Basic Books.

Blumenbach, Johann Friedrich. 1775. *De generis humani varietate nativa* [dissertation]. Göttingen: Typis Frid. Andr. Rosenbuschii.

———. 1776. *De generis humani varietate nativa liber* [first edition]. Göttingen: Vandenhoeck.

———. 1779. *Handbuch der Naturgeschichte*. Vol. 1 [first edition]. Göttingen: Johann Christian Dieterich.

———. 1780a. *Handbuch der Naturgeschichte*. Vol. 2 [first edition]. Göttingen: Johann Christian Dieterich.

———. 1780b. "Ueber den Bildungstrieb (Nisus formativus) und seinen Einfluß auf die Generation und Reproduction." *Göttingisches Magazin der Wissenschaften und Litteratur* 1 (5): 247–266.

———. 1781a. *De generis humani varietate nativa liber*, 2nd edn. Göttingen: Vandenhoek.

———. 1781b. *Über den Bildungstrieb und das Zeugungsgeschäfte*. Göttingen: Johann Christian Dieterich.

———. 1782. *Handbuch der Naturgeschichte*, 2nd edn. Göttingen: Johann Christian Dieterich.

———. 1788. *Handbuch der Naturgeschichte*, 3rd edn. Göttingen: Johann Christian Dieterich.

———. 1789a. *Über den Bildungstrieb*, 2nd edn. Göttingen: Johann Christian Dieterich.

———. 1789b. "Ueber Menschen=Racen und Schweine=Racen, vom Herrn Hofrath Blumenbach." *Magazin für das Neueste aus der Physik und Naturgeschichte* 6 (1): 1–13.

———. 1790. *Beyträge zur Naturgeschichte*. Vol. 1. Göttingen: Johann Christian Dieterich.

———. 1791. *Über den Bildungstrieb*, 3rd edn. Göttingen: Johann Christian Dieterich.

———. 1795. *De generis humani varietate nativa*, 3rd edn. Goettingen: Vandenhoek und Ruprecht.

———. 1797. *Handbuch der Naturgeschichte*, 5th edn. Göttingen: Johann Christian Dieterich.

———. 1798. *Über die natürlichen Verschiedenheiten im Menschengeschlechte: Nach der dritten Ausgabe und den Erinnerungen des Verfassers übersetzt, und mit einigen Zusätzen und erläuternden Anmerkungen herausgegeben von Johann Gottfried Gruber*. Leipzig: Breitkopf und Härtel.

———. 1799. *Handbuch der Naturgeschichte*, 6th edn. Göttingen: Johann Christian Dieterich.

———. 1806. *Beyträge zur Naturgeschichte*. Vol. 1, 2nd edn. Göttingen: Heinrich Dieterich.

Buffon, Georges-Louis Leclerc de, and Louis Jean-Marie Daubenton. 1749–1788. *Histoire naturelle, générale et particulière, avec la description du Cabinet du Roy*. 36 vols. Paris: Imprimerie royale.

———. 1749a. "Premier discours: De la manière d'étudier & de traiter l'Histoire Naturelle." In *Histoire naturelle générale et particulière*. Vol. 1, 1–62. Paris: Imprimerie royale.

———. 1749b. *Histoire naturelle générale et particulière*. Vol. 2. Paris: Imprimerie royale.

———. 1749c. *Histoire naturelle générale et particulière*. Vol. 3. Paris: Imprimerie royale.

———. 1766. *Histoire naturelle générale et particulière*. Vol. 14. Paris: Imprimerie royale.

———. 1771–1774. *Herrn von Buffons allgemeine Naturgeschichte. Eine freye mit einigen Zusätzen vermehrte Übersetzung nach der neuesten französ: Außgabe von 1769*. 7 vols. Berlin: Joachim Pauli Buchhändler.

———. 1777. *Histoire naturelle générale et particulière*. Vol. 33, Supplément. Vol. 4. Paris: Imprimerie royale.

Bunzl, Matti. 1996. "Franz Boas and the Humboldtian Tradition: From 'Volksgeist' and 'Nationalcharakter' to an Anthropological Concept of Culture." In *'Volksgeist' as Method and Ethic: Essays on Boasian Ethnography and the German Anthropological Tradition*, edited by George W. Stocking, 17–78. Madison, WI: University of Wisconsin Press.

Camper, Petrus. 1784. "Rede ueber den Ursprung und die Farbe der Schwarzen." In *Herrn Peter Campers saemmtliche kleinere Schriften die Arzney= Wundarzneykunst und Naturgeschichte betreffend*. Trans. J.F.M. Herbell. Vol. 1, 24–49. Leipzig: Crusius.

Carhart, Michael. 2008. *The Science of Culture in Enlightenment Germany*. Cambridge, MA: Harvard University Press.

Cassirer, Ernst. 1998. *Die Philosophie der Aufklärung*. Hamburg: Meiner.

Clifford, James, and George E. Marcus, eds. 1986. *Writing Culture: The Poetics and Politics of Ethnography*. Berkeley: University of California Press.

Dougherty, Frank William Peter. 1996. "Buffons Bedeutung für die Entwicklung des anthropologischen Denkens im Deutschland er zweiten Hälfte des 18. Jahrhunderts." In *Gesammelte Aufsätze zu Themen der klassischen Periode der Naturgeschichte/Collected Essays on Themes from the Classical Period of Natural History*, edited by Norbert Klatt, 70–88. Göttingen: Norbert Klatt.

Dreyfus, Hubert L., and Paul Rabinow. 1983. *Michel Foucault: Beyond Structuralism and Hermeneutics*, 2nd edn. with an afterword and interview with Michel Foucault. Chicago, IL: University of Chicago Press.

Duchet, Michèle. 1995. *Anthropologie et histoire au siècle des Lumières*. Paris: Albin Michel.

Fabian, Johannes. 1983. *Time and the Other: How Anthropology Makes Its Object*. New York: Columbia University Press.

Foucault, Michel. 1989. *The Order of Things: An Archaeology of the Human Sciences*. London: Routledge.

Hagner, Michael. 1999. "Enlightened Monsters." In *The Sciences in Enlightened Europe*, edited by William Clark, Jan Golinski, and Simon Schaffer, 175–217. Chicago, IL: University of Chicago Press.

Herder, Johann Gottfried. 1989. *Werke*, edited by Jürgen Brummack and Martin Bollacher. Vol. 6: *Ideen zur Philosophie der Geschichte der Menschheit*. Frankfurt am Main: Deutscher Klassiker Verlag.

———. 1994. *Werke*, edited by Jürgen Brummack and Martin Bollacher. Vol. 4: *Auch eine Philosophie der Geschichte zur Bildung der Menschheit*. Frankfurt am Main: Deutscher Klassiker Verlag.

Israel, Jonathan I. 2011. *Democratic Enlightenment: Philosophy, Revolution, and Human Rights 1750–1790*. Oxford: Oxford University Press.

Kant, Immanuel. [1798] 1995. *Werkausgabe*, edited by Wilhelm Weischedel. Vol. 12: *Anthropologie in paragmatischer Hinischt*. Frankfurt am Main: Suhrkamp.

Košenina, Alexander. 2008. *Literarische Anthropologie: Die Neuentdeckung des Menschen*. Berlin: Akademie Verlag.

Krauss, Werner. 1987. *Zur Anthropologie des 18. Jahrhunderts: Die Frühgeschichte der Menschheit im Blickpunkt der Aufklärung*. Berlin: Ullstein.

Lavater, Johann Caspar. 1775–1778. *Physiognomische Fragmente, zur Befoerderung der Menschenkenntniß und Menschenliebe*. 4 vols. Leipzig, Winterthur: Weidmanns Erben und Reich, und Heinrich Steiner und Compagnie.

Lepenies, Wolf. 1976. *Das Ende der Naturgeschichte: Wandel kultureller Selbstverständlichkeiten in den Wissenschaften des 18. und 19. Jahrhunderts*. Munich: Hanser.

———. 1980. "Naturgeschichte und Anthropologie im 18. Jahrhundert." In *Deutschlands kulturelle Entfaltung: Die Neubestimmung des Menschen*, edited by Bernhard Fabian, Wilhelm Schmidt-Biggemann, and Rudolf Vierhaus, 211–226. Munich: Kraus.

———. 1988. *NeubestimmungAutoren und Wissenschaftler im 18. Jahrhundert: Linné – Buffon – Winckelmann – Georg Forster – Erasmus Darwin*. Munich/Vienna: Hanser.

Lichtenberg, Georg Christoph. [1795] 1968–1992. "Geologische Phantasien (Franklins Geogenie)." In *Schriften und Briefe*. Vol. 3, 112–124.

———. 1968–1992. *Schriften und Briefe*, edited by Wolfgang Promies. 4 + 2 vols. Munich: Carl Hanser.

———. 1983–2004. *Briefwechsel*, edited by Ulrich Joost and Albrecht Schöne. 5 vols. Munich: Beck.

Marcus, George E., and Michael M. J. Fischer. 1986. *Anthropology as Cultural Critique: An Experimental Moment in the Human Sciences*. Chicago, IL: University of Chicago Press.

Meiners, C[hristoph]. 1785. *Grundriß der Geschichte der Menschheit*. Lemgo: Meyersche Buchhandlung.

———. 1793. *Grundriß der Geschichte der Menschheit*, 2nd edn. Lemgo: Meyersche Buchhandlung.

Montesquieu, [Charles Louis de Secondat de la Brède et de Montesquieu]. 1979. *De l'esprit des lois*. Vol. 1. Paris: Garnier-Flammarion.

Muthu, Sankar. 2003. *Enlightenment Against Empire*. Princeton: Princeton University Press.

Niekerk, Carl. 2004. "Man and Orangutan in Eighteenth-Century Thinking: Retracing the Early History of Dutch and German Anthropology." *Monatshefte* 96 (4): 477–502.

———. 2005. *Zwischen Naturgeschichte und Anthropologie. Lichtenberg im Kontext der Spätaufklärung*. Tübingen: Niemeyer.

———. 2009. "The Romantics and Other Cultures." In *The Cambridge Companion to German Romanticism*, edited by Nicholas Saul, 147–162. Cambridge: Cambridge University Press.

———. 2016. "The Problem of China: Asia and Enlightenment Anthropology (Buffon, de Pauw, Blumenbach, Herder)." In *China in the German Enlightenment*, edited by Bettina Brandt and Daniel L. Purdy, 97–117. Toronto: University of Toronto Press.

P[auw], [Cornelis] de. 1768–1770. *Recherches philosophiques sur les Américains, ou Mémoires intéressants pour servir à l'Histoire de l'Espèce humaine*. 3 vols. Berlin: George Jacques Decker.

———. 1769. *Philosophische Untersuchungen ueber die Amerikaner, oder wichtige Beitraege zur Geschichte des menschlichen Geschlechts: Erster Theil*. Berlin: Decker und Winter.

Platner, Ernst. 1772. *Anthropologie fuer Aerzte und Weltweise*. Erster Theil. Leipzig: Dyckische Buchhandlung.

Raynal, Abbé. 1770. *Histoire philosophique et politique des établissements et du commerce des Européens dans les deux Indes*. 6 vols. Amsterdam: [n.p.].

Schings, Hans-Jürgen, ed. 1994. *Der ganze Mensch: Anthropologie und Literatur im 18. Jahrhundert*. Stuttgart: Metzler.

Sloan, Phillip R. 1996. "Natural History, 1670–1802." In *Companion to the History of Modern Science*, edited by R.C. Olby, G.N. Cantor, J.R.R. Christie, and M.J.S. Hodge, 295–313. London: Routledge.

Stedman, J.G. 1796. *Narrative of a Five Years' Expedition Against the Revolted Negroes of Surinam in Guiana on the Wild Coast of South America from the Year 1772, to 1777*. Vol. 2. London: J. Johnson.

Todorov, Tzvetan. 1998. *Le jardin imparfait: La pensée humaniste en France*. Paris: Grasset.

Vermeulen, Han F. 2015. *Before Boas: The Genesis of Ethnography and Ethnology in the German Enlightenment*. Series: Critical Studies in the History of Anthropology. Lincoln, NE: University of Nebraska Press.

Zammito, John H. 2002. *Kant, Herder, and the Birth of Anthropology*. Chicago, IL: University of Chicago Press.

———. 2006. "Policing Polygeneticism in Germany, 1775: (Kames) Kant, and Blumenbach." In *The German Invention of Race*, edited by Sara Eigen and Mark Larrimore, 35–54. Albany: SUNY University Press.

Zantop, Susanne. 1997. *Colonial Fantasies: Conquest, Family, and Nation in Precolonial Germany, 1770–1870*. Durham: Duke University Press.

4 Climate change and creolization in French natural history, 1750–1795

E. C. Spary

In eighteenth-century France, the term "Creole" applied to anyone born under a different climate from their parents. Unlike today, it had no additional significance in terms of racial origin. The term "race" itself was frequently deployed loosely and even interchangeably with "species" in the period. Its original meaning, denoting a breeding stock, remained dominant into the nineteenth century. This essay will suggest the importance of taking that continuity seriously, in order to avoid a terminological confusion which has caused key transformations in early modern uses of the term to be overlooked.

Europeans described human variation in several different ways in the eighteenth century, drawing upon contemporary discussions of beauty, breeding, climate, and humoral theory. The undifferentiated application of the term "race" as currently understood to all early modern writing on other peoples demands replacement by a more historicist account of the plurality of these knowledge traditions invoked by philosophes when describing human variation. Preoccupation with the nobility of origins and its relationship with beauty led philosophes to rank and judge human peoples around the earth in ways that the modern reader finds uncomfortable, yet that very set of priorities, framed in the universalizing language of Enlightenment and civilization, could also prohibit commitment to essential natural differences separating Europeans from others. Furthermore, much of what is currently interpreted as "scientific racism" prior to 1800 was not a discussion of "races" in today's sense at all, but rather of Hippocratic arguments which explained variation between the peoples of the earth in climatic terms. Yet philosophes' medical debts have often remained invisible to historians – for example, discussions of "blood" could refer to one of the four medical humors, and most discussions of "race" were concerned with breeding and descent. Traditions of debate about medicine, aesthetics, and pedigree might align in writing about the peoples of the earth, but this was not necessarily the case; and even where they did, that alignment results in some very unexpected classifications of particular peoples. The trope of the "birth of scientific racism" in the late eighteenth century (Trouillot 2015, 78) is at best an oversimplification, and at worst obscures the role played by natural history in policies of discrimination against non-Europeans.

The writings of Georges-Louis Leclerc de Buffon, director of the royal natural history garden in Paris, continue to be seen as axiomatic to this Enlightenment redefinition of race (e.g., Dubois 2006, 4; Doron 2011, 76; Nelson 2010). Though Buffon and his fellow Burgundian, the physician and comparative anatomist Louis-Jean-Marie Daubenton, indeed defined the terms "race" and "species" more consistently than their contemporaries, the priority they accorded to climate as a cause of variation outweighed what stability they did accord to "race" as a natural category. Buffon's pronouncements on the *malleability* of races were, rather, what contemporaries took from his work. Several of his medical and scientific readers suggested that skin color changed over a relatively short time span. The debate over skin color or monogenism/polygenism needs to be framed within a broader contemporary discussion of degeneration – the departure of successive generations from the perfection of the original stock, which might be attributed to natural or social influences. Degeneration and the means to correct it became politically prominent issues in France from the 1770s onwards, as part of a broader call for national reform. Thus, "race" was discussed amid concerns about how social institutions and climate created the nation. Only by understanding the interplay between breeding and aesthetics on the one hand and Hippocratic models of medical geography on the other can we grasp the political nature of the contribution of philosophes like Buffon and his readers to this debate.

Creoles were interesting for many scientific and medical authors because they represented a state of flux between one climate and another. They were taken to be exemplary of the historical process that produced variation in the human species, one still ongoing in colonial situations, but also reversible. Both European and African bodies experienced radical physiological transformation on arrival in the colonies. The consequences of imbalance between the climate and the body could be profound: people changing climates suffered illnesses of varying severity, and frequently died. The medical management of creolization was thus a practical problem faced by all travelers, as medical handbooks reveal. The mutability of the notion of race implied by medical accounts of the creolization process stands in stark contrast to legal and customary classifications of colonial inhabitants on the basis of color. There was thus a tension between the fluidity of constitutions, taken for granted as a natural fact in metropolitan medicine and natural history, and skin color, a topic around which social, political, and juridical contestation clustered, particularly in the colonies. Political events rather than philosophical commitments, I argue, ultimately determined how and when the term "race" came to be applied in a more fixed sense, in contrast to the widespread medical view of human variation as climatogenic and relativistic.

A more historicist account of "race"

It is practically a trope that race assumed its modern role as a topos of scientific enquiry in the Enlightenment. Only recently has this statement been questioned, and even then often for apologetic or hagiographical purposes, apparently in an attempt to defend eighteenth-century naturalists against the charge of racism. Few scholars writing on the history of race have devoted much attention to the contemporary meanings and uses of the term (although see Hudson 1996; Colwill

1998; Doron 2011; Hocquet 2014; and, with reservations, Boulle 1988). Yet such attention reveals some unexpected features which undercut the presumption that "race" is a transhistorical theme from ancient times to the present day. In the sixth edition of the *Dictionnaire de l'Académie française*, published in 1835, the entry on race includes the familiar sense in which we use the term today:

> A multitude of men who originate from the same country, and resemble one another in facial features [and] exterior conformation. *The caucasian race. The mongol race. The malay race.*
>
> (*Dictionnaire* 1835)[1]

That is, the nineteenth-century definition is characterized by its *horizontal* focus upon variation across the earth's surface. By contrast, the fifth edition, published in 1798, offers instead what one might call a *vertical* definition of the term "race," common to many European languages from the Renaissance onwards, and which understood race in terms of historical origin:

> Line, all those who come from the same family. He is of a good race, of good race, of an illustrious or ancient race. He descends, he stems from a noble race, from a prosperous race. He is of the royal race [. . .]. Race is also used of domestic animals such as dogs, horses, etc. [. . .] One also says *This horse has race*, meaning, By his face and build he announces that he is from a good race.
>
> (*Dictionnaire* 1798)

Here there is no use of "race" to designate geographically separated human populations. In fact, the words "breed," "stock," or "descent," with their connotations of kinship, could be unproblematically substituted for every single use of the term "race," as in "the third race of kings" (Doron 2011). I venture to suggest that this is not true of many of today's uses of the term, and no longer true of the 1835 dictionary entry. We are thus talking about a set of meanings for the word "race" which have been lost, but which were standard for contemporary reading publics.

If the Enlightenment was such a key period in redefining the notion of human race in ways that allowed nineteenth-century racial theory to be supported by the natural sciences, why then is the new "horizontal" view invisible, even by the final years of the century? This was no artifact of the notorious slowness of dictionaries to reflect changes in usage: there is scant evidence that eighteenth-century French travelers, even those with close connections to metropolitan scientific institutions, systematically ordered the humans they encountered on their journeys into racial groups.[2] More typical was Voltaire's casual use of the term as an alternative to "species." For him, races were plural. In his history of Russia, for example, he remarked, "There are many more races of men than one thinks; those of the Samoyeds and the Hottentots seem to be the two extremes of our continent" (Voltaire 1756, 141; 1759–1763, I, 1). Decades later, the terms "race" and "species" were still being used in this looser sense, by scientifically instructed doctors like Charles Arthaud, traveling naturalists like Jean-Baptiste Leblond, or Republican priests like Henri Grégoire (McClellan 1992, 241; Leblond 1805, 5, 7; Grégoire 1826, 25–26). The fact that such key terms did not possess a stable agreed-upon meaning at any point throughout the period problematizes the secondary literature's preoccupation with the debate between monogenists and polygenists: for if a "species" was not a fixed thing, then statements about whether the human species was singular or plural surely fail to carry the explanatory, scientific, and moral significance historians often attribute to them. By the same token, the undifferentiated application of the term "race" to describe every sort of encounter between Europeans and others is problematized where "races" were not defined as fixed natural kinds. Only by the 1810s do racial classifications start to appear consistently in French voyage accounts, as travelers like François Péron began to record differences between Europeans and other peoples in quantitative terms, using a whole new battery of instruments invented for the purpose (Péron and Freycinet 1807–1816, I, passim). In short, my starting point is the suggestion that the main sense in which the term "race" is used today was not in common use until the early nineteenth century. Applying the term in this modern sense to all earlier debates without differentiation obscures the concerns of eighteenth-century authors, and implies that a scientifically supported racial theory was an inevitability, thus perpetuating a profoundly presentist attitude toward the investigation of the history of race itself. This in turn means that many of today's familiar associations between race, color, law, scientific truth, and prejudice cannot be assumed to have underpinned eighteenth-century French arguments.

Not all pejorative statements about other peoples were about race

The critique of other peoples in eighteenth-century French travel writings was primarily centered upon taste, beauty, and civilization. Old Regime naturalists existed in a world all too attentive to privilege, rank, and birth. Taste, deemed by most to be an acquired skill, allowed for social mobility and the display of social distinction. Outsiders to high society – like Buffon himself, ennobled in 1771 – were able to penetrate courtly circles if they could display good taste (Chaussinand-Nogaret 1985; Maza 1998; Shovlin 2006). Travel accounts mapped the global distribution of civility and its cognates, politeness and *honnêteté*, characteristics which at home differentiated the trustworthy, gentlemanly individual from the people (Pekacz 1999, 13–142; Bury 1996; Chartier 1987, 45–86; Muchembled 1998, 200–206; Revel 1989). This was the case regardless of whether travelers argued for or against the possibility of civilizing other peoples, and whether they celebrated European or non-European culture. If the term "race" did not undergo radical change in this period, the word "civilization" by contrast was a neologism, coined only in 1756 (Mazlish 2001, 293; Pagden 1988).

In this world, the importance of taste should not be underestimated – although often treated by historians as an epiphenomenal issue, it was a key mode of political action for the eighteenth-century French elite, insofar as it recapitulated the form taken by political power in the period (Burke 1992; also Bourdieu 1994). Travelers' reports of skin color cannot be necessarily read as a discourse about "race" at all. For example, French observers perceived the islanders of Tahiti as more like themselves than, say, the Poles or Germans were; the basis of comparison was not primarily skin color, genealogy, or geographical proximity but rather habits, manners, and approximation to antique corporeal ideals. Aesthetic judgments first applied to French ladies of fashion were extended to the Tahitians. Dark skin color detracted from beauty, but as many agreed, the Tahitian belles could be likened to suntanned French peasant women, and had a "European physiognomy." In fact, many on the Bougainville voyage held the Tahitians to be white and, much more importantly, to be beautiful. Conversely, the Pacharais of Tierra del Fuego, though paler, were deemed "short unfortunate degenerates of the human species, [. . .] meager, ugly, weak and vile" (quoted in Taillemite (1978), II, 224; also I, 318, 326; II, 395, 497). The cultivation and evaluation of beauty were an essential gentlemanly skill: a prime manner of demonstrating the possession of good taste was by displaying one's ability to judge what was beautiful in art, women, horses, or music. The twin weapons of taste and civility were deployed around the world in order to legitimate the standards upheld at home, and they were deemed by some to have universal applicability (Colwill 1998; Bindman 2002; Steintrage 2016, Chapter 6).

These were certainly deeply discriminatory discourses, which vilified certain peoples and exalted others according to Eurocentric standards, but it should be observed that they were not "racial" in either current or early modern senses of the term. Neither the question of kinship nor scientific arguments for dividing the global human population into subgroups figured in French discussions of Tahiti. If the Tahitians were positioned within a wider discussion about mankind, it was framed in terms of manners and social institutions, not pedigree. The politics of civilization and civility continued to predominate over concerns to classify peoples into a small number of overarching and naturally distinct races. Some decades later, the Republican abolitionist priest Henri Grégoire would describe discrimination against nonwhites as a *prejudice*; in using this term, he was specifically indicating that legislation on the basis of skin color had a political, not a natural, foundation, and that such differences would be abolished, not affirmed, by scientific progress (Grégoire 1826). His description of the colonial preference for whiteness as "skin nobility" pointedly evoked the social institutions swept away by the Revolution, but also the courtly origin of standards of beauty.

Unlike the majority of travel writers, Buffon explicitly set out to define races and their difference from true species. Natural species were individuals sharing similar characteristics which could produce fertile offspring only with one another, though they might produce infertile offspring with other, morphologically similar individuals. Natural races were groups sharing similar hereditary characteristics, but also capable of producing fertile offspring with individuals having different, also hereditary characteristics (Duchet 1971, Part II, Chapter 1; Boulukos 2008, Chapter 3; Harvey 2012, Chapter 5). Discussing the ass, Buffon underlined the similarity of body plan between quadrupeds and wondered whether this was grounds for claiming a common origin for all such beings, only to turn away from that heretical position to insist that there were intransgressible reproductive boundaries separating true species from one another – hence the infertility of mules (Lovejoy 1959; cf. Lyon and Sloan 1981). Thus, he differentiated between species and races because the former were the original, and therefore noble, products of Creation, while the latter were degenerations produced since Creation in consequence of the imperfect state of nature and society.

Buffon's attempt to define races in aesthetic terms makes sense only if he was using the term "race" as a synonym for "breed": a race was a group of similar individuals which bred *true to type*. The significance of this definition in terms of progeniture is obvious for a society where character, family descent, and property were closely aligned, and where beauty had long been represented as linked to nobility. Major discussions of racial formation occur, significantly, in Buffon's articles on the dog and horse, and they draw upon English and Spanish writing on the production of aesthetically pleasing domestic animals (*HN* 1749–1788, IV, 174–259, 280; V, 185–229; also Ritvo 1987; Doron 2011, 80–81). In man as in animals, Buffon's interest in naturally occurring races was reductionist: his goal was to identify the original stock types of a given species. For man, he suggested, "whiteness thus seems to be the primitive color of Nature, which climate, food and mores alter and change, even as far as yellowness, brownness or blackness" (*HN* 1749–1788, III, 502). Here, Buffon's account of the human varieties which most closely approached perfection was clearly grounded in a Eurocentric aesthetics, profoundly informed by concerns about beauty and civilization that he shared with contemporaries. The Lapp and Negro races were "two extremes [. . .] equally distant from the true and the beautiful" (quoted in Duchet 1971, 217; also Meijer 1999), but many or perhaps all races had subdivisions arising from their particular social or natural circumstances.

In Buffon's first major discussion of man, which appeared in the third volume of the *Histoire naturelle*, published in 1749, beauty accordingly plays a prominent role. This discussion seems at first glance profoundly muddled to a modern reader. Buffon uses the term "race" not only to refer to color ("the white race," "the black race") but also to cut across color boundaries entirely by asserting that "the black race" incorporated at least two distinct "races" defined on grounds other than color:

> It seems at first, on collating the testimony of travelers, that there are as many varieties in the black race as in the white; the blacks have, like the whites, their Tartars and their Circassians, those of Guinea are extremely ugly and have an unbearable smell, those of Soffala and Mozambique are beautiful and have no bad smell. It is thus necessary to divide the blacks into different races, and it seems to me that one can reduce them to two principal ones, the Negroes and the Kaffirs.
>
> (*HN* 1749–1788, III, 454)

A given "race" might in fact manifest a spectrum of colors: white, dusky, reddish, red or yellow copper, yellow, brown, and black.[3] As one might expect if a "race" described historical kinship, skin color was less important in differentiating between races than morphological characters and generalized family resemblance. This meant that race and geography did not map perfectly onto one another. Many African peoples, such as the Ethiopians and Moors, were in Buffon's view closer to the white race than to the black, and even black races must have started out white. That is, contrary to the claims of scholars like Doron (2011), Buffon did not produce a fixed, stable model of race at all, but rather a fluid and plural one.[4]

To the extent that Buffon's natural history was a political critique of his own society, it also constantly subverted these same notions of race rooted in breeding and birth, highlighting the power of external circumstance to mold physical and moral characteristics and so create or destroy "racial" differences. Buffon advanced strong arguments in support of the ancients' view that climate caused changes in human skin color – that is, the "races" he described were the historical product of living circumstances.

> It may take several centuries and the succession of a large number of generations for a white race to acquire the brown color by degrees and at last to become completely black, but it appears that with time a white people transported from the north to the equator could become brown and even completely black, especially if that same people were to change its customs and only use the productions of the hot country to which it had been transported for nourishment.
>
> (*HN* 1749–1788, III, 483)

Changes of climate, food, and government were the main causes of transformation in the human body. Color alone was insufficient to demarcate racial difference – Buffon appealed to plural characteristics, "the facial features, hair, skin, the smell of their bodies, their mores and their character" (*HN* 1749–1788, III, 453; Curran 2011, 107–111) as the grounds for separating peoples into distinct races. The local conditions of both nature and society acted reciprocally on individuals over time, giving rise to extremes of beauty and ugliness, savagery and civilization, within a single race.

Racial formation could be managed through science

Of all the influences upon bodies, Buffon claimed, "climate is the principal cause of variety in the human species" (*HN* 1749–1788, III, 519, 521). It was this foregrounding of climate, and the emphasis on external factors, both natural and social, as the cause of differentiation in the human species which would dominate Buffon's subsequent accounts of variation, racial formation, and human history. Boulle (1988, 224) views Buffon as having "systematized [the race concept] and given [it] the aura of unimpeachable scientific authority," but in fact it should be clear from the earlier examples that Buffon specifically never committed to race as a stable category. German and Dutch authors like Immanuel Kant, Johann Friedrich Blumenbach, and Cornelis de Pauw would diverge sharply from his views. Kant, in particular, vehemently rejected Buffon's claim that the species could be recognized by the capacity of its members to interbreed, on the basis that it allowed "for a plasticity of species that is in principle unlimited" (Shell 2006, 59; Zammito 2006, 41; Curran 2009, 176–177). But French authors, who accounted for the historical appearance of races primarily in medical geographical terms, continued to entertain a far more fluid and malleable definition.

A common contemporary view of nature was as the fallible executrix of an originally perfect divine plan. This view profoundly underpinned Buffon's thermoregulatory cosmology. In the more speculative writings he produced from the 1760s onwards, he welded his earlier, vulcanistic model of the origins of the earth to a meteorological view of how location determined physical and moral characteristics of human and animal bodies (Williams 1994). The revival of Hippocratism in European medical writings had focused scholarly attention on the ways that different climates affected the constitution; to the fore were Europeans' worries about the implications for their own bodies of the expansion of maritime empire. Buffon accounted for the historical emergence of differences in appearance around the earth's surface in the same terms, attending closely to the relationship between bodies and "airs, waters, places" (Earle 2012; Wheeler 2000; Niebyl 1971). This explains how, in his discussion of human variation, it was perfectly possible for two individuals to inhabit the same climatic zone and possess the same skin color, yet be of different "races" because they stemmed from different family stocks. In Buffon's cosmological speculations, the earth had originated as a blob of superheated matter knocked from the sun by a passing comet. Since that time it had progressively cooled. The physical features of both the earth and its living inhabitants emerged at different stages of this cooling process. Mankind had originated in the cool, temperate climates of Europe, so that civilization was the natural product of gentler climes. In hotter parts of the globe, human society typically existed in a state of indolence and despotism. In harsh conditions, the only animals to thrive were ferocious, untamable predators and asocial savages. Buffon famously proclaimed the Americas ripe for human improvement, which would eliminate the dangerous monsters still at large there, while bringing American "savages" into the fold of civilization (Vyverberg 1989; Spary 2000, Chapter 3). In any given location upon the earth's surface, humanity was thus in flux in Buffon's view, transforming as the climate changed.[5]

Such climatogenic change was often degenerative – in the literal and original sense that it produced the dwindling of a particular bloodline through the introduction of hereditary faults, malformations, and diseases. Even without human intervention, a stock subjected to climatic change would lose certain characteristics and acquire others. In a sinister thought experiment of 1766, Buffon proposed that

> To carry out an experiment on the change of color in the human species, it would be necessary to transport some individuals of that black race of

> Senegal to Denmark, where, man having commonly white skin, blond hair, blue eyes, the difference in the blood and the opposition of color is the greatest. It would be necessary to shut these Negroes in with their females, and scrupulously conserve their race without allowing them to interbreed; this is the only way to know how much time it would take to reintegrate the nature of man in this respect; and, similarly, how long it took to change [man's nature] from white to black.
>
> (*HN* 1749–1788, XIV, 314)

A major programmatic goal of the natural sciences was meteorological: through reason, man could bring about or preserve those climatic conditions which allowed civilization to flourish, postponing the eventual decline of the earth into an icy, lifeless desert. As Duchet (1971, 199) emphasizes, this was no linear theory of human descent or evolution. Instead, it was a prescription for the future management of society. Like the geoengineering solutions proposed in response to today's scientific debates about global warming, Buffon and other French naturalists proposed practical measures aimed at countering climatic deterioration, including the felling of the American forests. On his Burgundy estate, Buffon experimented extensively on animals, apparently seeking to gauge the relative power of climate, diet, domestication, and breeding to transform the physical and moral characteristics of races or breeds.

> It seems that the model of the beautiful and the good is dispersed over the whole earth, and that in each climate only a portion resides which is constantly degenerating, unless it is united with another portion taken from far away; so that in order to have good seed, beautiful flowers, etc., it is necessary to exchange seeds and never sow them in the same terrain which produced them; and, in the same way, in order to have beautiful horses, good dogs, etc., it is necessary to give foreign males to the females of the country, and reciprocally to the males of the country, foreign females; failing that, seeds, flowers, animals will degenerate, or rather take on such a strong hue of the climate, that matter will dominate over form and will seem to bastardize it [. . .] In mixing [. . .] the races, and above all in renewing them constantly with foreign races, the form seems to perfect itself, and Nature [seems] to revive herself and produce all that she can of the best.
>
> (*HN* 1749–1788, IV, 216–217)

Nelson (2010) describes such interventions as "racial engineering," but it should be noted that they addressed "perfection" in general, not "race" in particular. That is, "race" was interpreted in Hippocratic terms, as something both created and corrected by the state of the six non-naturals (Riley 1987; Jordanova 1979; Glacken 1967; Niebyl 1971). Knowledge about how breeding and climate interacted was important not because authors like Buffon were seeking to place race on a more scientific footing but because breeding appeared to be a technique for counteracting the consequences of climatic damage.[6] Natural variations in climate created homogeneous groups of living beings by the same mechanism that allowed men to produce new breeds or stabilize new characters in domestic animals (*HN* 1749–1788, XVII, 497). It followed that art and science might be used to minimize or reverse regional divergence produced by natural means, and return the species to its original perfection, or some approximation thereof.

Buffon's account of generation offered a mechanical model that explained how locale could produce variation in this way. Assimilative forces inserted molecules of living matter into a preexisting body plan or internal mold which characterized each species. Flaws were introduced into the design when inorganic molecules from surrounding inert matter were inadvertently introduced to the mold, imparting the particular chemical characteristics of a milieu to the body of the individual born there. Such differences were cumulative: over time, they also became inheritable, and were transmitted from parent to offspring (Sloan 1973; Winston 2005, 38; Quinlan 2007; Duchet 1971, 245). In this way, different climates eventually yielded breeds of men with distinctive physical characteristics. Even repeated experiences could become hereditary. The humped back characterizing the "secondary race" of African buffalo, for example, had originally appeared thanks to its use as a beast of burden (*HN* 1749–1788, XI, 328). For Buffon and his followers, their breeding experiments suggested that species regeneration was a practical problem that could be addressed by three means: interbreeding individuals from different locations, altering living conditions, and changing government (Spary 2000, Chapter 3; Peabody 1996, Chapter 4; Nelson 2010; 2013; Goldstein Sepinwall 2003; also Grégoire 1826, 46).

For Buffon, bad social and political institutions had effects upon bodies and minds that could be as profound as the local shortcomings of nature. Even local circumstances usually had their effect by means of diet and other aspects of lifestyle, which improved in a policed society as a matter of course, thanks to the arts and sciences. Because of the power of the social state to transform body and mind, man was less liable than animals to be affected by the localized effects of climate. And yet domestication or civilization, the third most significant cause of specific change after climate and diet, was a process which Buffon described in very ambivalent terms. Domestication was certainly a cause of civility and luxurious comfort, but it also produced degeneration from the natural state of liberty. Contemporaries sometimes treated domestication and slavery as synonymous; the problems of despotic governments, which made slaves of their subjects, were at the opposite end of the spectrum from the perils of the savage life, but both had material consequences for human and animal bodies. "One finds," said Buffon, "on all enslaved animals the stigmata of their captivity and the imprint of their irons" (*HN* 1749–1788, XIV, 317). The state of domesticated animals could best be illustrated by the figure of the human slave under despotic government. Buffon repeatedly cast humans in the position of despotic monarchs, with the power to transform the natural shape and character of animals: "One cannot doubt but that [man] is the author of all those enslaved races, as perfect to us as they are degenerate and vicious to Nature" (*HN* 1749–1788, XVII, 496; also IV, 169, 439; VIII, 282; XI, 17, 228–229; XVI, 252, 447; XXIII, 73). And yet in a passage on the dog, he argued that its nobility stemmed from its moral capacity to form a society with mankind (*HN* 1749–1788,

V, 185). Buffon's own political ambivalence about the legitimate limits of governance is clearly evident in these many discussions of slavery. In claiming that what enslaved individuals gained in gentleness and urbanity they lost in nobility and autonomy, the naturalist was echoing a long-standing debate over the decline of the nobility in France (Chaussinand-Nogaret 1985; Smith 1996, 2005; Doyle 2009). Most of his references to slavery in human populations were not about French colonialism (*HN* 1749–1788, II, 587; III, 381, 383, 412, 413, 418, 432, 434, 437; VII, 31); in a rare reference to the French colonial situation, he sharply criticized planters' violence toward African slaves (*HN* 1749–1788, III, 469–470). These highly political themes, metaphorically figured through discussions of animals, were clear to readers of Buffon. One of the closest of those readers, Jean-Jacques Rousseau, drew his claim that the subjects of French absolutist monarchy were degenerating because they resided under a despotic and luxurious regime direct from the pages of the *Histoire naturelle*. This political condition explained both the physical weakness and the moral bankruptcy of the French nation (Rousseau 1967–1971, II, 62, 72–76, 141–152, 173–176, 252). Both philosophes were here invoking a long-standing medical argument about the adverse effects of luxurious living upon the body (Thériault 1979; Emch-Deriaz 1992). While Doron (2011, 102) is correct to assert that "With Buffon, the whole vocabulary of nobility and breeding becomes part of natural history," then, he overlooks an entirely different dimension of Buffon's account of mutability in animal bodies which is far more radical, and which did not pass his eighteenth-century readers by.

Both Buffon and his readers agreed that good government could also perfect its subjects both physically and morally. Buffon's protégé Jean-Baptiste Leblond, for example, observed that the native of Bogotá, "civilized by the Spaniards, is no longer the same now that he lives in healthier, more spacious houses, is better dressed, and feeds himself with more varied and succulent foods." Formerly "dusky little men with red eyelids," these savages were now in the process of turning white, or more precisely European: "he comes proportionately closer to our constitution and color in the whiteness of his skin, and the development of his organs, as his race distances itself from the savage condition" (Leblond 1805, 39; also Buffon 2009, letter L619). Other commentators similarly enrolled the body in political critiques of absolutism or defenses of civilization. Claims that all Frenchmen were slaves underpinned Revolutionary calls for the moral and physical regeneration of the entire nation (Ozouf 1988; anon. 1789; Grégoire 1826, 26; Nelson 2013; Spary 2000, Chapter 5).

For eighteenth-century French readers, therefore, the slavery debate was partly or perhaps even largely reflexive. Both as scientific and as social phenomena, race and slavery were not necessarily synonymous in contemporary literature (Roach 1998). Slavery was a generalized condition, especially characteristic of despotic government; to describe its moral and physical effects was to participate in a wider debate about the illegitimacy of Old Regime governance. Insofar as slavery was discussed by naturalists and physicians, therefore, that discussion was not confined to a nonwhite population, while "racial" differences (in today's sense) were more usually seen as an effect than a cause of political and social differences.

Over the second half of the eighteenth century, these writings on human perfection as the product of rational labor increasingly contrasted with the brutal realities of colonial life, where a many-tiered social, legal, and penal hierarchy documented, classified, and legislated about minute variations in skin color. Such systems of color differentiation had existed in the French Antilles since the later seventeenth century, but they found little support in late eighteenth-century scientific or medical writings. The colonies formed a crucible for developing claims about the transmissibility of noble status through "blood" (Aubert 2004). If man was a "noble" being, so too were all those animal species that retained their primitive character, resisting the encroachments of climate (e.g., *HN* 1749–1788, IV, 109, 176, 239; VII, 26; IX, 4–10, 129–130; XVI, 76). But Buffon and others undercut such claims by invoking medical geography to reject the permanence of deviations from the original stock. Here views – and indeed personal experiences – of authors like Buffon echoed both recent developments in the French nobility, which intermarriage and ennoblement made one of the most open in Europe, and recent developments in colonial populations, where free people of color were surging into vivid political life at this time (Doyle 2009; Garrigus 2006).

Medical geography supported the non-fixity of races

Buffon's account of the mechanism of species degeneration or regeneration thus naturalized nobility and slavery, but also made nature and society interpermeable. It weakened rather than constituted a 'modern' racism grounded in nature, by implying that even naturally originating racial differences were mostly reversible with the application of reason. His meliorism has been studied as a philosophical and political issue (Sloan 1973; Winston 2005; Nelson 2010, 2013) but its links to medical geography, in which locale determined the physical and moral qualities of living bodies, have often gone unremarked in histories of race. Yet many authors writing on colonial health and creolism shared Buffon's broader Hippocratic framing. Elite travelers were advised to govern their bodies in accordance with medical precepts, deploying diet and other technologies of the self as a means of maintaining constitutional equilibrium when changing location (Pluchon 1991, 443–451; Sokalski 1986; Bory de Saint-Vincent 1804). Deacon (2000, 281) has termed such practices the "micro-management of environment." The rational management of health was crucial to civilization, which continued, in travel accounts circa 1800, to be heralded as the European accomplishment, achieved by constant labor upon the self. It was civilization which simultaneously elevated Europeans above other peoples and allowed them, through reason, to control the degenerative effects of nature and society, as in one 1801 medical guide for travelers:

> Without health, one does not truly have the use of one's intellectual faculties [. . .]. Now one cannot continue to enjoy health, if one is ignorant of the manner of conducting oneself at meals, during exercise, and in all the other parts

> of one's regimen. Occasions to neglect it appear more often at each step of a journey than at any other time.
>
> (Duplanil 1801, I, xi–xii)

At a distance from the metropolis, health was difficult to sustain. Travel was one of the most hazardous activities in Hippocratic terms, exposing travelers to new topographical conditions whose constitutional effects required countering with a tightly controlled regimen (Rupke 2000; Barrett 2000, 185–191; Harrison 1996).

The dangers of an imbalance between climate and body exceeded those resulting from most other changes in the non-naturals. Bodies were held to be formed by and for the climate of their birth. Joseph-Jacques Gardane, editor of the *Gazette de Santé* and specialist in Creole medicine, identified the cellular tissue as both the physical topos on which climate acted and the carrier for skin color. It was also the seat of illness resulting from "the abuse of the non-naturals, & above all [from] the intemperance of climates & seasons" (Gardane 1784, 2). Its effects were apparent in mental function, character, physical strength, and predisposition to illness, as well as in an inclination to particular forms of government; medical topography and regimen were thus fundamental to knowledge of "the origin of the customs & mores of different Peoples." "Like exotic plants," Gardane said of French Creole colonists, "they are born to parents foreign to the [climate] they inhabit" (Gardane 1784, 10). The legacy of a constitution undermined by the twin effects of climate and the luxurious plantation lifestyle was frequently degeneracy. Creoles' minds were swifter, yet more superficial and irritable; their bodies, weakened by transpiration and idleness, were narrow-chested and hunch-shouldered simulacra of the original European colonizers of the New World. Such views echoed Buffon's claim that the changes wrought in living bodies by exposure to differences in climate, diet, and social institutions could potentially occur "within a very small number of generations," and that they would affect young animals far more profoundly than adults (*HN* 1749–1788, IV, 217, quoted in Sloan 1973, 303).

The non-naturals were centrally implicated both in the original differentiation of peoples and in their distinctive physical and moral characteristics and customs. They were also responsible for the health of individuals and peoples, playing an instrumental role in the transformations suffered by colonial bodies. There was even a condition, "emigration fever," which formed an acknowledged stage in the constitution's adaptation to the new climate (Gardane 1784, 22; Bajon 1777, I, 22; Ledru 1810, I, 255–256; Lorry 1781, II, 285; Ward 2007, Chapters 4–5; Harrison 1996, 75; Morey Smith 2015). In describing this illness, Leblond explicitly referred to its effects on skin color:

> Whites originating from temperate zones or from the cold region of the tropics, coming to the hot region, undergo a change which shows itself in copious sweating, rashes, [and] intense itching [. . .]. This disease is a true naturalization which acclimatizes the individual; that naturalization is confirmed, if a certain dark pallor succeeds the red coloration of the skin.
>
> (Leblond 1805, 9)

Such climatogenic illness could be successfully managed by manipulating the non-naturals. Gardane outlined a complete program of self-preservation for Creoles traveling to Europe: incremental exposure to cooler weather, regular shipboard exercise, avoidance of hot humid air, and a diet of acidic vegetables. Dietary sobriety, above all, was to be perpetuated well after the Creole traveler's arrival in Paris (Gardane 1784, 35, 56–57, 100–101; Duplanil 1801, I, 157). Such health manuals responded to the growing volume of maritime traffic between the Americas and France which followed the end of the Seven Years' War in 1763, as improvements in shipping meant that the journey to the Caribbean islands reduced to just a few months.

Similarly, enslaved Africans were not put to work straightaway upon arrival in the American colonies, but were (in theory at least) subjected to a regimen of dieting and bathing to adjust their bodies to the new climate (Lorry 1781, II, 222). The choice of people of African origin as workers in the American colonies was sometimes explained by the claim that the African climate was closer to that of the Antilles than to that of Europe, so that African constitutions would suffer a milder shock from the transition to the New World than European ones (Leblond 1805, 8; Ledru 1810, II, 184–185). Colonial medical specialists remarked upon the many health problems suffered by first-generation European immigrants to the Antilles; by comparison, their offspring, born under a Caribbean sky, would be better acclimatized and more able to resist local disease. It was this perception of the transformed nature of people born in the colonies that made the category of Creole a unique and medically interesting one – in other words, for metropolitan authors and readers, the Creole body primarily exemplified a climatic, not a racial problematic.

Transforming the Creole

Certain prominent natural historical and medical authors writing in the late eighteenth century predicted that all Creole families, whether of European, African, indigenous, or mixed parentage, must end up dark-skinned through the effects of climate alone, even without the contributing and accelerating factors of interbreeding, education, or diet. Thus the net effect of scientific and medical debates over climate and descent in the second half of the eighteenth century, within the French empire, was indeed to produce natural explanations for human diversity, but in the very same breath to minimize the distinctions between people of different geographical origins, on the assumption that colonial differences would be progressively effaced under shared conditions of climate, society, government, and politics (Palmer 2016; for the English case, Boulukos 2008, Chapter 6, who describes this as a period of "ameliorationist convergence").

Buffon's *Histoire naturelle* had used the term "Creole" only infrequently, and never in connection with humans. Yet his account of climatogenic change was invoked to support medical accounts of the Creole body at the century's end, in the same period in which Creoles themselves were finding a new political and literary voice.[7] A striking and often overlooked feature of Gardane's work is that it is one of the few works to utilize the term "Creole" before the 1790s, and so to contribute to the construction of creolism as a medical category. During the French Revolution, the appellation "Creole" would become increasingly familiar in the metropolis, as a stream of publications and spokespeople poured forth from the colonies. Chief among the vocal representatives of creolism were individuals from the rich and populous Antilles. The defense of creolism as a virtuous and natural state was primarily conducted by Saint-Domingue's free population of color during the 1780s and 1790s (Garrigus 1997; also Debbasch 1967). Médéric-Louis-Élie Moreau de Saint-Méry, a Martinican Creole who had returned to Paris as a *parlementaire* before settling in Saint-Domingue, depicted black Antillean Creoles in a favorable light:

> The Creole negroes are born with physical & moral qualities, which give them a true right to superiority over those who have been transported from Africa, & this fact that here, domestication has embellished the species, while supporting a truth of the sublime Historian of nature [Buffon], might supply matter for doubting the excesses with which the masters' despotism has been reproached.
>
> (Moreau de Saint-Méry 1797–1798, I, 40)[8]

Just as French colonial societies reached the most prosperous and also most exploitative period of their existence, scientific and medical authors were at work to undermine rigid classificatory distinctions founded on color. In his *Description de Saint-Domingue*, probably written after he had fled the revolution in Haiti to set up a publishing house in Philadelphia, Moreau de Saint-Méry famously listed the complex classifications which had been established on the island in an attempt to pin down the proliferating intermediate conditions between white and black resulting from crosses between Europeans, Africans, and their descendants. He listed thirteen different classes based on the proportions of white and black "blood," from 1 to 128 parts, that a person contained.[9] These multiple appellations were in use because colonial opinion had it that there was "a line extending to infinity, [which] always separates the white descent from the other." Some historians (e.g., Garraway 2005) discussing Moreau de Saint-Méry's book stop at this point in the text, taking the classificatory table as evidence of the new scientific acceptance of fixity of race. But on the very next pages, Moreau de Saint-Méry underscores the *invalidity* of this same classificatory enterprise, sanctioned by "tradition" and "opinion" and enshrined in legislation (Boulle 1988, 233; Grégoire 1826, 9–12; Meadows 2000). The meticulous recording of degrees of difference and inheritance for each inhabitant of the colony was unscientific:

> It is the eye of prejudice which sees, & if it were to roam all over Europe, it would find, using this system, the wherewithal to create a nomenclature for color there too, for who has not observed, when travelling in that part of the world, pretty dark tints & characteristics which seem to belong to Africa? Surely there are some quadroons who are twice as white as any Spaniard or Italian.

> And if, when hardly a century and a half has gone by since the two colors began to mingle in the French Colony, there are already occasions on which one is reduced to doubt in the case of certain individuals, might not a few generations more bring about an absolute convergence where skin color is concerned, especially in a climate where the skin of Europeans themselves takes on a yellowish tone, when under its influence for a long time? That epoch might even be accelerated by particular circumstances, such as transplantation to a cold country, for example; it is known that the negro who inhabits France is less black there than in the Antilles.
>
> (Moreau de Saint-Méry 1797–1798, I, 87)

The Creole eye, returning to Europe, revealed the inherent prejudice of the color calculus; the eye of reason saw matters differently. Even so, Moreau de Saint-Méry's appeal to natural law over jurisdiction or custom did not preclude the expression of standard hierarchies based on degree of civilization, and he was opposed to political equality in the colony.

> The current truth is that the negro is in a true state of degeneration compared to the civilized European. That condition is such that it legitimates the claim that such degeneration, which is perhaps the work of centuries, would require further centuries before its effects disappeared completely.
>
> (Moreau de Saint-Méry 1797–1798, I, 62–63)

The response of this philosophe to legislation predicated upon skin color was thus to argue that the secure foundation for colonial hierarchies lay in social institutions, specifically not in fixed natural differences.[10] Even morphology was, in the view of both Moreau de Saint-Méry and Leblond, liable to transform under the influence of climate and social institutions:

> Black children born in our Colonies, who have the same physical upbringing & the same foods as in Africa, have, in general, a less flattened nose, less coarse lips and more regular features than African negroes. The nose elongates, the features soften, the yellow tint of the eyes abates proportionately as the generations become more distant from their primitive source, & these shades of difference are very noticeable. I have seen negroes with a very long aquiline nose, & this characteristic passing to all the individuals of the same family.
>
> (Moreau de Saint-Méry 1797–1798, I, 55)

Moreau de Saint-Méry was definitely no political egalitarian. But his reasons for supporting the superiority of whites and Creoles over people of African origin are very clear: they were founded upon arguments about civilization and aesthetics identical to those invoked much earlier in the century, not upon distinct "races" (in the modern sense), the natural reality of which he rejects – specifically by appealing to Buffon's arguments. Ironically, therefore, the universalizing aesthetic perspective of the philosophe's gaze rules out race as a stable natural category at the same time as it smuggles in inequality by another route.[11]

If Moreau de Saint-Méry represents the conservative extreme of the climatic argument, another former Saint-Domingue colonist, the grocer Siméon-Jérôme Ducoeurjoly, writing soon afterwards on the "cause and color of negroes," would offer a more radical position, but one which similarly emphasized the profoundly transformative effects of civilization and climate in tandem.[12] Skin color resulted from "the action of the air, the heat of the sun, and the quality of the exhalations with which the atmosphere is charged."[13] Ducoeurjoly invoked specific historical examples to argue that France's Antillean colonists were well on their way to blackness:

> The Saracens, Moors and Arabs who invaded West Africa in the seventh century and settled there, were white, dusky or yellow; after some generations they changed color and became as black as the earlier inhabitants [. . .]. When the Portuguese [. . .] began to form settlements [in Africa] towards the middle of the fourteenth century, they were not black at all; some settled in the Canaries and the Azores, the rest in the Cape Verde islands and on the African coast, close to the equator. The former have not undergone any change of color in the three centuries during which they have inhabited those islands, from father to son; the latter, closer to the equator, have become as black as the natives.

An out-and-out transformist, Ducoeurjoly also suggested, in a book specifically published to acknowledge Saint-Domingue's struggle for independence, that the flattened nose and thicker lips that French commentators identified with Negro status had originated as culturally specific practices of beauty which became hereditary: "the custom of squashing their noses, and pulling their lips to make them swell and thicken" was a "custom which finally decides nature to follow, willingly and in her own right, a course she was initially forced to adopt by means of art." And Ducoeurjoly concluded, "The color of negroes, as well as their particular traits, are thus purely extrinsic and accidental to them, and do not make them into a separate species of men" (Ducoeurjoly 1802, I, 17).

The model of human perfection continued, for many authors, to be the white European body; Ducoeurjoly's words betray the struggle of Europeans to accept otherness. Yet these arguments about the lability of bodily characteristics, and above all the quick pace of climatic transformation, had the political goal of minimizing "racial" difference by presenting race as an ephemeral, unstable categorization – certainly not the insuperable natural barrier it would become in scientific writings. Both Moreau de Saint-Méry and Ducoeurjoly continued to utilize the terms "race" and "species" interchangeably (Moreau de Saint-Méry 1797–1798, I, 39; Ducoeurjoly 1802, I, 17–18). Transformist views of race even appear to have become more widespread in the closing decades of the century. Writing of blackness in the 1760s, Claude-Nicolas Le Cat had invoked the history of European colonization to argue just the opposite of the authors quoted earlier: that climatic conditions had *not* affected the color of European settlers in Africa during their two centuries of residence there, and thus that color was an essential natural distinction between "negroes" and white Europeans (Le Cat 1787, 8–10; cf. also Shelford 2013). But during the Republican years, this transformist position was a fairly commonplace one, and it remained so, as Fogarty and Osborne (2003) and Harrison (1996) show, well into the nineteenth century in colonial policy and discourse.

Colonial and metropolitan opinion was thus divided over the significance of mixing races (in the old sense of "stock"), and the effects of changing climate. Many fleeing white planters doubtless shared the view that there was an insurmountable natural barrier between white and black in perpetuity. Ducoeurjoly's claims represented one political extreme in the debate over liberation and equality in these decades. Nature could be evoked quite differently on the opposing side of the debate.[14] A former director of the royal company which held a monopoly over the Senegal gum trade, for example, rejected the whole principle of creolism, asserting that black-skinned people were destined by Nature to inhabit torrid zones; he warned French readers to "preserve the races in their natural purity" and avoid the "mixture of black and white [which] is very dangerous for our population . . . it might degrade, corrupt and destroy it" (Durand 1802, I, 166; similarly, François Valentin de Cuillon 1802–1803, II, 294). Such fixist views almost certainly became more politically charged as the number, status, and power of free people of color in the French colonies increased after 1750 (Garrigus 1996; Dubois 2003; Palmer 2016, 33; Ghachem 2012; for the English case, Boulukos 2008, 141). Bowing to the planters' lobby, Louis XVI issued a royal declaration in August 1777 forbidding all blacks and people of color to enter the kingdom, even as servants, on the grounds that they would acquire habits of "independence" in France (Palmer 2016, 157). But not all planters were the same. Some enthusiastically embraced political liberty for their children: people like Arnoux, the Guadeloupe agronomist who freed his African slave and brought their three mulatto sons back to France to be reared according to Rousseauist precepts. The eldest died fighting in the Revolutionary army in 1792, shortly before the Convention Nationale abolished slavery in the French colonies (Archives Nationales (Paris), F^{10} 257, letter of 14 Thermidor, year II / 1 August, 1794; Palmer 2016 cites similar examples).[15] Others, like the colonial administrator Gabriel de Bory, invoked Buffon's writings in support of proposals to liberate all men of color and advocated breeding programs to produce "improved" mulattos to run the colonial administration (Nelson 2010, 1376–1377). Moreau de Saint-Méry (1797–1798, I, 69) too claimed that the proliferation of free people of color in recent decades stemmed from the desire of white fathers to liberate their children. These disagreements over race and its nature thus reflect stark divisions in French colonial society over the status of the *métis*, not scientific consensus over the existence of fixed natural groups within the human species; rather, views of race as a fixed natural category came at this time from those who profited by the exploitation of black labor.[16]

Natural historical and medical accounts of racial characteristics as climatogenic and highly mutable remained current in metropolises and colonies beyond 1800.

As Dubois (2004, Chapter 9) shows for the case of Guadeloupe, many people openly declared previously hidden slave ancestry during the 1790s under new equality legislation. The medical and natural historical literature discussed in this essay mirrored the chronology of the dispute over the classification of colonists according to color. The optimistic use of Buffon's writings to downplay differences – rather than to insist upon a fixed, scientifically determined notion of race – was at its height during the First Republic. After 1800, the brand-new science of anthropology, showcased in the recently founded *Société des Observateurs de l'Homme*, conferred a new heuristic importance and a new set of meanings upon race, which now became the focal point of both field research and metropolitan description (Stocking 1964; Chappey 2002; Staum 2000, 2003).

The shift this operated in readings of Buffon is clearly evident in the case of the radical cleric André-Pierre Ledru, who traveled to the Antilles between 1795 and 1797.[17] His manuscript travel diary portrayed colonial populations as increasingly creolized and homogenized. On some islands, such as Puerto Rico, Ledru observed, "the races are crossed to such an extent, that one usually only meets with dusky faces" (Ledru 1810, II, 163; Jangoux 2006). But when his manuscript was edited for publication in the 1800s, his editor, the naturalist Charles Sonnini de Manoncourt, inserted into the text a systematic classification of Antillean inhabitants into black, white, and crosses between the two. That is, one can date the moment at which Buffon's account of race was "officially" appropriated to underpin more fixist accounts of the differences within the human species fairly precisely: to the end of the Republic and the Haitian Revolution. It was this shift, I argue, that made it possible for nineteenth-century anthropologists to appropriate Buffon as the founder of their discipline (Blanckaert 1993).

It is no coincidence that Sonnini was also the editor of a new edition of Buffon's *Histoire naturelle* (*HN* 1798–1808). In the commentary he added to Ledru's text, he cited both Moreau de Saint-Méry's table of the intricate differentiations of Old Regime colonial legislation and a recent French translation of Blumenbach's *De generis humani varietate nativa* (Blumenbach 1804), to assert the scientific validity of racial classifications. Moreau de Saint-Méry's "line extending to infinity, [which] always separates the white descent from the other" – a line which the Creole author himself had rejected – now became a scientific instrument for the reliable identification and hierarchization of race in the Antilles. The introduction to the Blumenbach translation, written by the physician Frédéric Chardel, had invoked a humoral account of human variation, presenting skin color and morphological distinctions in terms very similar to those of Buffon, as infinitely nuanced around the globe, and as acquired characteristics originally produced by climate, eventually becoming hereditary. The word "race" appears only once in Chardel's introduction, and then in its early modern sense of "stock" (Chardel 1804). But, again, in editing Ledru, Sonnini ignored Chardel, and focused instead upon Blumenbach's table of crosses between races (Blumenbach 1804, 162–168; Ledru 1810, 163n1).

What had happened at this precise juncture to harden the category of race? Besides the institutional transformations accompanying the emergence of anthropology as a discipline, historians have pointed to the rapid takeover of the Parisian scientific world by the priorities of Napoleon Bonaparte. It was in 1803 that the future emperor would suppress the Class of Moral and Political Sciences at the National Institute for sheltering his political opponents. A personal antagonist of transformism, he also repudiated Jean-Baptiste Antoine Pierre de Lamarck, another of Buffon's protégés and interpreters and like them committed to a transformist view of nature (Outram 1984; Jordanova 1984; Burkhardt 1977; Corsi 1988). The battle of Vertières in November 1803 marked the definitive defeat of Napoleon's armies in their bid to regain control of the island of Saint-Domingue. The new Republic of Haiti had a constitution which, as Hörmann (2016, 185) notes, "stipulated that all Haitian citizens (irrespective of their physiological skin color) were to be considered 'black.'" That is, the same Republican color-blindness exhibited by the likes of Ducoeurjoly founded the new political order: blackness became a political, rather than a natural, category. In Paris during the Consulate, political opinion was divided over Haiti up to 1803 (Girard 2009). But published accounts reaching metropolitan reading publics after 1803 emphasized the "carnage and ruin" and the unnaturalness of the sudden "reversal of racial power" that had accompanied the birth of the Republic of Haiti. These were tales fit to sway even ardent abolitionists like Alexander von Humboldt (Hörmann 2016; Young Lee 2010; Rebok 2009; on the American reception, Clavin 2007). At a time of increasing reaction against Republican extremism in government, equality in the colonies could henceforth be recaptured as a threat. Redefining race aided this process: even natural truth was coordinated around the new political climate.

Notes

1 All translations are the author's own, except where otherwise stated.
2 For one early attempt by the traveler François Bernier to identify distinct races across the globe, which was, however, virtually never mentioned in eighteenth-century works, cf. Boulle (2003).
3 Very similar remarks were made by Pierre Louis Moreau de Maupertuis; see Curran (2009, 158–159) and Sloan (1973, 300).
4 This pluralism is recognized by Hocquet (2014, 22–24).
5 Nelson's claim that scholarship on degeneration "has focused too narrowly on theories of climate" (2010, 1371) does not sufficiently frame Buffon's writings against his overall cosmology; contemporaries clearly grasped that his remarks referred to medical geography.
6 *HN* (1749–1788, XI, 295–296). My lengthy discussion of these issues (Spary 2000, Chapter 3) has apparently escaped Nelson's attention (2010, 1372n25).
7 The chronology of the decline of this view has been well documented by Harrison (1996) for the British case, and mirrors the French situation.
8 On Saint-Domingue as a center for scientific enquiry in the late Old Regime, see especially McClellan (1992; 2000) and Dubois (2006). On the attempts of Creoles to argue for supremacy over persons born in Africa, see Saint-Louis (2006).
9 The entire discussion, including quotations, draws upon Moreau de Saint-Méry (1797–1798, I, 71–86).

10 Formal distinctions between degrees of African and European heredity were enshrined in law by the 1780s, but their implementation was haphazard because of the power of Creole élites (Peabody 1996; Garrigus 1996, 21).
11 Dubois (2006, 6) precisely pinpoints this manoeuver in which "universal claims were intertwined with justifications [. . .] based on the incapacity of certain 'others' to enjoy their natural rights."
12 I take Ducoeurjoly's name from Archives Nationales (Paris), Minutier Central, étude LXV/333, apprenticeship of 23.4.1761.
13 All quotes are from Ducoeurjoly (1802, I, 14–16), partly paraphrasing Du Bos, *Réflexions critiques sur la poésie* (4th ed.: 1740, II, 253–254).
14 Nelson (2010, 1389) notes that "even as modern ideas of race were coming into being, both sides of the dialectic of race were already present."
15 On such families, see especially Palmer (2016).
16 Nelson (2010; 2013) emphasizes the importance of assessing the debate over race from the colonial standpoint.
17 On Ledru, see *BHV* (1816–1819, IV, 165–166).

Bibliography

Anon. 1789. *L'esclavage anéanti, et la résurrection de la France, aux représentans de la nation.* Paris: n. p.

Arnold, David. 1996. "Introduction: Tropical Medicine Before Manson." In *Warm Climates and Western Medicine: The Emergence of Tropical Medicine, 1500–1900*, edited by David Arnold, 1–19. Amsterdam: Rodopi.

Aubert, Gabriel. 2004. " 'The Blood of France': Race and Purity of Blood in the French Atlantic World." *The William and Mary Quarterly*, 3rd series, 61 (3): 439–478.

Bajon, Bertrand. 1777. *Mémoires pour servir a l'histoire de Cayenne, et de la Guiane Françoise*. 2 vols. Paris: Grangé.

Barrett, Frank A. 2000. *Disease & Geography: The History of an Idea.* Toronto: Geographical Monographs.

BHV. 1816–1819. *Biographie des hommes vivants.* 5 vols. Paris: Michaud and Antoine Boucher.

Bindman, David. 2002. *Ape to Apollo: Aesthetics and the Idea of Race in the 18th Century.* London: Reaktion Books.

Blanckaert, Claude. 1993. "Buffon and the Natural History of Man: Writing History and the 'Foundational Myth' of Anthropology." *History of the Human Sciences* 6 (1): 13–50.

Blumenbach, Johann F. 1804. *De l'Unité du genre humain et de ses variétés, ouvrage précédé d'une lettre à Joseph Banks, baronet et président de la Société Royale de Londres.* Paris: Allut.

Bory de St.-Vincent, Jean-Baptiste. 1804. *Voyage dans les quatre principales iles des mers d'Afrique, fait par ordre du gouvernement.* 3 vols. Paris: Buisson.

Boulle, Pierre H. 1988. "In Defense of Slavery: Origins of a Racist Ideology in France." In *History From Below: Studies in Popular Protest and Popular Ideology*, edited by Frederick Krantz, 219–246. Oxford: Basil Blackwell.

———. 2003. "François Bernier and the Origins of the Modern Concept of Race." In *The Color of Liberty: Histories of Race in France*, edited by Sue Peabody and Tyler E. Stovall, 11–27. Durham: Duke University Press.

Boulukos, George. 2008. *The Grateful Slave: The Emergence of Race in Eighteenth-Century British and American Culture*. New York: Cambridge University Press.

Bourdieu, Pierre. 1994. *Distinction: A Social Critique of the Judgement of Taste*. London: Routledge.

Buffon, Georges-Louis Leclerc de. 2009. "Correspondance." URL: www.buffon.cnrs.fr/correspondance/.

Burke, Peter. 1992. *The Fabrication of Louis XIV*. New Haven: Yale University Press.

Burkhardt, Richard W. 1977. *The Spirit of System: Lamarck and Evolutionary Biology*. Cambridge, MA: Harvard University Press.

Bury, Emmanuel. 1996. *Littérature et politesse: L'invention de l'honnête homme 1580–1750*. Paris: Presses Universitaires de France.

Chappey, Jean-Luc. 2002. *La Société des Observateurs de l'Homme (1799–1804): Des anthropologues au temps de Bonaparte*. Paris: Société des études robespierristes.

Chardel, Frédéric. 1804. "Discours préliminaire du traducteur." In *De l'Unité du genre humain et de ses variétés, ouvrage précédé d'une lettre à Joseph Banks, baronet et président de la Société Royale de Londres, by Johann Friedrich Blumenbach*, i–x. Paris: Allut.

Chartier, Roger. 1987. *Lectures et lecteurs dans la France de l'Ancien Regime*. Paris: Seuil.

Chaussinand-Nogaret, Guy. 1985. *The French Nobility in the Eighteenth Century: From Feudalism to Enlightenment*. Cambridge: Cambridge University Press.

Clavin, Matt. 2007. "Race, Rebellion, and the Gothic: Inventing the Haitian Revolution." *Early American Studies* 5 (1): 1–29.

Colwill, Elizabeth. 1998. "Sex, Savagery and Slavery in the Shaping of the French Body Politic." In *From the Royal to the Republican Body: Incorporating the Political in Seventeenth- and Eighteenth-Century France*, edited by Sara E. Melzer and Kathryn Norberg, 198–223. Berkeley: University of California Press.

Corsi, Pietro. 1988. *The Age of Lamarck: Evolutionary Theories in France 1790–1830*. Berkeley: University of California Press.

Curran, Andrew. 2009. "Rethinking Race History: The Role of the Albino in the French Enlightenment." *History and Theory* 48 (3): 151–179.

Deacon, Harriet. 2000. "The Politics of Medical Topography: Seeking Healthiness at the Cape During the Nineteenth Century." In *Pathologies of Travel*, edited by Richard Wrigley and George Revill, 279–297. Amsterdam: Rodopi.

Debbasch, Yves. 1967. *Couleur et liberté. Le jeu du critère ethnique dans un ordre juridique esclavagiste, I: L'affranchi dans les possessions françaises de la Caraïbe (1635–1833)*. Paris: Dalloz.

Dictionnaire. 1798. "Dictionnaire de l'Académie Françoise." 2 vols. URL: http://artfl-project.uchicago.edu/.

———. 1835. "Dictionnaire de l'Académie Française." 3 vols. URL: http://artfl-project.uchicago.edu/.

Doron, Claude-Olivier. 2011. "Race and Genealogy: Buffon and the Formation of the Concept of 'Race'." *Humana Mente* 22: 75–109.

Doyle, William. 2009. *Aristocracy and Its Enemies in the Age of Revolution*. Oxford: Oxford University Press.

Dubois, Laurent. 2003. "Inscribing Race in the Revolutionary French Antilles." In *The Color of Liberty: Histories of Race in France*, edited by Sue Peabody and Tyler E. Stovall, 95–107. Durham: Duke University Press.

———. 2004. *A Colony of Citizens: Revolution and Slave Emancipation in the French Caribbean, 1787–1804*. Chapel Hill: University of North Carolina Press.

———. 2006. "An Enslaved Enlightenment: Rethinking the Intellectual History of the French Atlantic." *Social History* 31 (1): 1–14.

Duchet, Michèle. 1971. *Anthropologie et histoire au siècle des lumières: Buffon, Voltaire, Rousseau, Helvétius, Diderot*. Paris: François Maspero.

Ducoeurjoly, Siméon-Jérôme. 1802. *Manuel des habitans de Saint-Domingue, contenant un précis de cette Ile, depuis sa découverte*. 2 vols. Paris: Arthus-Bertrand.

Duplanil, Jean-Denis. 1801. *Médecine du Voyageur; ou Avis sur les Moyens de conserver la santé, et de remédier aux accidens et aux maladies auxquels on est exposé dans les Voyages, tant par terre que par mer*. 3 vols. Paris: Moutardier.

Durand, Jean-Baptiste-Léonard. 1802. *Voyage au Senegal, ou Memoires historiques, philosophiques et politiques sur les découvertes, les établissemens et le commerce des Européens dans les mers de l'Océan atlantique, depuis le Cap-Blanc jusqu'à la rivière de Serre-Lionne*. . . 2 vols. Paris: Henri Agasse.

Earle, Rebecca. 2012. *The Body of the Conquistador: Food, Race and the Colonial Experience in Spanish America, 1492–1700*. Cambridge: Cambridge University Press.

Emch-Dériaz, Antoinette. 1992. *Tissot: Physician of the Enlightenment*. New York: Peter Lang.

Fogarty, Richard, and Michael A. Osborne. 2003. "Constructions and Functions of Race in French Military Medicine, 1830–1920." In *The Color of Liberty: Histories of Race in France*, edited by Sue Peabody and Tyler E. Stovall, 206–236. Durham: Duke University Press.

Gardane, Joseph-Jacques. 1784. *Des Maladies des Créoles en Europe*. Paris: Valade.

Garraway, Doris. 2005. "Race, Reproduction, and Family Romance in Moreau de Saint-Méry's Description . . . de la Partie Française de l'Isle Saint-Domingue." *Eighteenth-Century Studies* 38 (2): 227–246.

Garrigus, John D. 1996. "Colour, Class and Identity on the Eve of the Haitian Revolution: Saint-Domingue's Free Coloured Elite as *Colons Américains*." *Slavery & Abolition* 17 (1): 20–43.

———. 1997. " 'Sons of the Same Father': Gender, Race, and Citizenship in French Saint-Domingue, 1760–1792." In *Visions and Revisions of Eighteenth-Century France*, edited by Christine Adams, Jack R. Censer, and Lisa J. Graham, 137–153. University Park, PA: Pennsylvania State University Press.

———. 2006. *Before Haiti: Race and Citizenship in French Saint-Domingue*. New York: Palgrave Macmillan.

Ghachem, Malick W. 2012. *The Old Regime and the Haitian Revolution*. Cambridge: Cambridge University Press.

Girard, Philippe. 2009. "Napoléon Bonaparte and the Emancipation Issue in Saint-Domingue, 1799–1803." *French Historical Studies* 32 (4): 587–618.

Glacken, Clarence J. 1967. *Traces on the Rhodian Shore: Nature and Culture in Western Thought from Ancient Times to the End of the Eighteenth Century*. Berkeley: University of California Press.

Goldstein Sepinwall, Alyssa. 2003. "Eliminating Races, Eliminating Difference: Blacks, Jews, and the Abbé Grégoire." In *The Color of Liberty: Histories of Race in France*, edited by Sue Peabody and Tyler E. Stovall, 28–41. Durham: Duke University Press.

Grégoire, Henri. 1826. *De la Noblesse de la peau, ou du préjugé des blancs contre la couleur des Africains et celle de leurs descendans noirs et sang-mêlés*. Paris: Baudouin frères.

Harrison, Mark. 1996. " 'The Tender Frame of Man': Disease, Climate and Racial Difference in India and the West Indies, 1760–1860." *Bulletin of the History of Medicine* 70: 68–93.

Harvey, David Allen. 2012. *The French Enlightenment and Its Others: The Mandarin, the Savage, and the Invention of the Human Sciences*. New York: Palgrave Macmillan.

HN. 1749–1788. "Georges-Louis Leclerc de Buffon, Louis-Jean-Marie Daubenton, Philibert Gueneau de Montbeillard and Gabriel-Louis Bexon." In *Histoire Naturelle, Générale et Particulière, avec la description du cabinet du roy*. 36 vols. Paris: Imprimerie du Roi. URL: www.buffon.cnrs.fr/.

———. 1798–1808. "Georges-Louis Leclerc de Buffon, Louis-Jean-Marie Daubenton, Philibert Gueneau de Montbeillard, Gabriel-Louis Bexon, and Charles Sonnini de Manoncourt." In *Histoire Naturelle, Générale et Particulière* [. . .] . *Nouvelle édition*. 127 vol. Paris: F. Dufart.

Hocquet, Thierry. 2014. "Biologization of Race and Racialization of the Human: Bernier, Buffon, Linnaeus." In *The Invention of Race: Scientific and Popular Representations*, edited by Nicolas Bancel, Thomas David, and Dominic Thomas, 17–32. New York: Routledge.

Hörmann, Raphael. 2016. "'A Very Hell of Horrors'? The Haitian Revolution and the Early Transatlantic Haitian Gothic." *Slavery & Abolition* 37 (1): 183–205.

Hudson, Nicolas. 1996. "From 'Nation' to 'Race': The Origin of Racial Classification in Eighteenth-Century Thought." *Eighteenth-Century Studies* 29 (3): 247–264.

Jangoux, Michel. 2006. "Nicolas Baudin par son contemporain André Pierre Ledru: une autre perception du capitaine naturaliste." In *Relations savantes: voyages et discours scientifiques*, edited by Sophie Linon-Chipon and Daniela Vaj, 107–120. Paris: Presses de l'Université Paris-Sorbonne.

Jordanova, Ludmilla. 1979. "Earth Science and Environmental Medicine: The Synthesis of the Late Enlightenment." In *Images of the Earth: Essays in the History of the Environmental Sciences*, edited by Ludmilla Jordanova and Roy Porter, 119–146. Chalfont St. Giles: British Society for the History of Science.

———. 1984. *Lamarck*. Oxford: Oxford University Press.

Le Cat, Claude-Nicolas. 1787. *Traité de la couleur de la peau humaine en général, de celle des negres en particulier, et de la métamorphose d'une de ces couleurs en l'autre, soit de naissance, soit accidentellement*. Amsterdam: n. p.

Leblond, Jean-Baptiste. 1805. *Observations sur la Fièvre Jaune, et sur les Maladies des Tropiques*. Paris: Barrois.

Ledru, André-Pierre. 1810. *Voyage aux Iles de Ténériffe, La Trinité, Saint-Thomas, Sainte-Croix et Porto-Ricco*. 2 vols. Paris: Arthus-Bertrand.

Lorry, Anne-Charles. 1781. *Essai sur les alimens, pour servir de commentaire aux livres diététiques d'Hippocrate*. 2 vols. Paris: Didot jeune.

Lovejoy, Arthur O. 1959. "Buffon and the Problem of Species." In *Forerunners of Darwin: 1745–1859*, edited by Bentley Glass, Owsei Temkin, and William L. Straus, Jr., 84–113. Baltimore: Johns Hopkins University Press.

Lyon, John, and Phillip R. Sloan, eds. 1981. *From Natural History to the History of Nature: Readings from Buffon and His Critics*. Notre Dame: University of Notre Dame Press.

Maza, Sarah. 1998. "Luxury, Morality and Social Change: Why There Was No Middle-Class Consciousness in Pre-Revolutionary France." In *The French Revolution: Recent Debates and New Controversies*, edited by Gary Kates, 113–130. New York: Routledge.

Mazlish, Bruce. 2001. "Civilization in a Historical and Global Perspective." *International Sociology* 16: 293–300.

McClellan, James E. III. 1992. *Colonialism and Science: Saint Domingue in the Old Regime*. Baltimore: Johns Hopkins University Press.

———. 2000. "L'Historiographie d'une académie coloniale: le 'Cercle des Philadelphes' (1784–1793)." *Annales Historiques de la Révolution Française* 320: 77–88.

Meadows, R. Darrell. 2000. "Engineering Exile: Networks and the French Atlantic Community, 1789–1809." *French Historical Studies* 23 (1): 67–102.

Meijer, Miriam C. 1999. *Race and Aesthetics in the Anthropology of Petrus Camper (1722–1789)*. Amsterdam: Rodopi.

Moreau de Saint-Méry, Médéric-Louis-Élie. 1797–1798. *Description topographique, physique, civile, politique et historique de la partie française de l'Isle Saint-Domingue*. 2 vols. Philadelphia: Moreau de Saint-Méry.

Morey Smith, Sean. 2015. "Seasoning and Abolition: Humoural Medicine in the Eighteenth-Century British Atlantic." *Slavery & Abolition* 36 (4): 684–703.

Muchembled, Robert. 1998. *La Société policée. Politique et politesse en France du XVIe au XXe siècle*. Paris: Éditions du Seuil.

Nelson, William. 2010. "Making Men: Enlightenment Ideas of Racial Engineering." *American Historical Review* 115 (5): 1364–1394.

———. 2013. "Colonizing France: Revolutionary Regeneration and the First French Empire." In *The French Revolution in Global Perspective*, edited by Suzanne Desan, Lynn Hunt, and William Max Nelson, 73–85. Ithaca: Cornell University Press.

Niebyl, Peter H. 1971. "The Non-Naturals." *Bulletin of the History of Medicine* 45: 486–492.

Outram, Dorinda. 1984. *Georges Cuvier: Vocation, Science and Authority in Post-Revolutionary France*. Manchester: Manchester University Press.

Ozouf, Mona. 1988. "La Révolution française et l'idée de l'homme nouveau." In *The French Revolution and the Creation of Modern Political Culture. II: The Political Culture of the French Revolution*, edited by Colin Lucas, 213–232. Oxford: Pergamon Press.

Pagden, Anthony. 1988. "The 'Defense of Civilization' in Eighteenth-Century Social Theory." *History of the Human Sciences* 1: 33–45.

Palmer, Jennifer L. 2016. *Intimate Bonds: Family and Slavery in the French Atlantic*. Philadelphia: University of Pennsylvania Press.

Peabody, Sue. 1996. *'There Are No Slaves in France': The Political Culture of Race and Slavery in the Ancien Régime*. New York: Oxford University Press.

Pekacz, Jolanta T. 1999. *Conservative Tradition in Pre-Revolutionary France: Parisian Salon Women*. New York: Peter Lang.

Péron, François, and Louis-Claude de Saulces de Freycinet. 1807–1816. *Voyage de découvertes aux Terres Australes. Histoire*. 2 vols. Paris: Imprimerie Impériale.

Pluchon, Pierre. 1991. *Histoire de la colonisation française. I: Le Premier empire colonial, des origines à la Restauration*. Paris: Fayard.

Quinlan, Sean. 2007. *The Great Nation in Decline: Sex, Modernity and Health Crises in Revolutionary France c. 1750–1850*. Aldershot: Ashgate.

Rebok, Sandra. 2009. "La Révolution de Haïti vue par deux personnages contemporains: le scientifique prussien Alexander von Humboldt et l'homme d'état américain Thomas Jefferson." *French Colonial History* 10: 75–95.

Revel, Jacques. 1989. "The Uses of Civility." In *A History of Private Life. III: Passions of the Renaissance*, edited by Roger Chartier, 167–205. Cambridge, MA: The Belknap Press of Harvard University Press.

Riley, James C. 1987. *The Eighteenth-Century Campaign to Avoid Disease*. New York: Palgrave Macmillan.

Ritvo, Harriet. 1987. *The Animal Estate: The English and Other Creatures in the Victorian Age*. Cambridge, MA: Harvard University Press.

Roach, Joseph. 1998. "Body of Law: The Sun King and the Code Noir." In *From the Royal to the Republican Body: Incorporating the Political in Seventeenth- and Eighteenth-Century France*, edited by Sara E. Melzer and Kathryn Norberg, 113–130. Berkeley: University of California Press.

Rousseau, Jean-Jacques. 1967–1971. *Oeuvres complètes*. 3 vols. Paris: Seuil.

Rupke, Nicolaas, ed. 2000. *Medical Geography in Historical Perspective*. London: Wellcome Trust Centre for the History of Medicine.

Saint-Louis, Vertus. 2006. "Le Surgissement du terme 'africain' pendant la révolution de Saint-Domingue." *Ethnologies* 28 (1): 147–171.

Shelford, April G. 2013. "Race and Scripture in the Eighteenth-Century French Caribbean." *Atlantic Studies* 10 (1): 69–87.

Shell, Susan M. 2006. "Kant's Conception of a Human Race." In *The German Invention of Race*, edited by Sara Eigen and Mark Larrimore, 55–72. Albany: SUNY University Press.

Shovlin, John. 2006. *The Political Economy of Virtue: Luxury, Patriotism, and the Origins of the French Revolution*. Ithaca: Cornell University Press.

Sloan, Phillip R. 1973. "The Idea of Racial Degeneracy in Buffon's Histoire Naturelle." *Studies in Eighteenth-Century Culture* 3: 293–321.

Sloan, Phillip R., and John Lyon. "Introduction." In *From Natural History to the History of Nature: Readings from Buffon and His Critics*, edited by John Lyon and Phillip R. Sloan, 1–32. Notre Dame: University of Notre Dame Press.

Smith, Jay M. 1996. *The Culture of Merit: Nobility, Royal Service, and the Making of Absolute Monarchy in France, 1600–1789*. Ann Arbor: University of Michigan Press.

———. 2005. *Nobility Reimagined: The Patriotic Nation in Eighteenth-Century France*. Ithaca: Cornell University Press.

Sokalski, Alexander A. 1986. "'Impressions Africaines': The Chevalier de Boufflers in Senegal, 1785–1787." *Studies in Eighteenth-Century Culture* 16: 251–270.

Spary, Emma C. 2000. *Utopia's Garden: French Natural History from Old Regime to Revolution*. Chicago, IL: University of Chicago Press.

Staum, Martin S. 2000. "The Paris Geographical Society Constructs the Other, 1821–1850." *Journal of Historical Geography* 26 (2): 222–238.

———. 2003. *Labelling People: French Scholars on Society, Race, and Empire 1815–1848*. Ithaca: Cornell University Press.

Steintrage, James A. 2016. *The Autonomy of Pleasure: Libertines, License, and Sexual Revolution*. New York: Columbia University Press.

Stocking, George. 1964. "French Anthropology in 1800." *Isis* 55: 134–150.

Taillemite, Étienne. 1978. *Bougainville et ses compagnons autour du monde 1766–1769, Journaux de navigation*. 2 vols. Paris: Imprimerie Nationale.

Thériault, Serge A. 1979. *Jean-Jacques Rousseau et la médecine naturelle*. Montréal: L'Aurore.

Trouillot, Michel-Rolph. 2015. *Silencing the Past: Power and the Production of History*. Boston: Beacon Press.

Valentin de Cuillon, François. 1802–1803. *Examen de l'esclavage en général, et particulièrement de l'esclavage des nègres dans les colonies françaises de l'Amérique*. 2 vols. Paris: Desenne and Maradan.

Voltaire, François-Marie Arouet de. 1756. *Essay sur l'Histoire Générale, et sur les Moeurs et l'Esprit des Nations, depuis Charlemagne jusqu'à nos jours*. Geneva: Cramer.

———. 1759–1763. *Histoire de l'Empire de Russie sous Pierre le Grand*. 2 vols. n.p.: n.p.

Vyverberg, Henry. 1989. *Human Nature, Cultural Diversity, and the French Enlightenment*. New York: Oxford University Press.

Ward, Candace. 2007. *Desire and Disorder: Fevers, Fictions, and Feeling in English Georgian Culture*. Lewisburg: Bucknell University Press.

Wheeler, Roxann. 2000. *The Complexion of Race: Categories of Difference in Eighteenth-Century British Culture*. Philadelphia: University of Pennsylvania Press.
Williams, Elizabeth A. 1994. *The Physical and the Moral: Anthropology, Physiology, and Philosophical Medicine in France, 1750–1850*. New York: Cambridge University Press.
Winston, Michael E. 2005. *From Perfectibility to Perversion: Meliorism in Eighteenth-Century France*. New York: Peter Lang.
Young Lee, Paula. 2010. "The Curious Affair of Monsieur Martin the Bear." *Journal for Eighteenth-Century Studies* 33 (4): 615–629.
Zammito, John H. 2006. "Policing Polygeneticism in Germany, 1775: (Kames) Kant, and Blumenbach." In *The German Invention of Race*, edited by Sara Eigen and Mark Larrimore, 35–54. Albany: SUNY University Press.

5 Blumenbach's collection of human skulls

Wolfgang Böker

In the course of more than half a century, Johann Friedrich Blumenbach put together an "apparatus" in support of his anthropological research of which the main part consisted of some 240 human skulls. Essential was less the numerical size of the collection than its diversity in terms of geography and the age of the individual items. The collection covered all parts of the world that during Blumenbach's lifetime were accessible, and contained specimens from Egyptian antiquity to Blumenbach's own days. Some 600 skulls were added after 1840 to what is known as the "Blumenbachsche Schädelsammlung." Here I describe the development of the skull collection up to the year of Blumenbach's death, adding an overview of the posthumous history of the collection. Emphasis is placed on the first decade of the acquisition of skulls. The question is addressed at what times and in which ways the material proved formative of, and significant for, Blumenbach's work.

Blumenbach's collection of human skulls

A distinction must be made between the history of Blumenbach's collection of human skulls and that of Göttingen University's "Academic Museum" (*Academisches Museum*), managed by him. The museum was established in 1773, when the university purchased the vast private collection of Christian Wilhelm Büttner, professor of natural history at Göttingen, for academic teaching. It therefore could build on an existing collection and the museum was a public institution (Plesker 2015). By contrast, the skull collection was started from scratch by Blumenbach himself and was his private property, "which the university has no more claim to than to my legwear," as he told Georg Forster in May 1792 (Dougherty 2012, 158). Accordingly, the skull collection is not mentioned in Johann Stephan Pütter's quasi-official chronicle of Göttingen University (Pütter 1788; Pütter and Saalfeld 1820; Pütter, Saalfeld and Oesterley 1838). Blumenbach was well aware of the fact that in starting a specialized collection for the natural history of man, he created something different from earlier "anatomical cabinets" (Blumenbach 1790b, 3–4; Nutz 2009, 257–259).

The skull collection was only part – however, a central part – of a more comprehensive assemblage of human remains, which from 1790 was referred to by

Blumenbach as his "apparatus anthropologicus" (Blumenbach 1790b, 5).[1] Other important and continuously expanded areas of this "apparatus" were hair, tissue, and portraits (Blumenbach 1790a, 74–75). This is apparent from Blumenbach's manuscript catalogues of his collection,[2] which document its growth between 1795 and 1840, and from the use of characteristic portraits to visualize different varieties of mankind in his *Abbildungen naturhistorischer Gegenstände* (Blumenbach 1796; 1810). Tissue preparations from Blumenbach's time are no longer extant. This makes the skulls the only part of the collection that still exists in its entirety, kept in a single place.

An important feature of the "apparatus anthropologicus" was a meticulous documentation of the provenance of its items and the manner of their acquisition. Blumenbach archived his correspondence with the donors as well as other documents that could certify the authenticity of his specimens; he did this separately for each skull in numbered envelopes (Blumenbach 1790b, 5, 1806, 63–65; English translation in Bendyshe 1865, 301–302). It seems, however, that this system was abandoned after Blumenbach's death, because many of the letters which accompanied the incoming skulls are no longer kept with the skull collection today, but are part of Blumenbach's manuscripts in Göttingen's University Library (Dougherty 2006–2015, list of letters/provenances). Among the Blumenbach papers are also letters and reports that his donors had received from others and forwarded to Blumenbach in order to document the provenance of skulls (e.g., Dougherty 2013, 179–180 letter nr. 1099, 300 fn 15; Dougherty 2015, 340–341 letter nr. 1638). Other sources for the history of the skull collection are four extant catalogues[3] by Blumenbach, covering the period from 1793 to 1840, and publications by Blumenbach, especially the *Decades collectionis suae craniorum diversarum gentium illustratae* (Blumenbach 1790b; 1793; 1795a; 1800; 1808b; 1820) and their German "summaries" in the *Göttingische gelehrte Anzeigen*.[4] Moreover, a late nineteenth-century inventory of the collection (Spengel 1877) is extant and provides additional information.

Development of the skull collection, 1775–1840

It seems that before 1784 Blumenbach possessed only three original crania: a skull from a Göttingen churchyard, mentioned for the first time in 1775;[5] another from an ossuary in Murten/Switzerland, given to Blumenbach by a former student in 1778;[6] and a third of an ancient Egyptian mummy, bought from a Gdansk (Danzig) merchant, who passed through Göttingen in 1779 (Blumenbach 1790b, 13). Only this third specimen became part of the "official" skull collection, documented by Blumenbach in his manuscript catalogues and publications (Blumenbach 1790b, 13; 1795b, XXIX nr. 23). In 1800, Blumenbach retrospectively dated the beginning of his skull collecting back to around 1780 (Blumenbach 1800, 3).

In 1784, Christian Friedrich Michaelis, upon his return from the United States, gave Blumenbach the skulls of a North American native and of an African slave who had died in New York.[7] Blumenbach mentioned this in a letter of September 24, 1784, to Pieter Camper, also asking if Camper could provide him with the skull of a "Hottentot" (Dougherty 2007, 190 letter nr. 306). This appears to indicate the beginning of a quest for skulls of non-European ethnic groups (Dougherty 2007, XIV) and from precisely defined parts of the world.[8]

Two catalogues, both dating from 1794 and together listing eighty-three complete skulls with the respective year of acquisition, allow a reconstruction of the annual growth of the collection during the years 1784 to 1795 (Table 5.1). This is also the decade of Blumenbach's most intensive study of skull morphology and its significance for physical anthropology (see ahead).

Two later catalogues, covering the period 1795 to 1836, only very rarely give the acquisition date for the listed skulls.[9] Some information about the further growth of the collection can be derived, however, from contemporaneous publications (Table 5.2).[10]

Within the first ten years of his collecting activity Blumenbach acquired one third of the skulls that he was to possess at the end of his life more than forty years later. Fifty-seven of the first eighty-three skulls, however, were obtained from a single donor, the Russian state counselor Georg Thomas von Asch. Apart from these, by 1795, Blumenbach had obtained only twenty-six of a total of ca. eighty-three skulls – that is, around one eighth. Taking into account this, his success as a collector increased in later years, when his collection grew by ca. forty skulls per decade. Many of the skulls were sent to Göttingen by Blumenbach's former students, whose number, of course, grew over the years (Segala 2013, 31).

Table 5.1 Blumenbach's skull collection between 1778 and 1794. Sources: SUB Göttingen, Cod. Ms. Blumenbach I, nr. 1 and nr. 2.

Year	*Number*	*Total*
1778	1	1
1779	1	2
1780	0	2
1781	0	2
1782	0	2
1783	0	2
1784	2	4
1785	2	6
1786	3	9
1787	3	12
1788	2	14
1789	10	24
1790	12	36
1791	14	50
1792	10	61
1793	7	68
1794	8	76
Date unknown	8	83

Table 5.2 Blumenbach's skull collection in 1806, 1817, and 1840. Sources, for 1806: Blumenbach (1808a, 200 [printed version of a lecture given in 1806]); cf. Tantini (1812, 48), who saw 130 skulls in 1807. For 1817: SUB Göttingen, Cod. Ms. Blumenbach I, nr. 4: total of numbered items when the catalogue was started in 1817; among these three complete mummies. For 1840: report in preparation of the purchase of Blumenbach's collections for Göttingen University, mentioning "about 229" skulls (Reich and Gehler 2012, 174); cf. a slightly different number given in 1856, referring to 1840: "245 ganze Schädel und Schädelfragmente, sowie eine ägyptische und eine Guanchenmumie" (Wagner 1856, 235).

Year	*Number*
1806	134
1817	149
1840	229 (or 245)

State institutions and public funds did not play a direct part in the building of the skull collection, although they did in the acquisition of other specimens for the Göttingen Academic Museum. In 1781, for example, the Hanoverian government instructed a member of its staff in London to look for objects from the Islands in the Pacific for the Göttingen Museum and funded their purchase; also, in 1799, it acquired Johann Reinhold Forster's collection of similar objects (Urban 1998). Even under French rule the museum received gifts from the government of the Kingdom of Westphalia (Blumenbach 1809). By contrast, we do not know of any official instructions to diplomats to look for skulls on behalf of Blumenbach or of any financing of expeditions for this purpose. It was only indirectly that Blumenbach could benefit from state institutions – for example, through Asch and Blumenbach's British patron Joseph Banks, the president of the Royal Society, and through some of his students, who belonged to influential families or became themselves highly ranking civil servants later in their careers. Moreover, there is no evidence that Blumenbach obtained non-European skulls by purchase (except for the one he bought in 1779), nor that he brought back skulls from his own travels to Switzerland, London, and Paris. To a large extent the development of the skull collection depended upon Blumenbach's circle of acquaintances – for example, his students – and therefore did not follow a systematic plan.

The skull collection after 1840

During Blumenbach's lifetime, the skull collection was not kept in the building of the Academic Museum, as previously assumed (Ehlers 1901, 404), but privately in his Göttingen home on the street now called "Neustadt" (Nawa 2010, 43 fn 170; Dougherty 2007, 202, letter nr. 310 fn 3). It could be studied there at his invitation by passing visitors, such as Franz Joseph Gall and Francesco Tantini (Klatt 2013, 36–53; Tantini 1812, 48). After Blumenbach's death on January 22, 1840, his private collections, including the skulls, were bought by the Kingdom of Hanover and became part of the university's Academic Museum (Reich and

Gehler 2012). The skulls thus became accessible to the scholarly public without restrictions. Rudolph Wagner, Blumenbach's successor as professor of physiology and zoology at Göttingen University, became responsible for the anthropological and zoological collection, which included the skulls (Wagner 1860, 166–167). In 1842, these were transferred to Wagner's newly founded Institute of Physiology (Wagner 1856, 234–235, English version in Bendyshe 1865, 348; Ehlers 1901, 434; Nawa 2010, 45). In the mid-1850s, interest from inside and outside Göttingen in the collection increased, in part due to the engagement of Wagner (Wagner 1856, 241, English version in Bendyshe 1865, 353–354). On the initiative of Karl Ernst von Baer and Wagner, those who worked on the collection and other anthropologists convened at Göttingen for a congress on September 24, 1861. This date can be considered the founding moment of German biological anthropology as an independent academic discipline (Hoßfeld 2005, 87).

After Wagner's death the skull collection was transferred to the university's Institute of Anatomy and is kept today in its "Zentrum Anatomie" (Wagner 1890, 37; Schultz and Reich 2013, 50–51). The skulls no longer serve the study of human morphological diversity, in part as a consequence of the racist abuse of anthropology under National Socialism. Moreover, research into the biological diversity of humankind – Blumenbach's "generis humani varietas nativa" – is today primarily based on genetics. In recent decades, Blumenbach's skull collection has been rediscovered for its value as "biohistorical documents" that can be applied to forensic anthropology and medicine, to the history of medicine, and to paleopathology, archaeology, and ethnology (Schultz 2012).

The skull collection and Blumenbach's research on the natural history of mankind

When Blumenbach was studying the natural morphological diversity of the human species for his dissertation *De generis humani varietate nativa*, he had almost no human skulls or other relevant anatomical specimens at his disposal, not even in the Academic Museum, with its more than 12,000 specimens. Among these were only three complete skulls.[11] Blumenbach himself possessed one human skull at the time, and not until 1778 and 1779 did he obtain two other skulls. Thus in the first (1775/1776) and second (1781) editions of his dissertation he relied almost exclusively on published descriptions and images of skulls. In fact, in 1776, he made it a matter of discussion whether the shape of the skull could be used to define "varieties" of humankind. He concluded that information based on reports and pictures was too inconsistent and contradictory to define human variety, as shown by the fact that the one skull from the Göttingen churchyard he possessed resembled images of North American skulls "as like as two peas in a pod" (Blumenbach 1776, 63). Moreover, he thought that it was impossible to determine "natural" shapes of skulls, because in most cases skulls were subject to lifetime deformation, due to ways of living and disfigurement for the purpose of matching cultural norms of beauty ("omnem fere capitis formae in diversis populis differentiam, unice fere vitae generi et arti tribuendam esse," Blumenbach 1776, 68;

English version in Bendyshe 1865, 121). In the second edition of *De generis humani varietate nativa*, Blumenbach made some changes to this line of argument (mainly because of the newly introduced fifth variety), but came to the same conclusion (Blumenbach 1781, 88). Originally, he therefore rejected skull shape as a criterion of human variety, using instead skin color, facial features, and color of hair (Dougherty 2007, XII–XIII; Nutz 2009, 260; Vermeulen 2015, 372). On this basis he recognized four main varieties of the human species, successively located in Europe and the neighboring parts of Asia and Africa, central and southern Asia, sub-Saharan Africa, and America.

In his *Handbuch der Naturgeschichte* (Blumenbach 1779, 63–64) and in the second edition of his dissertation (Blumenbach 1781, 51–52, 93) Blumenbach introduced a major modification of the schema of four main varieties by adding a fifth, located in Southeast Asia, the southern Pacific, and Australia and called the "Malayan" variety from 1795 on (Blumenbach 1795b, 286). This expanded classification was not based on the arrival of skulls from these parts of the world, because Blumenbach tried to obtain Pacific skulls only after 1787. Instead, the new, enlarged categorization resulted from a more careful ("curatius," "propius") reexamination of the sources he already had used in 1775/1776 – that is, visual representations and reports in the literature ("postquam autem in Asiae orientalis et Americae diuersas gentes curatius inquisiui easque vt ita dicam propius spectaui"; Blumenbach 1781, 51; cf. Dougherty 2007, XII–XIII). The same appears to be true for two minor adjustments *within* the five-varieties scheme, made slightly later and probably also not triggered by the examination of actual skulls.[12]

To repeat, in 1784 Blumenbach was given the skulls of a North American native and an African, and in the following years, he revised his opinion on the significance of skull morphology for the study of the "varietas nativa" of humans. Since the early 1780s, osteology had been a focus of Blumenbach's studies. He announced lectures on osteology for the first time in 1777 (Anon. 1777, 263), and from 1781 on he worked on a textbook of osteology (Dougherty 2006, 227, letter nr. 145), which was published in 1786. More than one third of its 480 pages dealt with the bones of the skull (Blumenbach 1786, 93–264). Obviously, Blumenbach was aware of the special "semantic density" (Nutz 2009, 261) of this part of the skeletal system. Already in this textbook he expressed his confidence that the "natural diversity of human races" could be demonstrated by features of the skull (Blumenbach 1786, 85–90). In substantiation of this idea, he described and compared the prominent characteristics of three non-European skulls in his possession, using a "beautiful European skull" as "model and standard of [their] diversity" (Blumenbach 1786, 86 fn **; cf. Dougherty 2007, XIV fn 55). This foreshadowed Blumenbach's later notion of a morphologically neutral skull shape representing the original type, as distinct from skulls with more prominent features which are derivative variations. Apparently, the arrival from 1784 on of non-European skulls had an impact on Blumenbach's anthropological studies by sparking a new assessment of skull morphology and its significance in defining the natural variety of the human species.

On December 12, 1789, when Blumenbach's collection had grown to twenty-four skulls (see earlier), he gave a lecture about ten of them at the Göttingen Academy (Societät der Wissenschaften) (Blumenbach 1790c) and published a leaflet with annotated copperplate engravings of these crania, the first *Decas collectionis suae craniorum* (Blumenbach 1790b). In the introduction he programmatically formulated new methodological guidelines for his future anthropological research.[13] First, it should no longer be based on reports and random pictures, but on real bone and tissue material. To this end and as mentioned earlier, he had set up an anthropological collection of different types of specimens: skulls, fetuses, hair, and tissue preparations, as well as reliable portraits, drawn from life.[14] This diversity of documentary material reflected Blumenbach's belief that the complexity of organic bodies – for which he coined the term "Gesamthabitus" – could be grasped only by a multidimensional, "natural" method, without artificial abstraction – that is, without reduction to one feature or one type of evidence.[15] Second, the skulls of Blumenbach's collection made it now possible to use skull morphology as the main – but not the only – basis for defining variety within the human species.[16]

The substantially revised edition of *De generis humani varietate* of 1795 was based on the new approach. This edition also contained pictures of human skulls, which the earlier editions did not. To sum up: by 1795 Blumenbach possessed eighty-three skulls, among them, since 1794, the five "model skulls"; he now used them to visualize the five varieties of mankind, already defined fifteen years before. In the new edition the text was preceded by an index of Blumenbach's anthropological collection, referred to as "sup[p]ellex anthropologica" (Blumenbach 1795b, XXI–XLIV). It listed eighty-two skulls[17] and other items and was meant to substantiate Blumenbach's empirically based studies and inferences.

Just as important as the new methodology were two innovations in Blumenbach's description of the five main varieties of humankind, which for the first time were given names. First, each variety was defined on the basis of a characteristic skull shape. Second, between these individual shapes a formal and "genetic" relationship was established. This means that Blumenbach used cranial features not only to give his typology an additional, new foundation (Dougherty 2007, XIV; Nutz 2009, 259) but also, more importantly, to describe the five varieties of humankind as a continuous spectrum of forms (Figure 5.1). For this purpose, he defined a "neutral" original form, characterized by the absence of any projecting parts – namely, the skull of the "Caucasian" variety. All other forms were seen as derived from the original form, showing changes in either cheekbones or jaws and resulting in two "extreme" forms: the "Aethiopian" skull on the one side and the "Mongolian" on the other, each representing an alternative direction of change. Blumenbach visualized this by a view of three skulls from above (Figure 5.2) and termed this method of comparison "norma verticalis" (Blumenbach 1795b, 203–205).

Between the original form and the extremes Blumenbach placed two transitional forms, the "Malayan" and the "American."[18] He claimed that his interpretation was based primarily on the shape of the skulls, using skin color as additional evidence.[19]

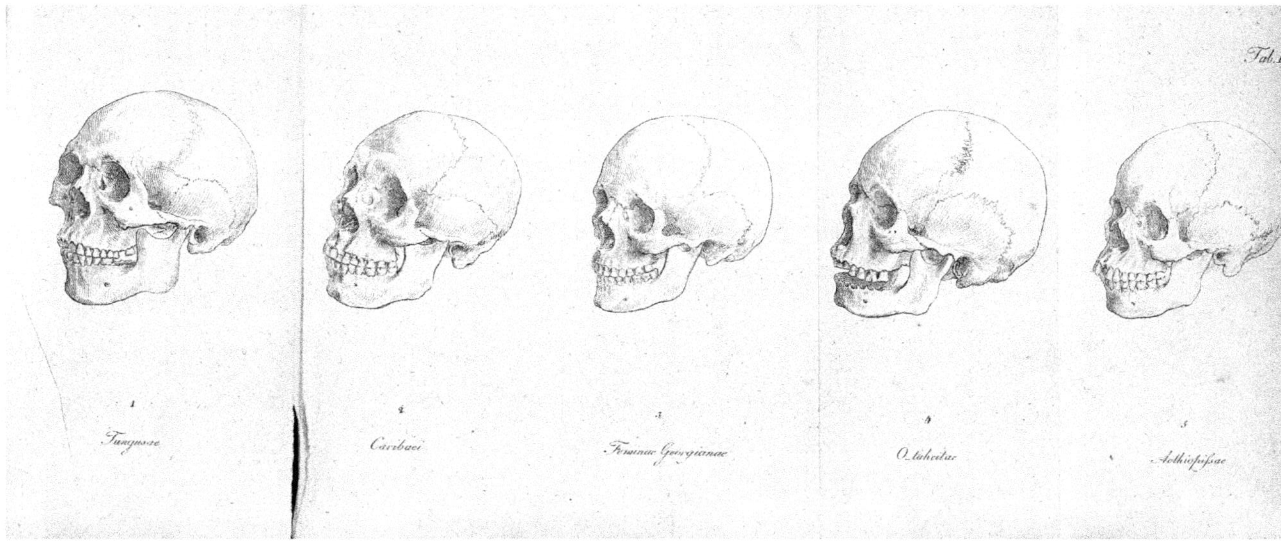

Figure 5.1 Blumenbach's five model skulls in semi-profile, aligned horizontally. Copperplate engraving (from Blumenbach 1795b, Tabula II).

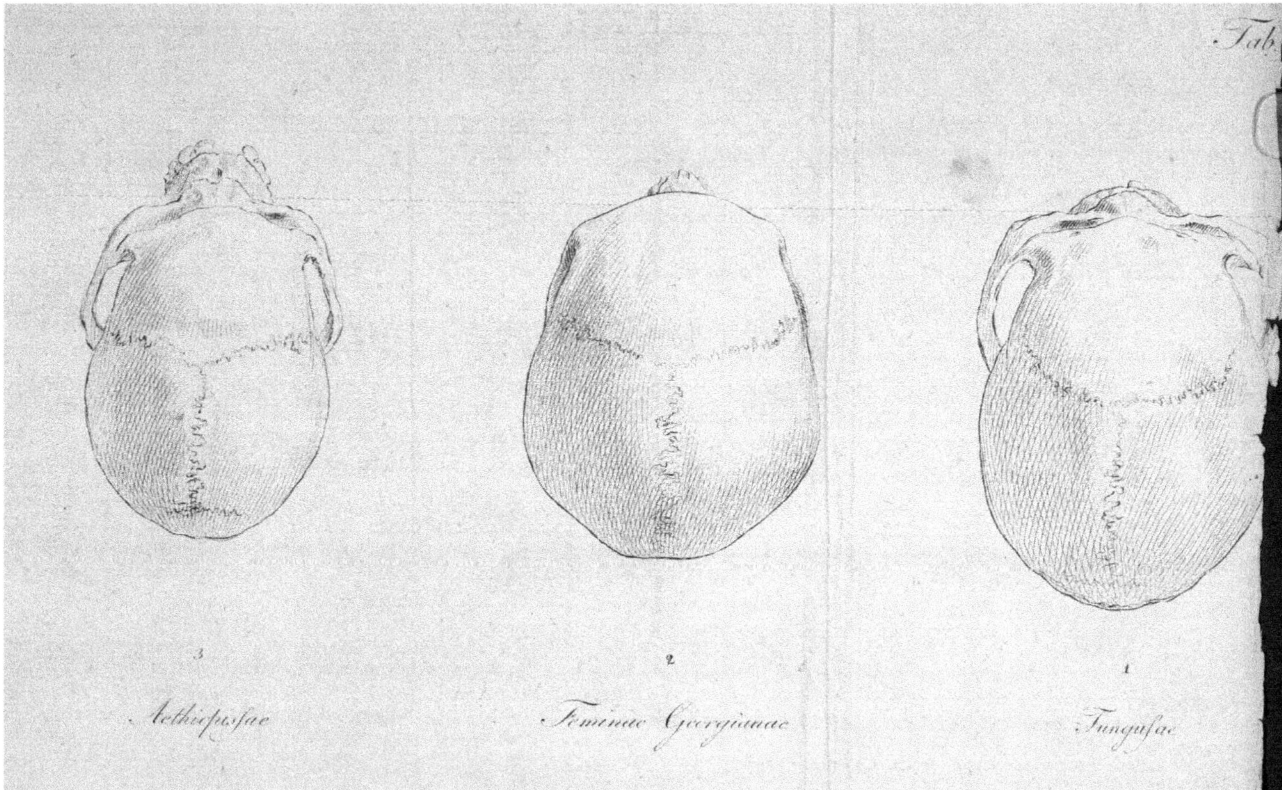

Figure 5.2 Three of Blumenbach’s five model skulls, viewed from above. Copperplate engraving (from Blumenbach 1795b, Tabula I).

This crucial innovation of Blumenbach's concept was visualized by the copperplate engraving of Figure 5.1, showing the five model skulls aligned horizontally with the "Caucasian" skull in the center and the "Mongolian" and the "Aethiopian" on the two opposite ends. Skull morphology allowed arranging the varieties that so far had been described separately and individually as a linear continuum of related forms, thus underlining the central idea of Blumenbach's anthropological research: the unity of humankind.

In addition to the acquisition of non-European skulls by Blumenbach, two of his colleagues, working in the same field, may have contributed to his turn toward skull morphology (Nutz 2009, 260–263). Already in the 1760s, Camper had formed a small collection of skulls, among these eight non-European specimens, and had used them to study ethnic differences. And in 1784, Samuel Thomas Soemmerring had introduced a novel, strictly "anatomical and physiological argumentative strategy" (Nutz 2009, 261–262) in his anthropological treatise about "the bodily difference between the Negro and the European" (Soemmerring 1784). In his critical reception of Soemmerring's results, Blumenbach may have felt prompted to adopt this method. Yet it seems unlikely that he "discovered skulls as crucial evidence for the natural history of man only at the suggestion of Camper and Soemmering" (Nutz 2009, 263).

Concluding observations

The interpretation of the diversity of human skull morphology presented by Blumenbach in the 1795 edition of *De generis humani varietate* and in the first three volumes of *Decades craniorum* (Blumenbach 1790b, 1793, 1795a) represented the zenith and to a certain degree the end of his conceptual skull work. In the following years, Blumenbach's focus was no longer on developing scientific interpretations of the incoming specimens but on enlarging and administering the collection. After 1795, the number of skulls on display in Blumenbach's house grew by an average of four skulls per year, and Blumenbach published another thirty-five engravings of them, albeit at a much reduced frequency compared to the period between 1790 and 1795 (Blumenbach 1800, 1808b, 1820, 1828). The new skulls were used to confirm the classification of 1795 and did not trigger a revision or supplementation with subvarieties. No tabular or statistical techniques for presenting and interpreting large numbers of specimens were introduced, either in publications or in private manuscript catalogues of the collection. In the post-1795 descriptions of skulls, Blumenbach did not change his "cranioscopical" method (Karliczek and Jank 2010, 59) of describing the features of individual skulls and matching them with the "Gesamthabitus" of a variety. He stuck to his rejection of the unsatisfactory craniometrical methods suggested by Camper and Louis-Jean-Marie Daubenton, which he had first expressed in 1790 and reaffirmed more than three decades later (Blumenbach 1790b, 7–10, 1815, 20 fn *, 1824, 20 fn **). However, he also did not try to develop more refined measuring procedures for the purpose of grasping the complexity of skull morphology.

Blumenbach's manuscript catalogues of the skull collection confirm this observation. The earliest of them, dating from 1793, is an unstructured list of items according to their geographical (and ethnical) provenance, whereas the second, started in 1794/1795, arranges them in groups according to topics and didactic principles.[20] This represented a comprehensive grasp and conceptualization of the empirical material. During the following two decades new skulls were fitted into the categories established in 1795 and given an "identification number." When in 1817 Blumenbach started a new catalogue, the categories were not amended but merged into more general groups and supplemented by only two new categories – namely, for mummies and ancient skulls. Later entries became increasingly unclear, and new skulls were no longer marked with identification numbers.

An analysis of the early phase of Blumenbach's skull collecting invalidates the widely held belief that the skulls provided the empirical basis for Blumenbach's famous classification of five "varieties" of humans. Blumenbach developed this concept well before he possessed any skulls from parts of the world other than his own – that is, Europe. A growing number of original skulls at his disposal subsequently convinced him of the significance of skull morphology for physical anthropology. More importantly, he turned skull morphology into crucial evidence for his thesis of the fundamental unity of all the varieties of mankind.

Notes

1 Cf. also the title of the manuscript catalogue of his collection "Catalogus meiner Schedelsammlung u. des übrigen dazu gehörigen anthropologischen Apparats" (1817), Staats- und Universitätsbibliothek (SUB) Göttingen, Cod. Ms. Blumenbach I, nr. 4.

2 SUB Göttingen, Cod. Ms. Blumenbach I, nr. 3 and nr. 4.

3 SUB Göttingen, Cod. Ms. Blumenbach I, nr. 1–4 and nr. 4 Anhang. Digital versions of these catalogues are available online via the website of the project "Johann Friedrich Blumenbach – Online" (www.blumenbach-online.de).

4 For Blumenbach's skull publications see Kroke (2010). A searchable online version with access to digital versions of all publications by Blumenbach is available via the website of the project "Johann Friedrich Blumenbach – Online".

5 "Ipse cranium, vetustum satis, praeterita aestate e sepulcreto vrbis erutum, coram habeo" (Blumenbach 1775, 63). According to the publication date of this first edition of Blumenbach's dissertation, the text was finished by the end of 1775 and printed within the context of the graduation process, as shown by the date of the oral exam (September 16, 1775) on its title page. This date was omitted from the title page of the otherwise identical and better known edition of 1776 (Blumenbach 1776). The skull from the Göttingen churchyard was not listed in the printed catalogue of the skull collection (Blumenbach 1795b, XXI–XXXIV). It also does not appear in any of the manuscript catalogues of the collection. It can be identified, however, by a label on one of the extant skulls of the collection (Spengel 1877, 10, "Nr. 56 [= Nr. des Göttinger Katalogs 259]").

6 SUB Göttingen, Cod. Ms. Blumenbach I, nr. 1, fol. 1, unnumbered entry (between nr. 34 and nr. 36); not in Blumenbach 1795b, and in the other manuscript catalogues, but (without number) in a list of skulls on display in Blumenbach's home, SUB Göttingen, Cod. Ms. Blumenbach I, nr. 4 Anhang, ("Vierter Schrank, zweyte Reihe"). See Spengel (1877, 16, "Nr. 88 [= Nr. des Göttinger Katalogs 295]"). About the donator Samuel Wyss see Dougherty (2007, XIII).

7 SUB Göttingen, Cod. Ms. Blumenbach I, nr. 1, fol. 7 nr. 59, and fol. 8 nr. 65.

8 For example, letter to Joseph Banks, June 20, 1787 (Dougherty 2010, 109–110, letter nr. 464); letter to Johannes Loretz, July 7, 1791 (Dougherty 2012, 76–78, letter nr. 691); letter from Georg Thomas von Asch to Blumenbach, Aug. 30 (Julian date: Aug. 16), 1785: "Keine Bemühungen sollen mir zu schwer sein, die verlangte [sic] Schedel von Asiatischen Völkern Ihnen zu verschaffen" (Dougherty 2007, 312–313, letter nr. 375); letter from Friedrich Scholl to Blumenbach, after October 20, 1786: "Einen *Cretin* Kopf zu bekommen wird schwehr halten doch seye versichert daß ich alles mögliche anwenden werde um einen zu erhalten" (Dougherty 2010, 50, letter nr. 427; emphasis in original).

9 SUB Göttingen, Cod. Ms. Blumenbach I, nr. 3 and nr. 4.

10 A scientific database of the skull collection will be developed by the project "Johann Friedrich Blumenbach – Online" and provide more detailed information.

11 The original catalogue, written by Blumenbach, and a duplicate are kept in Göttingen University's Institute of Social and Cultural Anthropology (*Institut für Ethnologie*) (Urban 2001, 98 fn 4). An online version of the catalogue is available via www.blumenbach-online.de. In the section "Der Mensch" are listed: "16. Ein Türken Schedel," "17. der Schedel eines Franzosen aus der Schlacht von Hoechstaedt," "18. Ein aufgesägter Schedel ohne calvaria," "19. Ein seitwärts aufgesägter Schedel mit den präparierten Fortsätzen der dura mater," "20. Der Schedel eines 8 monatlichen Foetus," "21. Eine einfache Calvaria von ungewöhnlicher Dicke," and three lower jaws, four bone preparations, two skin preparations, and several preparations of internal organs and vessels.

12 Whereas in the first and second editions of his *Handbuch der Naturgeschichte* (Blumenbach 1779, 63; Blumenbach 1782, 60), Blumenbach had classified the inhabitants of the North American Arctic as belonging to the first (since 1795 "Caucasian") "varietas," he classified them as belonging to the second (since 1795 "Mongolian") in the third edition (Blumenbach 1788, 61). Skulls of indigenous people from Arctic America ("Eskimos"), sent by Johann Loretz, arrived in Göttingen only in 1794 (cf. SUB Göttingen, Cod. Ms. Blumenbach I, nr. 4, fol. 8, nrs. 25 and 26). And in the 1795 edition of *De generis humani varietate nativa*, Blumenbach reclassified the inhabitants of Finland and Lapland in the same way (Blumenbach 1795b, 290–292). Skulls from Lapland arrived in Göttingen only in 1800; cf. a letter from Carl Peter Thunberg to Blumenbach, November 15, 1800 (Dougherty 2012, 522–523, letter nr. 1347). Since 1790, however, there was a skull from Finland in Blumenbach's collection, and two others since 1792; cf. SUB Göttingen, Cod. Ms. Blumenbach I, nr. 1, fol. 2, nrs. 9–11.

13 Nutz (2009, 260) refers only to the illustrations of the *Decades* as part of Blumenbach's strategy of verification ("Beglaubigungsstrategie") and does not seem to take into account the introduction of the first *Decas*. He thus gives the impression that it was not before 1795 that Blumenbach used skull morphology for his classification of human varieties.

14 "Inprimis vero eo vitio laborare libellum [i.e., *De generis humani varietate nativa*, 1776] facile agnoui, quod in eo conficiendo fere vnice ex libris sapere mihi licuerit, ita vt maxima pars eorum quae de hominum varietatibus continebat, non nisi ex itinerariorum farragine compilata, verbo ex charta in papyrum translata esset.

Nec studio ergo nec sumptibus peperci vt huic defectui quantum licebat mederer, vtque subsidia mihi compararem *ex ipsa natura petita*, quibus varietates gentium rite dignosci possunt, crania viz., embryones, pilorum specimina, aliasque corporis partes huic scopo inseruientes. Denique vero et imaginum ad viuum curate factarum copiam" (Blumenbach 1790b, 4; emphasis in original).

15 "Verum in vniuersum et in hac historiae naturalis parte vt in reliquis *naturalem* quam vocant methodum, quae longe plurimos corporis organici characteres simul sumtos complectitur, artificiali qua singularem tantum characterem a reliquis abstractum

pro fundamento systematico ponunt, praeferendam puto: ideoque in ordinandis et describendis variarum gentium craniis neque DAUBENTONII, neque CAMPERI, neque DVRERI regulas vnice adhibui sed vniuersum eorum habitum indagaui" (Blumenbach 1790b, 9; emphasis in original). For Blumenbach's concept of "Gesamthabitus" cf. Dougherty (1985).

16 "Craniorum inquam quibus ad gentilitias varietates distinguendas et definiendas nulla alia humani corporis pars aptior videtur, cum caput osseum [. . .] stabilitati suae maximam conformationis et partium relatiuae proportionis varietatem iunctam habeat, vnde characteres nationum certissimas desumere licet" (Blumenbach 1790b, 5). In the following paragraphs Blumenbach established methodological principles for the study of skulls. However, in defining the five main human varieties Blumenbach never abandoned referring to skin color, facial traits (nose, lips, eyes), and the structure of hair (Blumenbach 1790b, 11; Blumenbach 1795b, 287–296; Blumenbach 1830, 56–58).

17 Excluded from the list were two skulls of uncertain origin. Blumenbach, however, already included an ancient Roman cranium (Blumenbach 1795b, XXXI nr. 38), promised by Stefano Borgia on April 25, 1795, but arriving in Göttingen only in May 1797; cf. Blumenbach's letter to Borgia, May 6, 1797 (Dougherty 2013, 127–128, letter nr. 1062).

18 "*Caucasiam* [varietatem] ob caussas infra enarrandas pro primigenia habendam primo loco posui. Haec utrinque in bina ab invicem remotissima et diversissima extrema abiit, hinc nempe in *Mongolicam*, illinc in *Aethiopicam*. Medios vero inter istam primigeniam et hasce extremas varietates locos tenent reliquae binae: *Americana* nempe inter Caucasiam et Mongolicam. *Malaica* inter eandem istam Caucasiam et Aethiopicam" (Blumenbach 1795b, 286–287; emphasis in original).

19 "*Primo* enim loco eam venustissimam ut vidimus [. . .] *cranii formam* prae se fert ista stirps, ex qua, tanquam, ex figuratione media et primigenia, reliquae utrinque usque ad ultima bina extrema (hinc scilicet Mongolicum, illinc contra Aethiopicum) simplicissima gradatione fluunt. *Tum* vero eadem albo *colore* est, quem itidem pro primitivo generis humani habere licet, cum ex eo, ut supra exposuimus [. . .] facilis in fuscum degeneratio sit, longe difficilior contra ex fusco, [. . .] in candidum" (Blumenbach 1795b, 304; emphasis added).

20 SUB Göttingen, Cod. Ms. Blumenbach I, nr. 2, which is basically identical with the printed catalogue in the third edition of *De generis humani varietate* (Blumenbach 1795b).

Bibliography

Anon. 1777. "Göttingen." *Göttingische Anzeigen von gelehrten Sachen 1777* (33. Stück, 17. März), 257–264.

Bendyshe, Thomas, trans. 1865. *The Anthropological Treatises of Johann Friedrich Blumenbach* [. . .]. London: Published for the Anthropological Society, by Longman, Green, Longman, Roberts, & Green.

Blumenbach, Johann Friedrich. 1775. *De generis humani varietate nativa.* Goettingae: Typis Frid. Andr. Rosenbuschii.

———. 1776. *De generis humani varietate nativa liber.* Goettingae: Apud Viduam Abr. Vandenhoeck.

———. 1779. *Handbuch der Naturgeschichte.* Teil 1. Göttingen: Bey Johann Christian Dieterich.

———. 1781. *De generis humani varietate nativa liber.* Editio altera longe auctior et emendatior. Goettingae: Apud Viduam Abr. Vandenhoeck.

———. 1782. *Handbuch der Naturgeschichte*. Zweyte durchgehends verbesserte Ausgabe. Göttingen: Bey Johann Christian Dieterich.

———. 1786. *Geschichte und Beschreibung der Knochen des menschlichen Körpers*. Göttingen: Bey Johann Christian Dieterich.

———. 1788. *Handbuch der Naturgeschichte*. Dritte sehr verbesserte Ausgabe. Göttingen: Bey Johann Christian Dieterich.

————. 1790a. *Beyträge zur Naturgeschichte*. Erster Theil. Göttingen: Bey Johann Christian Dieterich.

———. 1790b. *Decas collectionis suae craniorum diversarum gentium illustrata*. Göttingen: Apud Ioann. Christ. Dieterich.

———. 1790c. "Göttingen." *Göttingische Anzeigen von gelehrten Sachen 1790* (3. Stück, 4. Januar), 25–29.

———. 1793. *Decas altera collectionis suae craniorum diversarum gentium illustrata*. Göttingen: Apud Ioann. Christ. Dieterich.

———. 1795a. *Decas tertia collectionis suae craniorum diversarum gentium illustrata*. Göttingen: Apud Ioann. Christ. Dieterich.

———. 1795b. *De generis humani varietate nativa. Editio tertia*. Goettingae: Apud Vandenhoek et Ruprecht.

———. 1796. *Abbildungen naturhistorischer Gegenstände*. Heft 1. Göttingen: Dieterich (2nd ed., 1810).

———. 1800. *Decas quarta collectionis suae craniorum diversarum gentium illustrata*. Göttingen: Apud Joann. Christian. Dieterich.

———. 1806. *Beyträge zur Naturgeschichte*. Erster Theil, zweyte Ausgabe. Göttingen: Bey Heinrich Dieterich. (English translation in Bendyshe 1865).

———. 1808a. "Jo. Frid: Blumenbachii decas quinta collectionis suae craniorum diversarum gentium illustrata. D. XXV. Aug. MDCCCVI." *Commentationes Societatis Regiae Scientiarum Gottingensis* XVI: 199–216.

———. 1808b. *Decas quinta collectionis suae craniorum diversarum gentium illustrata*. Göttingen: Apud Henricum Dieterich.

———. 1809. "Göttingen." *Göttingische gelehrte Anzeigen* 1809 (93. Stück, 12. Juni), 921–922.

———. 1815. *Handbuch der vergleichenden Anatomie*. Zweite verbesserte und vermehrte Auflage. Göttingen: Bey Heinrich Dieterich.

———. 1820. *Decas sexta collectionis suae craniorum diversarum gentium illustrata*. Göttingen: Apud Henricum Dieterich.

———. 1824. *Handbuch der vergleichenden Anatomie*. Dritte verbesserte und vermehrte Auflage. Göttingen: In der Dieterichschen Buchhandlung.

———. 1828. *Nova pentas collectionis suae craniorum diversarum gentium illustrata*. Göttingen: Sumtibus Dieterichianis.

———. 1830. *Handbuch der Naturgeschichte*. Zwölfte rechtmäßige Ausgabe. Göttingen: In der Dieterich'schen Buchhandlung.

Dougherty, Frank William Peter. 1985. "Johann Friedrich Blumenbach und Samuel Thomas Soemmerring. Eine Auseinandersetzung in anthropologischer Hinsicht?" In *Samuel Thomas Soemmerring und die Gelehrten der Goethezeit*, edited by Gunter Mann and Franz Dumont, 35–56. Stuttgart: Gustav Fischer.

———. 2006. *The Correspondence of Johann Friedrich Blumenbach. Rev., augm. and ed. by Norbert Klatt*. Vol. 1 (1773–1782). Vol. 2 of Brosamen zur Blumenbach-Forschung. Göttingen: Klatt.

———. 2007. *The Correspondence of Johann Friedrich Blumenbach. Rev., augm. and ed. by Norbert Klatt*. Vol. 2 (1783–1785). Vol. 3 of Brosamen zur Blumenbach-Forschung. Göttingen: Klatt.
———. 2010. *The Correspondence of Johann Friedrich Blumenbach. Rev., augm. and ed. by Norbert Klatt*. Vol. 3 (1785–1790). Vol. 4 of Brosamen zur Blumenbach-Forschung. Göttingen: Klatt.
———. 2012. *The Correspondence of Johann Friedrich Blumenbach. Rev., augm. and ed. by Norbert Klatt*. Vol. 4 (1791–1795). Vol. 5 of Brosamen zur Blumenbach-Forschung. Göttingen: Klatt.
———. 2013. *The Correspondence of Johann Friedrich Blumenbach. Rev., augm. and ed. by Norbert Klatt*. Vol. 5 (1796–1800). Vol. 6 of Brosamen zur Blumenbach-Forschung. Göttingen: Klatt.
———. 2015. *The Correspondence of Johann Friedrich Blumenbach. Rev., augm. and ed. by Norbert Klatt*. Vol. 6 (1801–1805). Vol. 7 of Brosamen zur Blumenbach-Forschung. Göttingen: Klatt.
Ehlers, Ernst Heinrich. 1901. "Göttinger Zoologen." In *Festschrift zur Feier des hundertfünfzigjährigen Bestehens der Königlichen Gesellschaft der Wissenschaften zu Göttingen*. [Vol. 3:] Beiträge zur Gelehrtengeschichte Göttingens, 391–494. Berlin: Weidmannsche Buchhandlung.
Hoßfeld, Uwe. 2005. *Geschichte der biologischen Anthropologie in Deutschland: von den Anfängen bis in die Nachkriegszeit*. Stuttgart: Franz Steiner.
Karliczek, André, and Marlen Jank. 2010. "Quantifizieren, Typisieren, Hierarchisieren? Peter Camper und der Winkel der Natur." In *Natur im Kasten. Lichtbild, Schattenriss, Umzeichnung und Naturselbstdruck um 1800* [exhibition catalogue], edited by Olaf Breidbach, Kerrin Klinger, and André Karliczek, 58–78. Jena: Ernst-Haeckel-Haus.
Klatt, Norbert. 2013. "Johann Friedrich Blumenbach und Franz-Joseph Gall – Eine Begegnung." In *Kleine Beiträge zur Blumenbach-Forschung*, edited by Norbert Klatt. Vol. 5, 7–58. Göttingen: Klatt-Verlag.
Kroke, Claudia. 2010. *Johann Friedrich Blumenbach: Bibliographie seiner Schriften. Unter Mitarbeit von Wolfgang Böker und Reimer Eck*. Vol. 2 of Schriften zur Göttinger Universitätsgeschichte. Göttingen: Universitätsverlag.
Nawa, Christine. 2010. "Zum 'öffentlichen Gebrauche' bestimmt: Das Academische Museum Göttingen." *Göttinger Jahrbuch* 58: 23–62.
Nutz, Thomas. 2009. "Varietäten des Menschengeschlechts." In *Die Wissenschaft vom Menschen in der Zeit der Aufklärung*. Berlin: Böhlau Verlag.
Plesker, Nadine. 2015. "Das Königlich Academische Museum in Göttingen." In *Tempel der Kunst*, edited by Bénédicte Savoy, 413–442. Vienna: Böhlau Verlag.
Pütter, Johann Stephan. 1788. *Versuch einer academischen Gelehrten-Geschichte von der Georg-Augustus-Universität zu Göttingen. Zweyter Theil von 1765 bis 1788*. Göttingen: Vandenhoeck und Ruprecht.
———, and Friedrich Saalfeld. 1820. *Versuch einer academischen Gelehrten-Geschichte von der Georg-Augustus-Universität zu Göttingen. Dritter Theil von 1788 bis 1820*. Hannover: Helwingsche Hofbuchhandlung.
———, Friedrich Saalfeld, and Georg Heinrich Oesterley. 1838. *Versuch einer academischen Gelehrten-Geschichte von der Georg-Augustus-Universität zu Göttingen*. Vierter Theil von 1820 bis zur ersten Säcularfeier der Universität im Jahre 1837. Göttingen: Vandenhoeck und Ruprecht.
Reich, Mike, and Alexander Gehler. 2012. "Der Ankauf der Privatsammlung von J. F. Blumenbach (1752–1840) durch die Universität Göttingen." *Philippia* 15 (3): 169–187.

Schultz, Michael. 2012. "Die Blumenbachsche Schädelsammlung: Eine historische Schädelsammlung als Quelle interdisziplinärer Forschung." In *Dinge des Wissens: Die Sammlungen, Museen und Gärten der Universität Göttingen*, edited by Georg-August-Universität Göttingen, 106–116. Göttingen: Wallstein-Verlag.

———, and Mike Reich. 2013. "Blumenbachsche Schädelsammlung." In *Die Sammlungen, Museen und Gärten der Universität Göttingen*, edited by Georg-August-Universität Göttingen, 50–51. Göttingen: Universitätsverlag.

Segala, Marco. 2013. "Auf den Schultern eines Riesen. Arthur Schopenhauer als Student Johann Friedrich Blumenbachs." In ". . . die Kunst zu sehn." In *Arthur Schopenhauers Mitschriften der Vorlesungen Johann Friedrich Blumenbachs (1809–1811)*, edited by Jochen Stolberg and Wolfgang Böker. Vol. 3 of Schriften zur Göttinger Universitätsgeschichte, 13–40. Göttingen: Universitätsverlag.

Soemmerring, Samuel Thomas. 1784. *Über die körperliche Verschiedenheit des Mohren vom Europäer*. Mainz: n. p.

Spengel, Johann Wilhelm. 1877. *II. Göttingen. Die von Blumenbach gegründete Anthropologische Sammlung der Universität Göttingen: Aufgenommen im Jahre 1874*. Braunschweig: Vieweg. Vol. 2 of *Die anthropologischen Sammlungen Deutschlands*.

Tantini, Franceso. 1812. "Descrizione del gabinetto antropologico del [. . .] Blumenbach." In *Opuscoli scientifici*, edited by Franceso Tantini. Vol. 1, 43–64. Pisa: Nestri.

Urban, Manfred. 1998. "The Acquisition History of the Göttingen Collection." In *James Cook: Gifts and Treasures from the South Sea. The Cook/Forster Collection Göttingen*, edited by Brigitta Hauser-Schäublin and Gundolf Krüger, 56–85. New York: Prestel.

———. 2001. "Die Völkerkundliche Sammlung." In *"Ganz für das Studium angelegt": Die Museen, Sammlungen und Gärten der Universität Göttingen*, edited by Dietrich Hoffmann and Kathrin Maack-Rheinländer, 91–98. Göttingen: Wallstein.

Vermeulen, Han F. 2015. *Before Boas: The Genesis of Ethnography and Ethnology in the German Enlightenment*. Lincoln: The University of Nebraska Press.

Wagner, Hermann.1890. "Physiologisches Institut." In *Chronik der Georg-Augusts-Universität für das Rechnungsjahr 1889–1890: Mit Rückblicken auf die früheren Jahrzehnte 1837–1890*, 36–37. Göttingen: Dieterichsche Universitätsbuchdruckerei.

Wagner, Rudolph. 1856. "Die anthropologische Sammlung des Physiologischen Instituts." *Nachrichten von der Georg-Augusts-Universität und der Königl. Gesellschaft der Wissenschaften zu Göttingen*. Vom Jahr 1856. Nr. 14 (6. Okt. 1856), 233–242 (English translation in Bendyshe 1865, 347–356).

———. 1860. "Physiologisches Institut." *Nachrichten von der Georg-Augusts-Universität und der Königl: Gesellschaft der Wissenschaften zu Göttingen*. Vom Jahr 1860. Nr. 16 (28. Mai 1860), 165–175.

———. 1861. "Ueber eine im September d. J. stattgehabte Zusammenkunft einiger Anthropologen zum Zwecke gemeinsamer Besprechungen." *Nachrichten von der Georg-Augusts-Universität und der Königl. Gesellschaft der Wissenschaften zu Göttingen*. Vom Jahr 1861. Nr. 19 (27. November 1861), 313–321.

6 Blumenbach's theory of human races and the natural unity of humankind

Thomas Junker

Introduction

The Göttingen physician and naturalist Johann Friedrich Blumenbach was one of the founders of biological anthropology. He first presented his research program for the new science in his dissertation *De generis humani varietate nativa* (1775). The extended and significantly revised third edition published in 1795 was translated into several languages and made his ideas accessible to a greater audience. To a large extent Blumenbach's impact was a consequence of the fact that he took up a political debate that was passionately discussed in the eighteenth century, and presented a scientific solution for it: is it possible to prove that all humans belong to the same biological species considering the empirically observable "natural variety"? If this should be the case, how can the differences between the geographic varieties – in the human species as well as in other animals – be explained?

Racial geometry

Two decades ago Stephen Jay Gould added a short chapter to the second edition of his *The Mismeasure of Man* (1996) dealing with the historical impact of Blumenbach's anthropological work. On the one hand, Gould emphasized that Blumenbach had been a staunch opponent of slavery and racism. On the other, he criticized him for indirectly having promoted these practices by designing a hierarchical "racial geometry": "Blumenbach meant well, but ended up affirming racial hierarchy by way of geometry and aesthetics, not by any overt viciousness" (Gould 1996a, 49). Gould based his conclusion mainly on quotations from the last chapter of Blumenbach's book *On the Natural Variety of Mankind* (3rd ed., 1795). As additional proof he presented an illustration that was allegedly from the *Anthropological Treatises* (1865), a collection that included the English translation of some of Blumenbach's works (see Figure 6.1).

In 1998 I pointed out that the illustration was a forgery, or at least a distortion, fabricated to give Gould's argument credibility: the original nonhierarchical ordering of skulls in Blumenbach's book (this volume, Figure 5.1) had been rearranged in a hierarchical way. By looking at the illustration the readers of *The Mismeasure*

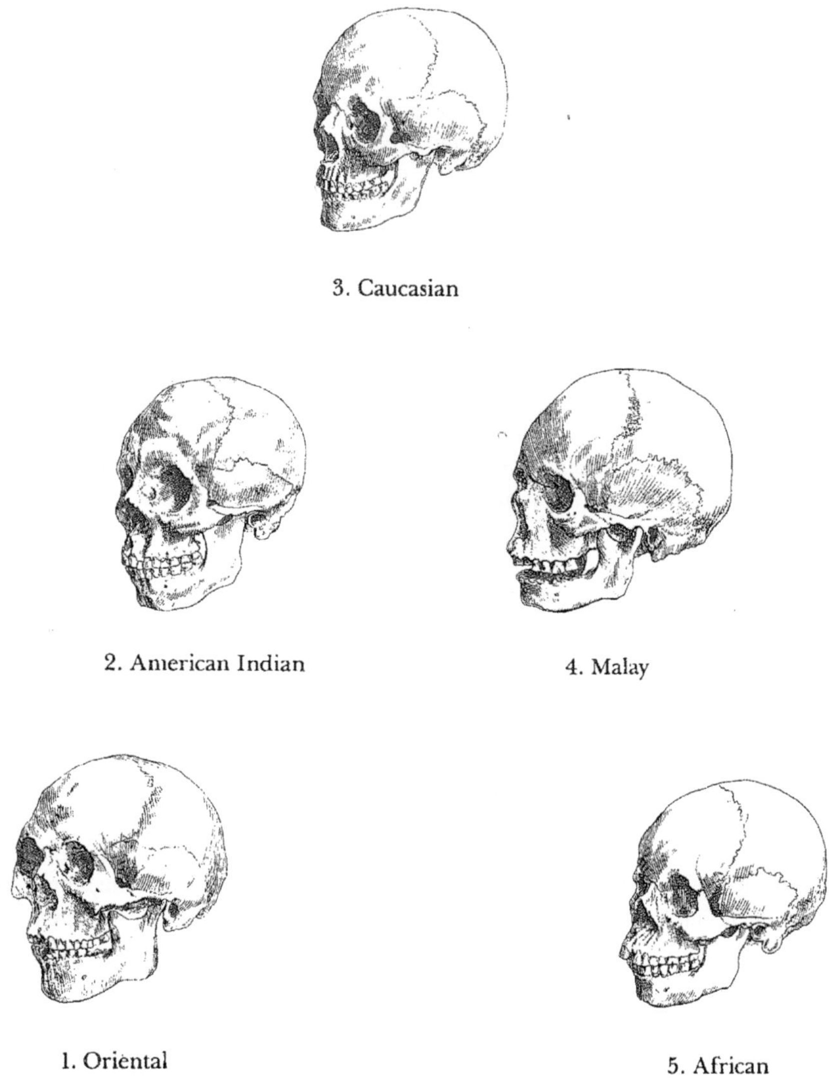

Figure 6.1 The illustration from Gould's *The Mismeasure of Man* (1996b, 409) comes with the caption "Blumenbach's racial geometry with two lines of 'degeneration' extending out through intermediary stages from a central Caucasian 'ideal.' From *Anthropological Treatises*, J. F. Blumenbach 1865."

of Man should intuitively understand what Blumenbach had thought – or, should we say, what Gould claimed Blumenbach had thought?

In his response to my criticism Gould admitted that the illustration was falsely attributed to Blumenbach. However, he saw no need to revise his claim that

Blumenbach had "changed the mental geometry of human order to a scheme that has promoted conventional racism ever since" (Gould 1996b, 405). He contended that his argument rested "entirely upon Blumenbach's text" and that the fabricated figure remained "superfluous and additional to my intent and analysis" (Gould 1998, 502; see Junker 1998). Both Gould's interpretation and his illustration have been taken up by several authors since then (Quintyn 2010, 22; Sussman 2014, 21).

Was Gould right after all when he argued that Blumenbach paved the way for nineteenth-century racism? Or did he misrepresent and misunderstand Blumenbach's intentions, results, and historical influence?

Blumenbach's one long argument

First I will argue that Blumenbach's classification of human varieties can be fully understood only when this aspect of his work is put in perspective. In his scientific publications he emphasized again and again that the so-called racial differences within the human species are superficial and transient, rendering any classification more or less arbitrary:

> All national differences in the form and colour of the human body [. . .] run so insensibly, by so many shades and transitions one into the other, that it is impossible to separate them by any but very arbitrary limits.
>
> (Blumenbach 1825, 35–36)

The view that the subdivision of the human species into varieties is of secondary importance also shaped the structure of *On the Natural Variety of Mankind*. The introductory material – "Letter to Sir Joseph Banks," "Index of the Anthropological Collection," and "Explanation of the Plates" – covers fourteen pages (English ed., 1865). The main bulk of the text adds up to 115 pages and is divided into four sections: "I. On the difference between man and other animals. II. On the causes and ways by which animals degenerate universally. III. On the causes and ways in which mankind have degenerated in particular. IV. Five principal varieties of mankind, one species."

Section IV, which presents Blumenbach's classification of human races, and on which Gould based his conclusion, is the shortest, with only thirteen pages. Furthermore it should be noted that the book is not a collection of various topics nor does the last section serve as a summary. The book rather constitutes one long argument, with the purpose of answering an overarching question: do the geographical populations of humankind belong to a single species? That is, are they varieties (races) or are they different species? In Blumenbach's words,

> Now we come nearer to the primary object of the whole treatise, for we are to inquire of what kind and how great is the natural diversity which separates the races and the multifarious nations of men; and to consider whether the origin of this diversity can be traced to degeneration, or whether it is not so

> great as to compel us rather to conclude that there is more than one original species of man.
>
> (Blumenbach [1795] 1865, 188)

The focus of Blumenbach's book is the question of monogenism versus polygenism and not the construction of a racial classification. But insofar as the geographical varieties were regarded as separate species by other authors their origin and status had to be discussed. But why should it be important whether there is only one human species or several of them? In Blumenbach's publications three aspects are mentioned: science, religion, and politics.

With respect to the first aspect, the question of how to properly classify the human genus had been of scientific interest since the founder of modern biological systematics, Carl Linnaeus, had included humans in his *Systema Naturae* and had recognized two human species: *Homo sapiens* and *Homo troglodytes* (the "cave man"). For *Homo sapiens* Linnaeus listed six varieties: *Homo ferus*, *americanus*, *europaeus*, *asiaticus*, *afer* (African), and *monstrosus* (1758–1759, I, 20–24). Blumenbach had already discarded the *Homo troglodytes*, the wild and monstrous man, as "fabulous imaginations with which the Natural History of the human species has been burdened" (Blumenbach 1825, 38); but which status should be given to Linnaeus's four geographical varieties?

Second, the thesis that there is more than one human species could be seen as a religious heresy. This made polygenism attractive for freethinkers like Voltaire:

> But there were people, among them the tolerance preacher Voltaire, who considered the reputation of the catechism as a great thorn in their flesh. This might have been, but they should not have overturned the results of a reasonable historiography and in addition of physiology, physics, chemistry and so on for this reason.
>
> (Gruber 1798, VII–VIII [transl. TJ]; see Voltaire [1734] 1989, 422–423)

Third, the unity of humankind was of political importance insofar as monogenism could serve as an argument against slavery:

> However there were still means to wake up the slave traders from their slumber. [. . .] Among the various scholars from different nations who tried to defend the unity of mankind, [. . .] among us Hofrath Blumenbach came forward.
>
> (Gruber 1798, VIII–IX [transl. TJ])

Blumenbach's attempt to infer legal and moral rules from the biological unity of humankind (which still had to be proven) was not undisputed. Georg Forster, for example, had argued that the common origin did not prevent "degenerated Europeans" from ruling "over their fellow white men as despotically as they rule over negroes." Forster considered the existence of several "original human tribes" probable, but was a firm opponent of slavery at the same time (Forster 1786, 161–165 [transl. TJ]).

Nevertheless Blumenbach was convinced that a scientific proof of the affiliation of all humans to the same biological species would carry more weight than a simple moral appeal, in order to wake up "the slave traders from their slumber." But how could the unity of mankind (the "one species" hypothesis) be proven scientifically?

What is a species?

In the pre-Darwinian world view of the eighteenth century the biological species could be straightforwardly defined by postulating an independent origin for each species – either through divine creation or through spontaneous generation. In addition it was assumed that each species had an inner cohesion and temporal stability that prevented major departures from the original form as well as sterility when matings of individuals between different species occurred.

As the French natural historian Georges-Louis Leclerc de Buffon wrote,

> There is in Nature a general prototype of each species on which every individual is modelled [. . .]. The first animal, the first horse, for example, was the exterior model and the interior mold [moule intérieur], on which all horses that are born, all those that exist, and all those that will be born are formed.
> (Buffon 1753, 216 [transl. TJ])

Blumenbach adopted Buffon's species concept with minor changes, replacing the "moule intérieur" with a "Bildungstrieb" (formative drive): "The *Formative Impulse* [. . .] preserves the equally determinate form and habit of all the individual species of organized beings" (Blumenbach 1825, 12).

The definition of the species by its separate origin and "general prototype," however, did not solve the practical problem of how to determine whether two given individuals belong to the same or to different species: "Now we come to the real difficulty, which is to set forth the characters by which, *in the natural world*, we may distinguish mere varieties from genuine species" (Blumenbach [1795] 1865, 188). To solve this problem Buffon had suggested two criteria – similarity and fertility: "The species is nothing else than a constant succession of similar individuals that can reproduce together" (Buffon 1753, 386 [transl. TJ]).

At a first glance the criterion "fertility" ("Buffon's rule") had the great advantage that it seemed to settle the question unequivocally through experimental proof. The philosopher Immanuel Kant, for example, was sure that in this way the unity of humankind had been established once and for all:

> Buffon's rule: that animals which generate between them fertile young (whatever the difference in bodily form they may possess) belong to one and the same physical genus [= species], must be looked upon as the general definition for a natural genus in animals [. . .]. According to this definition all humans on the wide earth belong to one and the same natural genus [= species].
> (Kant [1775] 1977, 11 [transl. TJ])

If Kant was right and Buffon's rule provided all the evidence needed, why did Blumenbach devote so much time and energy to find and collect further proof? Because the situation was not as clear-cut as it seemed: (1) domestication leads to a highly artificial way of life that may have distorted the natural mating behavior: "But, as in the domestic animals which man has subdued, this character seemed ambiguous and uncertain, on account of the enslaved life they lead." (2) For wild animals, on the other hand, practical difficulties hindered the necessary experiments: "For, in the first place, what very little chance is there of bringing so many wild animals [. . .], to that test of copulation?" (Blumenbach [1795] 1865, 189; see Roger 1989, 410–418).

And (3) the "test of copulation" had to be applied not only to the geographical populations of humankind but also to humans and apes. The second part of the experiment was rejected by Blumenbach on aesthetic and moral grounds:

> Not indeed that horrid stories are wanting of the union of men with brutes, [. . .] still we have never known any instance related on good authority of any such connexion being fruitful, or that any hybrid has ever been produced from the horrid union of beast and man.
>
> (Blumenbach [1795] 1865, 201–202)

If "the principle sought to be deduced from copulation is not sufficient to define the idea of species and its difference from variety" (Blumenbach [1795] 1865, 189–190) the additional evidence provided by the criterion of similarity becomes relevant. Blumenbach was convinced that it is possible to discriminate between varieties and species on the basis of the degree of the observed differences within a species and between different species: since species have separate origins and different directions of the formative drive we observe major differences. Varieties on the other hand have a common origin. The direction of their formative drive can be modified through external causes over several generations (= degeneration), but this will cause only minor differences:

> We say that animals belong to one and the same species, if they agree so well in form and constitution, that those things in which they do differ may have arisen from degeneration. We say that those, on the other hand, are of different species, whose essential difference is such as cannot be explained by the known sources of degeneration.
>
> (Blumenbach [1795] 1865, 188)

Starting from these assumptions Blumenbach attempted to prove the common origin and species unity of all humans by comparing the "difference between man and other animals" (Section I) with the difference between human varieties (Section III and IV):

> He who means to write about the variety of mankind, and to describe the points in which the races of men differ from each other in bodily constitution,

> must first of all investigate those differences which separate man himself from the rest of the animals.
>
> (Blumenbach [1795] 1865, 163)

With regard to the first part of his argument, Blumenbach lists several "external characters by which Man is distinguished, not only from animals in general, but from the Apes which most closely resemble him": "The erect position; The broad, flat pelvis; The two hands; The regular and close set rows of teeth" (Blumenbach [1795] 1865, 164). Interestingly the human female has further bodily modifications that set her apart from other animals:

> The female, beside the peculiar form of the bosom in the bloom of life, possesses two other characters of distinction from the male, and from all other animals, viz. a periodical discharge of blood during a certain number of years; and a particular part in the sexual organs, the absence or destruction of which forms a physical sign of loss of virginity, and which, at least as far as regards its form and position, has not been remarked in any other animal.
>
> (Blumenbach 1825, 35)

In addition Blumenbach mentions differences in "the mental faculties of man": "excepting the sexual propensity, he presents few traces of instinct, and of the mechanical kind none [. . .]. On the other hand, he is exclusively in possession of reason, and of speech." And there are behavioral differences:

> Man is, in himself, a defenceless, helpless, creature. [. . .] This necessity of assistance, and his numerous urgent wants, prove the natural destination of man for *social connexion*. [. . .] [It remains to be seen if] man is destined elsewhere for monogamy, as well as in Europe.
>
> His residence and his diet are both unrestricted.
>
> (Blumenbach 1825, 35)

Differences between the varieties of mankind

In a second step Blumenbach listed the differences between human varieties: the color of the skin, the hair, and the iris of the eye, the face and form of skulls, the teeth, ears, breasts, genitals, legs, feet and hands, stature, and morbific affection. For all these differences he claimed that they could have been caused by degeneration. In conclusion Blumenbach sums up his one long argument: there is only one human species and the differences between human populations are superficial.

> Thus too there is with this that insensible transition by which as we saw the other varieties also run together, and which, compared with what was discussed in the earlier sections of the book, about the causes and ways of degeneration, and the analogous phenomena of degeneration in the other domestic animals, brings us to that conclusion, which seems to flow spontaneously

> from physiological principles applied by the aid of critical zoology to the natural history of mankind; which is, *That no doubt can any longer remain but that we are with great probability right in referring all and singular as many varieties of man as are at present known to one and the same species.*
>
> (Blumenbach [1795] 1865, 275–276)

Gould's argument revisited

Gould would probably not have objected to my reconstruction of Blumenbach's general argument regarding the unity of the human species. In his chapter on "racial geometry" he arrived at a similar conclusion, praising Blumenbach for his egalitarian ideas, and emphasizing that he was "the man most committed to human unity." But how could he at the same time have "promoted conventional racism ever since" (Gould 1996b, 405)?

To explain this obvious contradiction Gould suggested that Blumenbach's text cannot be taken at face value but should be read at a deeper level, unveiling "unconscious presupposition" and a hidden agenda that were not spelled out but "implied." On a superficial level Blumenbach may have argued against the inequality of human geographical populations, but at the same time he presented a completely different "implied mental picture of human diversity" – a picture not about unity and equality but about hierarchy and "linear ranking by putative worth" leading "from a Caucasian ideal to least desirable Oriental and African endpoints" (Gould 1996a, 49; 1996b, 405, 410; 1998, 502).

The problem with Gould's argument is that Blumenbach to my knowledge nowhere writes about "a Caucasian ideal" or about "least desirable Oriental and African endpoints." This would be surprising indeed, because he explicitly rejected this kind of typological reasoning over and over again in his publications. Let me give just one quotation where he referred to the notion of an aesthetic ideal and discarded it:

> As to the physiognomy of the negro, the difference no doubt is astonishing if you put an ugly negro (and there are ugly negroes as well as ugly Europeans) exactly opposite the Greek ideal. But this is precisely to offend against one of the rules given above. [. . .] I can, on the contrary, declare that amongst the negroes and negresses whom I have been able to observe attentively, and I have seen no small number of them, [. . .] all are more or less different from one another, and through all sorts of gradations run imperceptibly into the appearance of men of other kinds up to the most pleasing conformation.
>
> (Blumenbach [1806] 1865, 306)

Beauty

Since Blumenbach did not openly argue for a hierarchical ranking, Gould had to rely on two lines of circumstantial evidence to make his case: Blumenbach's five-race system and his reference to beauty. In his response to my criticism

Gould reaffirmed this claim with the words "I both base and rest my case upon his words" (Gould 1998, 503), followed by this quote from Blumenbach:

> I have allotted the first place to the Caucasian, for the reasons given below, which make me esteem it the primeval one. This diverges in both directions into two, most remote and very different from each other; on the one side, namely, into the Ethiopian, and on the other into the Mongolian. The remaining two occupy the intermediate positions between that primeval one and these two extreme varieties; that is, the American between the Caucasian and Mongolian; the Malay between the same Caucasian and Ethiopian.
> (Blumenbach [1795] 1865, 264–265)

In this quotation, however, there is no reference to beauty and nothing that implies a linear ranking or a hierarchy. The phrase "the first place" just refers to Blumenbach's idea that the Caucasian variety is the original form and that the other varieties have departed from this starting point into different directions and in different degrees.

As mentioned earlier Blumenbach thought that every biological species has a single point of departure and an original form, and that species are not immutable but can change over time within rather narrow limits through environmental influences: "Among the many causes of degeneration, the principal are the influence of climate and of food; and in man and animals, the mode of life" (Blumenbach 1825, 16).

This change, called "degeneration," was thought to depend on the amount of time elapsed and on the differences in the environment encountered by the individuals during their migrations:

> It is allowable to suppose that the people dispersed through the various parts of the world have, according to the differences in the degree and duration of the influence of climate and other causes of degeneration, either deviated still more from the form of the primary race, or approximated more closely to it.
> (Blumenbach 1825, 37 fn)

Based on this idea Blumenbach had to assume that an original human form had existed or might still exist that had not changed through degeneration. Why did he think that the Caucasian variety had conserved the original features of the prototype?

> For in the first place, that [Caucasian] stock displays, [. . .] the most beautiful form of the skull from which, as from a mean and primeval type, the others diverge by most easy gradations on both sides to the two ultimate extremes (that is, on the one side the Mongolian, on the other the Ethiopian). Besides, it is white in colour, which we may fairly assume to have been the primitive colour of mankind, since, as we have shown above, it is very easy for that to degenerate into brown, but very much more difficult for dark to become white.
> (Blumenbach [1795] 1865, 269)

According to Blumenbach the Caucasians can be identified as the original human form because their skulls have the most beautiful form and represent a "mean type" and because their skin color is white. But these characteristics result from the alleged prototypal nature of the Caucasian variety and do not explain why it has this status in the first place.

The main reason, not explicitly mentioned by Blumenbach, can be found in the biblical legends. According to this narrative Noah's ark was thought to have landed at Mount Ararat (near the Caucasus) after the deluge and the repopulation of the earth by humans (and other animals) started from this area. Blumenbach's understanding of degeneration on the other hand suggested that populations remaining close to the original habitat of the species would closely resemble the original forms, because degeneration was thought to be caused by the environmental change that happened during migration. This assumption could also explain why Noah would "in all respects [be] a second Adam" if it was accepted that the timespan between Adam and Noah had been rather short and that the environment had remained fairly constant (Coleridge [1819] 1990, 4548).

Is Blumenbach ranking the human varieties? As a matter of fact, he says that the Caucasian variety has the most beautiful form of the skull and some of the people still living in the Caucasus area, the Georgians, are the most beautiful race of men. It should be noted that Blumenbach (and other eighteenth-century authors) were "not intending, by this word [degeneration], the modern sense of deterioration, but the literal meaning of departure from an initial form of humanity at the creation (*de* means 'from' and *genus* refers to our original stock)" (Gould 1996b, 407). But even then the word had slightly negative connotations, in the sense that any modification of the original form (created by God) through natural causes would tend to be a deterioration, not an improvement. This reminds us of the observation that the modern word "evolution" stands for neutral biological change, but often comes with a positive connotation in the sense of progress.

To sum up: on the one hand, Blumenbach invoked beauty primarily as a substantiation of his notion that the Caucasian variety is the original human form, and not as a criterion to rank human varieties in general. On the other hand, his references to greater or lesser beauty particularly of the skulls create a certain amount of one-sidedness and favoritism. Insofar Gould's claim is not completely beside the point, but he exaggerated to the point of caricature. It is probably fair to say that Blumenbach was ambivalent; he was committed to an egalitarian view of human populations, but could not completely overcome the notions that the Europeans were privileged and that degeneration had a slightly negative touch.

Five races

For Gould another feature of the Blumenbach's ideas had even more disastrous consequences – his five-race system:

> By moving from the Linnaean four-race system to his own five-race scheme, Blumenbach radically changed the geometry of human order from a geographically based model without explicit ranking to a double hierarchy of

> worth, oddly based upon perceived beauty and fanning out in two directions from a Caucasian ideal.
>
> (Gould 1996b, 403)

Gould claimed that Blumenbach added a fifth race to enable the transformation from a geographical to a hierarchical model:

> The four-race system contained no appropriate group, and could therefore not be transformed into the new geometry of a pinnacle with two symmetrical limbs leading to maximal departure from ideal form. But invention of a fifth racial category for forms intermediate between Europeans and Africans would complete the new geometry – and Blumenbach *therefore added the Malay race*, not as a minor factual refinement, but as the enabler of a thorough geometric transformation in theories (mental pictures) about human diversity.
>
> (Gould 1996b, 411; emphasis added)

How did Blumenbach himself explain the addition of a fifth race? In the introductory letter that precedes the third edition of his book on the *On the Natural Variety of Mankind* the reason is clearly stated. The letter is to Joseph Banks, who took part in James Cook's first expedition to the South Sea. As Blumenbach explains, in the times when Linnaeus wrote his *Systema Naturae* "only four parts of the terraqueous globe and its inhabitants were known." But now,

> after your three-years' voyage round the world, illustrious Sir, when a more accurate knowledge of the nations who are dispersed far and wide over the islands of the Southern Ocean had been obtained by the cultivators of natural history and anthropology, it became very clear that the Linnaean division of mankind could no longer be adhered to.
>
> (Blumenbach [1795] 1865, 150)

After having divided "all mankind into four varieties" in the first edition of his book, he "was compelled to give up that division, and to place in its stead the following five varieties, as more consonant to nature" (Blumenbach [1781] 1865, 99 fn). Linnaeus had known four geographic regions and accordingly identified four races. After having discovered the islands of the Southern Ocean five geographic regions were known and it made sense to speak of five races.

I don't want to deny the possibility that after having moved from four to five races, Blumenbach appreciated the fact that this model had symmetrical features. Symmetry, however, is not hierarchy.

Mental geometries

I have always wondered why Gould was so sure that his interpretation of what Blumenbach had "implied" was correct, even after it turned out that the visualized representation had been manipulated and Gould had to rely on the circumstantial evidence gained from a selective reading of Blumenbach's text. So, why was Gould so sure that the mental geometry he envisioned was the correct one?

One possibility is that he had actually seen a similar figure – not in Blumenbach's own publications, but somehow connected to Blumenbach. This might very well be the case, as the following example demonstrates. The *Table Talk* of the English poet Samuel Taylor Coleridge contains an illustration that displays exactly what Gould had envisioned, accompanied by the appropriate words: "Blumenbach's scale of dignity may be thus figured" (Figure 6.2).

Coleridge's visualization is particularly significant, since he had attended Blumenbach's lectures at Göttingen in the year 1799 (Levere 1981, 16–20). The *wedge*, however, is only one of several possible interpretations of Blumenbach's ideas. The first American edition of Coleridge's *Table Talk*, for example, contains a modified version: a vertical ordering (Figure 6.3).

As these two versions demonstrate, the visualization of a theoretical concept is more than a simple translation. It must be seen rather as a creative process in its own right. But insofar as illustrations have the important function to epitomize an argument or interpretation "in a useful way (especially for the nonprofessional readers)" (Gould 1998, 502), it makes sense to contrast the wedge and the vertical line with another, hopefully more appropriate visualization.

If we take the insight seriously that "degeneration" first of all meant neutral change from an ancestral form that serves as the prototype (the Caucasian variety), then the implied mental geometry takes the shape of a family tree of the human species, with the ancestral form at the root of the tree and branches extending from it (Figure 6.4).

In Blumenbach's system with five human varieties we would expect four independent lines of descent that branch out from the ancestral form – that is, the extreme varieties (African and Mongolian) stem directly from the Caucasian form and not from the middle forms (Malay and American), because the extreme forms are geographically closer to the origin and the middle forms are further away. The

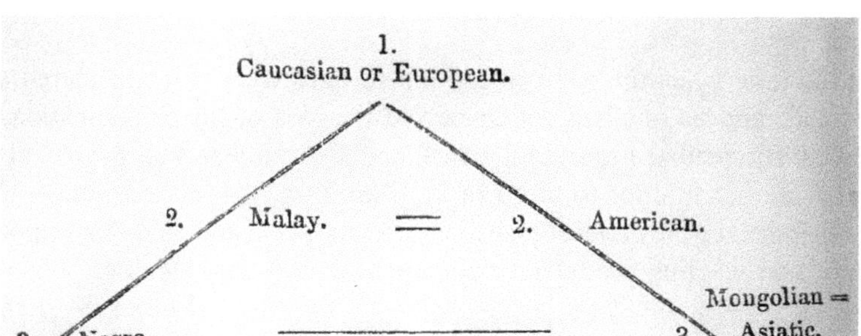

Figure 6.2 "Blumenbach's scale of dignity" (from Coleridge 1835a, 55–56). My thanks to John Michael for pointing this out to me.

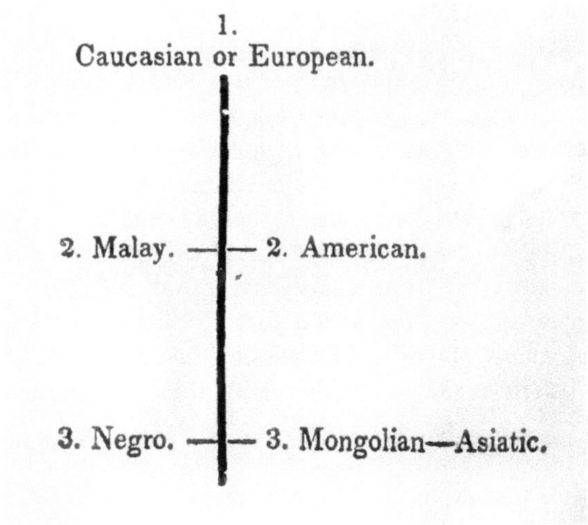

Figure 6.3 The modified version of "Blumenbach's scale of dignity" (from Coleridge 1835b, 60).

family-tree mental geometry of Blumenbach's theories is my interpretation. But I would argue that it has as much validity as Gould's hierarchical wedge – probably more.

Nevertheless Blumenbach used neither a hierarchical wedge nor a vertical line, nor a family tree for his publication, but a horizontal line (this volume, Figure 5.1).

Why did Blumenbach visualize his ideas in this way? Because he was truly committed to the unity and equality of all human populations? This may very well be the case.

Did Blumenbach's contemporaries see it that way? Some of them did not. But some did. Let me give just two examples. In biographical reminiscences by Karl Friedrich Heinrich Marx we read,

> At the time when the negroes and the savages were still considered as half animals, and no one had yet conceived the idea of the emancipation of the slaves, Blumenbach raised his voice, and showed that their psychical qualities were not inferior to those of the European, that even amongst the latter themselves the greatest possible differences existed, and that opportunity alone was wanting for the development of their higher faculties.
>
> (Marx [1840] 1865, 9)

A second example is the title of the French translation of Blumenbach's book by Frédéric Chardel. It reads, "On the unity of the human genus and on its varieties" (Figure 6.5).

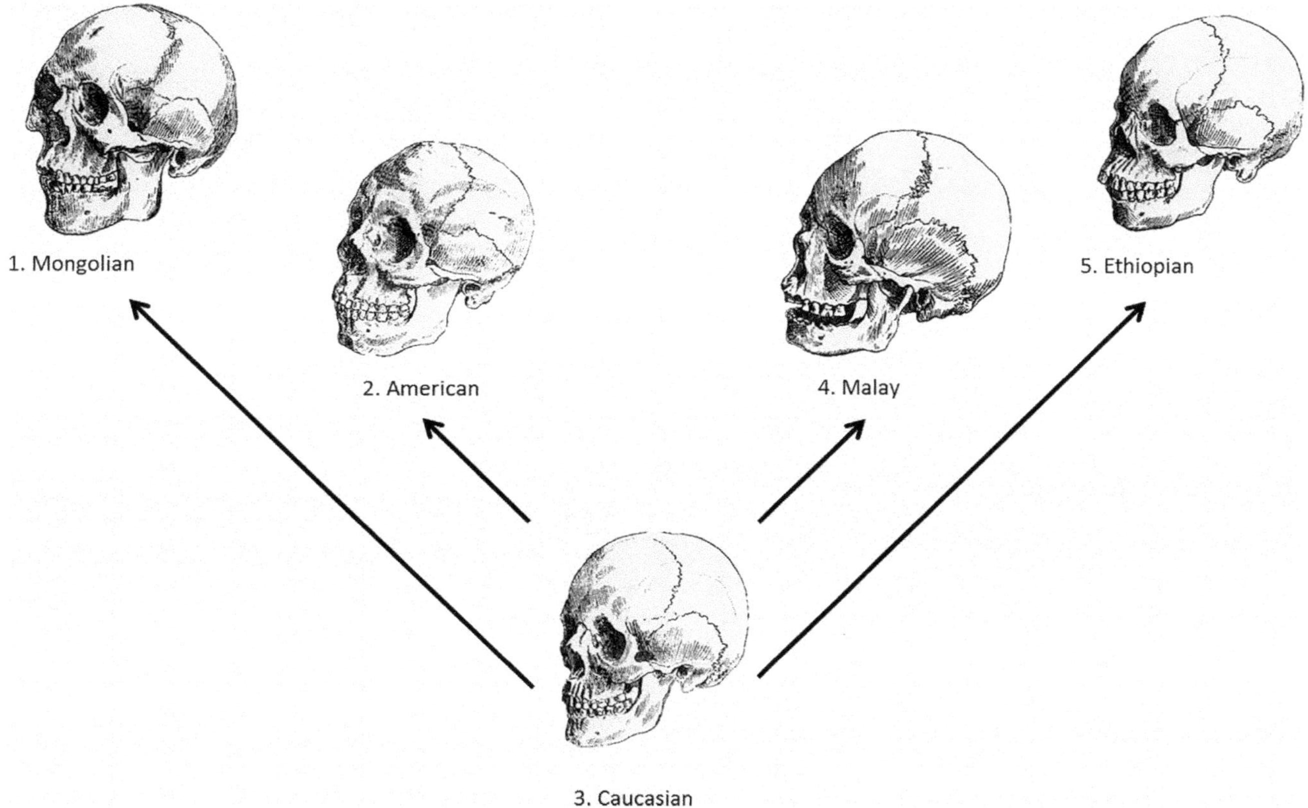

Figure 6.4 Blumenbach's racial geometry can be visualized as a family tree with an ancestral form and several varieties ascending from it. The degree of modification is indicated by the vertical distance.

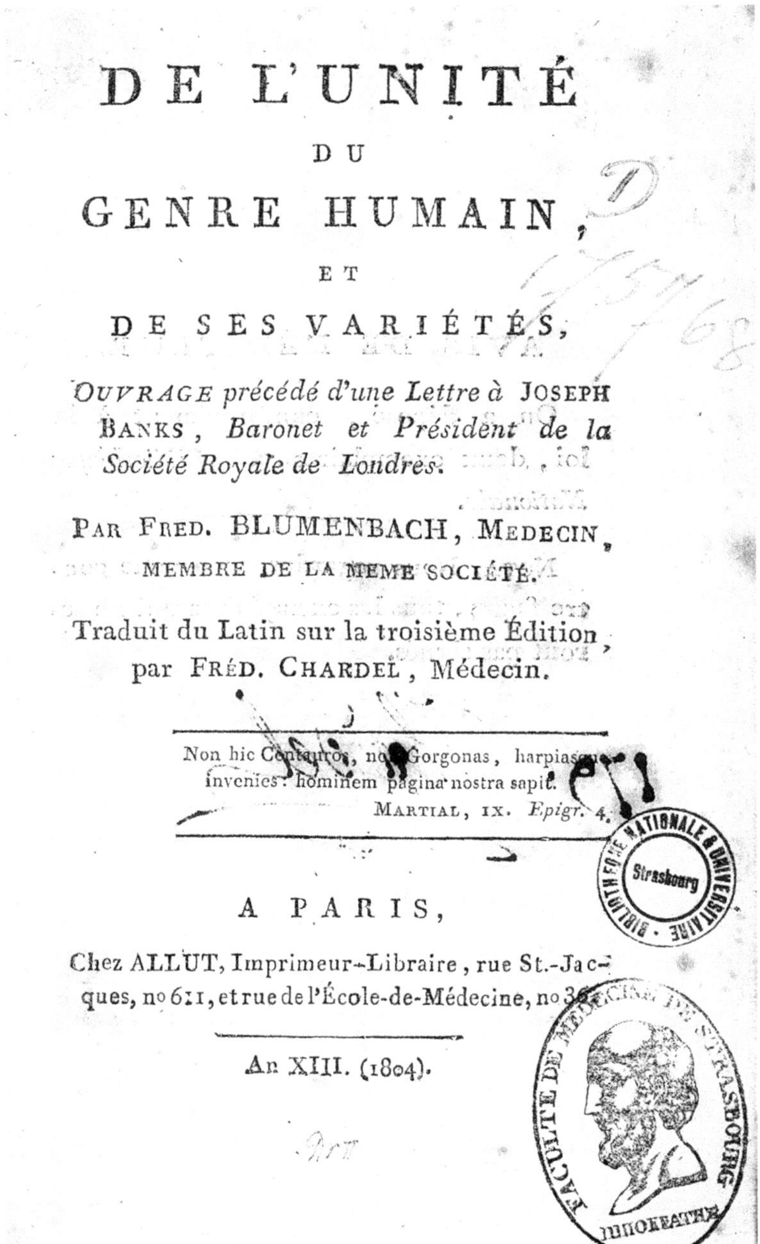

DE L'UNITÉ

DU

GENRE HUMAIN,

ET

DE SES VARIÉTÉS,

OUVRAGE précédé d'une Lettre à JOSEPH BANKS, *Baronet et Président de la Société Royale de Londres.*

PAR FRED. BLUMENBACH, MEDECIN, MEMBRE DE LA MEME SOCIÉTÉ.

Traduit du Latin sur la troisième Édition, par FRÉD. CHARDEL, Médecin.

Non hic Centauros, non Gorgonas, harpiasque invenies: hominem pagina nostra sapit.

MARTIAL, IX. *Epigr.* 4.

A PARIS,

Chez ALLUT, Imprimeur-Libraire, rue St.-Jacques, no 611, et rue de l'École-de-Médecine, no 36.

An XIII. (1804).

Figure 6.5 Title page of the French translation of Blumenbach's *De generis humani varietate nativa* (from Blumenbach 1804).

The title of the French edition actually reflects Blumenbach's intentions much better than the Latin, German, and English titles. If the French title would have been used for the other editions as well, some of the misunderstandings and misrepresentations of Blumenbach's ideas might have been prevented.

Bibliography

Blumenbach, Johann Friedrich. 1775. *De generis humani varietate nativa.* Göttingen: Rosenbusch.

———. 1795. *De generis humani varietate nativa.* Editio tertia. Praemissa est epistola ad virum Josephum Banks. Göttingen: Vandenhoek et Ruprecht.

———. 1798. *Über die natürlichen Verschiedenheiten im Menschengeschlechte.* Nach der dritten Ausgabe und den Erinnerungen des Verfassers übersetzt, und mit einigen Zusätzen und erläuternden Anmerkungen herausgegeben von Johann Gottfried Gruber. Leipzig: Breitkopf und Härtel.

———. 1804. *De l'unité du genre humain, et de ses varietés, Ouvrage précédé d'une Lettre à Joseph Banks.* Traduit du Latin sur la troisième Edition, par Frédéric Chardel. Paris: Allut.

———. 1825. *A Manual of the Elements of Natural History.* Translated from the tenth German edition by R. T. Gore. London: W. Simpkin & R. Marshall.

———. 1865. *The Anthropological Treatises of Johann Friedrich Blumenbach,* edited by Thomas Bendyshe. London: Published for the Anthropological Society, by Longman, Green, Longman, Roberts, & Green.

Buffon, Georges. 1753. *Histoire naturelle générale et particulière, avec la description du cabinet du roi.* Vol. 4. Paris: l'Imprimerie Royale.

Coleridge, Samuel Taylor. 1835a. *Specimens of the Table Talk of the Late Samuel Taylor Coleridge.* 2 vols. London: John Murray.

———. 1835b. *Specimens of the Table Talk of the Late Samuel Taylor Coleridge.* 2 vols. New York: Harper & Brothers.

———. 1990. *The Notebooks of Samuel Taylor Coleridge.* Vol. 4, 1819–1826, edited by Kathleen Coburn and Merton Christensen. London: Routledge.

Forster, Georg. 1786. "Noch etwas über die Menschenraßen." *Der Teutsche Merkur* (October and November): 57–86, 150–166.

Gould, Stephen Jay. 1996a. *The Mismeasure of Man.* Rev. and exp. edn. New York: Norton.

———. 1996b. "Racial Geometry." In Gould 1996a, 401–412.

———. 1998. "On Mental and Visual Geometry." *Isis* 89: 502–504.

Gruber, Johann Gottfried. 1798. "[Vorwort]." In Blumenbach 1798, V–XIV.

Junker, Thomas. 1998. "Critiques and Contentions: Blumenbach's Racial Geometry." *Isis* 89: 498–501.

Kant, Immanuel. [1775] 1977. "Von den verschiedenen Racen der Menschen." In *Schriften zur Anthropologie, Geschichtsphilosophie, Politik und Pädagogik 1.* Werkausgabe, Bd. 11, 7–30. Frankfurt am Main: Suhrkamp.

Levere, Trevor H. 1981. *Poetry Realized in Nature: Samuel Taylor Coleridge and Early Nineteenth-Century Science.* Cambridge: Cambridge University Press.

Linnaeus, Carl. 1758–1759. *Systema naturae per Regna tria naturae, secundum classes, ordines, genera, species. . .,* 10th edn. 2 vols. Holmiae: Laurentius Salvius.

Marx, Karl Friedrich Heinrich. 1865. "Memoir of J. F. Blumenbach." In Blumenbach 1865, 1–45.

Quintyn, Conrad B. 2010. *The Existence or Non-Existence of Race? Forensic Anthropology, Population Admixture, and the Future of Racial Classification in the U.S.* Youngstown, NY: Teneo Press.
Roger, Jacques. 1989. *Buffon: un philosophe au Jardin du Roi*. Paris: Fayard.
Sussman, Robert Wald. 2014. *The Myth of Race: The Troubling Persistence of an Unscientific Idea*. Cambridge, MA: Harvard University Press.
Voltaire. [1734] 1989. "Traité de métaphysique." In *Les Œuvres complètes de Voltaire: The Complete Works of Voltaire*, edited by W. H. Barber and Ulla Kölving. Tome 14, 415–503. Oxford: The Voltaire Foundation/ Taylor Institution.

7 A defense of human rights

Blumenbach on albinism

Renato G. Mazzolini

"In Spain we call them albinos," wrote the chronicler Bartolomé Leonardo de Argensola in 1609 (Argensola 1609, 71). He was referring to descriptions by travelers of certain inhabitants of the Moluccan Islands, who, unlike the majority of the population, had white skin "like the Germans" and eyes that were extremely sensitive to sunlight. From the sixteenth century onwards, in fact, numerous European travelers in Africa, Asia, and the Americas reported seeing white-skinned individuals begotten by persons of color. They found the phenomenon bewildering, and diverse terms were used to denote those individuals: *Leucoœthiopes* was the learned Latin term, whence derived those of national languages, such as White Moors, *Weisse Mohren*, *Mori-bianchi*, *Maures blancs*, and *Negres blancs*, as well as derogatory terms like *Dondos* and *Kakerlakken*, and finally the term of Spanish and Portuguese descent, *Albinos*, which at the end of the eighteenth century was incorporated into scientific terminology by all the European languages.

The question of how to consider and classify albinos was much debated.[1] Still in the mid-eighteenth century various authors – Voltaire, for example – could not conceive that blacks could generate white children and argued that albinos were a distinct human race (Voltaire 1748; Voltaire 1765, 4–5; Voltaire 1768, 119–120; see also Abanime 1975). Linnaeus, in the tenth edition of *Systema naturae*, classified them together with orangutans among the troglodytes (cave-dwellers) and wrote that they belonged to a genus very distinct from *Homo sapiens* (Linnaeus 1758–1759, I, 24). Maupertuis suggested instead that albinos could be begotten by blacks, but suffered from a hereditary disease (Maupertuis 1745, 149–152).

Blumenbach was still a student when, on February 23, 1775, he sent his first letter to the celebrated Albrecht von Haller, informing him of the dissertation that he was writing at the University of Göttingen, where many years previously Haller had been a professor of anatomy, physiology, and botany. Blumenbach told Haller that for some time he had been collecting material on the diversities and anomalies of the human race, and that his purpose was to "defend the rights of humanity" and to refute the ridiculous confusion between an authentic ape like the orangutan and "a white moor" (Dougherty and Klatt 2006–2015, I, 14). Again in that letter, Blumenbach said that he had noticed "a true White Moor of our climate" among his acquaintances. Haller replied that "White Moors" were suffering from a disease and did not constitute a distinct human variety

(Dougherty and Klatt 2006–2015, I, 16). This was an important confirmation of what Blumenbach would argue in his dissertation: albinos should not be confused with monkeys; rather, they were humans afflicted by a disease. Blumenbach, therefore, did not share the opinion of either Voltaire or Linnaeus; and it is understandable that before expressing himself publicly he wanted to know the opinion of Haller.

More interesting from a general historical point of view is Blumenbach's claim concerning "a true White Moor of our climate" – that is, the European climate. In fact, all the descriptions of albinos from the sixteenth century until the 1770s concerned individuals living in Africa, Asia, and the Americas, but not in Europe. This failure to recognize that there were albinos also in Europe may seem absurd. Yet, if we ask two simple questions concerning the sources which between 1640 and 1809 dealt with albinism, or referred to cases of individuals that we now recognize as suffering from albinism, the answers are clear. What sources indicate the presence of albinos in Africa, America, and Asia? What sources do so also in Europe? Figure 7.1 illustrates the results.

An analysis of 123 texts published between 1640 and 1809 shows that before the 1770s no author had realized that the condition which afflicted certain individuals in Africa, America, and Asia was also present in Europe. In particular, out of eighty-one texts published before 1784, only four stated that albinism was also present in Europe, while seventy-seven saw no connection at all between albinism in non-European countries and the albinism which certainly existed also in Europe. After some scholars had claimed in the 1780s that there were cases of albinism in Europe, one notes a significant increase in this assertion.

As said, in his dissertation of 1775, the young Blumenbach did not accept the Linnaean classification that relegated albinos to troglodytes. He instead affirmed

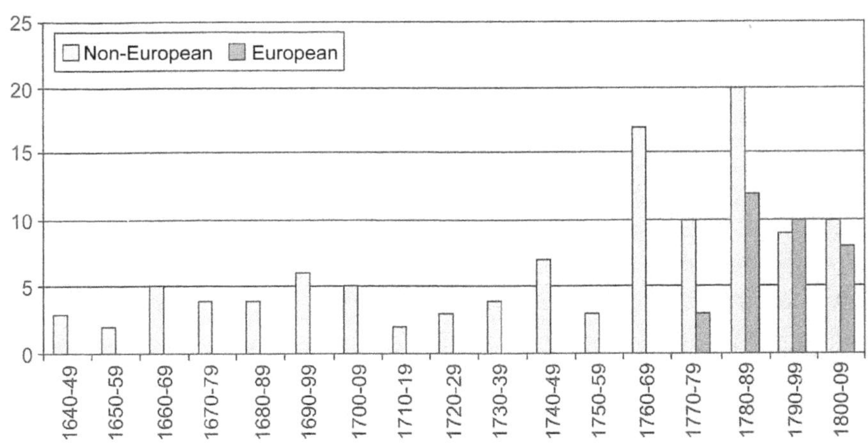

Figure 7.1 Histogram of 123 texts mentioning albinos, published between 1640 and 1809 (from Mazzolini 2006). Gray bars indicate texts describing albinos in Africa, America, and Asia, black bars for texts that describe albinos also in Europe.

their fully human nature, arguing that they were suffering from a disease. He countered the claims of scholars who maintained that albinos were weak and stupid by proposing an interpretation of the testimonies of travelers which demonstrated that, in fact, albinos were able to work at night, and that their mental faculties were absolutely not affected by the disease. He rebutted the depictions by travelers and commentators of albinos as marginalized by, or even expelled from, society by citing cases in which they occupied prestigious roles at certain African courts. He also asserted that the disease was already known to the ancients, and he adduced cases, including that of a young Saxon friend, which suggested that albinos existed in Europe as well. But, he concluded, it was still necessary to demonstrate the universality of the disease and its endemic or sporadic nature (Blumenbach 1775, 81–86, 88–91).

In 1776 the amateur geologist and scholar of alpine glaciers Marc-Théodore Bourrit published a work in which he recounted that, on passing close to the *glacier des Bois* on Mont Blanc, near Chamonix, he examined two young albinos living in that valley. "They were two orphaned children," he wrote;

> the older child was twelve years old, the younger one was ten: their hair, eyebrows, and eyelashes were white as milk, their complexion had the same whiteness, but they had eyes like those of white rabbits, and they were so weak that they could hardly move.
>
> (Bourrit 1776, 7–18; see also Bourrit 1785, 118)

At night, however, they were able to see and move better than during the day. Moreover, they correctly answered the questions that were put to them.

Blumenbach was greatly impressed by the description of the two young Savoyard albinos and mentioned them in 1781 in the second edition of *De generis humani varietate* (Blumenbach 1781, 121). In 1783, during a journey to Switzerland, he went to Chamonix for the purpose of examining them and reported the results in a paper entitled *De oculis Leucaethiopum et iridis motu commentatio*, read on October 9, 1784, to the Göttingen Royal Society of Sciences. This paper, published in 1786 in the proceedings of the Academy, was divided into two distinct parts: one dealt with the eyes of the albinos, the other with movements of the iris (Blumenbach 1786).

Blumenbach reported that the skin of the two boys was white, except for an unusual redness of the face. Their hair, ranging in color from white to yellow, resembled goat wool. The color of their very mobile irises was similar to those of white rabbits "pale violet with red in the middle," while the pupils were "almost bright red." Late at night they could see no better than other men, but at twilight or in pale moonlight they saw much better. They could not stand sunlight, that of a candle, or the glare of sun-reflecting snow. Both boys were myopic, and had sufficient intelligence and abilities to read and write. Several of their family members had brown hair. Contrary to the claim of Cornelis de Pauw that albinos lived only in the torrid zones of the tropics, it could now be demonstrated that they lived in glacial regions as well (Blumenbach 1786, 33).

Having described the Savoyard albino brothers, Blumenbach emphasized the importance of the eye's black pigment for vision. He pointed out, for instance, that in constructing telescopes and microscopes we imitate nature by painting the inner walls of the optical tubes black. He thus considered the absence of this pigment in the eyes of rabbits and white mice to be indicative of a disease. With respect to his 1775 dissertation, Blumenbach changed his opinion concerning the weaknesses of albinos. In fact, in order to substantiate the notion that albinism debilitated individuals, he argued that the white breeds of domestic animals were weaker than the colored ones. There was, he argued, a correlation among eye color, skin color, and cutaneous appendages, such as hair, beard, and bristles. Also testifying to this sympathetic correlation was the common experience that people with blue eyes have very light-colored skin, while people with brown eyes and brown hair have darker skin. He suggested that this depended on the similarity among certain inner anatomical structures of the skin and eye (Blumenbach 1786, 38–40). He then dwelt at length on the way in which the color of the iris could be artificially modified or appear different according to the intensity of light on it. Blumenbach concluded by conjecturing that the eyes of albinos must be almost entirely devoid of black pigment, and that the disease from which they suffered must be congenital and sometimes hereditary.

In 1784 Francesco Buzzi, an eye surgeon at the *Ospedale Maggiore* of Milan, furnished the anatomical proof that Blumenbach had lacked (Buzzi 1784; German translation Buzzi 1798). A peasant aged about thirty died at the *Ospedale Maggiore* in January 1783. He was an albino, and Buzzi obtained permission to perform a dissection. He was unable to isolate the Malpighian layer in the epidermis, and concluded that it was either invisible or nonexistent. He also demonstrated the absence of pigment in the posterior surface of the iris and choroid. To prove that the rosy color of the pupil depended on rays reflected from the ocular fundus and its lateral blood vessels, Buzzi performed a series of experiments on white and black rabbits. He then discussed various issues pertaining to the vision of albinos, claiming that they were not myopic, because "in a moderate light they can see clearly from afar and close by." This he had ascertained with albinos to whom he had given spectacles "with lenses of various focuses," and who had reacted like all other individuals with good eyesight (Buzzi 1784, 87). Their problems of vision were due solely to the intensity of light, which is why Buzzi correctly defined them as suffering from heliophobia (photophobia). He then denied that they were nyctalopic, as Haller had claimed, "because they see well, and do so both during the day and at night, with the sole difference that strong light dazzles them for the reasons already stated" (Buzzi 1784, 87) – that is, the absence of black pigment in the choroid and the back of the iris.

Blumenbach had concluded by induction that albinism is a disease of the eye pigmentation correlated to the skin and its adnexa; Buzzi did so with an anatomical demonstration carried out on a European albino and physiological experiments on rabbits. The main difference between the two authors was that according to Blumenbach albinism could be hereditary, while according to Buzzi it could not. In retrospect, we can say that Blumenbach was right. He reached his conclusion through careful analysis of the evidence provided by travelers; and as a scholar of human natural history, he conceived human populations in terms of successions of generations and sought to identify recurrences among them. A second difference that distinguished them was the analysis of the vision of albinos. According to Blumenbach, they were myopic; according to Buzzi, they were not. Once again in retrospect, we can say that, as far as this specific topic is concerned, Buzzi was right, for subsequent research has shown that in many cases albinos are not myopic but suffer only from heliophobia (photophobia). In this case Buzzi was helped in his analysis by his profession as optometrist. Finally, neither of them made a particular contribution to the understanding of nystagmus, but both of them were correct in maintaining that albinism is a pigmentary disorder that can occur among all peoples on earth.

In the second volume of his *Voyages dans les Alps* of 1786, the Genevan naturalist Horace-Benedict de Saussure devoted several pages to a discussion of albinism and described the two Savoyard albinos discovered by Bourrit. In those same pages he set forth the results of the research conducted by Buzzi and Blumenbach, passing very flattering judgment on the investigations carried out by Buzzi and far less flattering judgment on those of Blumenbach because they were not as direct (Saussure 1779–1796, II, 480–495). De Saussure's text reverberated widely through the learned circles of Europe; indeed, in 1790 the *Encyclopaedia Britannica* published an anonymous extract from it entitled *A Description of the Two Albinos of Europe* (Anonymous 1790). Europeans suddenly discovered, or better *saw*, the albinos living among them. Diverse scholars informed Blumenbach that they had seen an albino in Denmark, or in the city of Würzburg, or in Gotha, or in the Harz Mountains, and he made this information public in the periodical that he edited (Kutzer 1990, 210–211, 215).

The anthropological tradition inaugurated by Blumenbach stressed the universality of albinism, since it was an indirect proof of the unity of the human species. "Albinos are to be found among all five human races," wrote Christoph Girtanner in 1792 (Girtanner 1796, 240). In 1806, his words were echoed by Wilhelm Liebsch (a pupil of Blumenbach): "albinos are to be found among all five human races" (Liebsch 1806–1808, I, 290). The Englishman William Lawrence, who dedicated his anthropological work to Blumenbach, wrote in 1819, "So far is this variety from being peculiar to the Negro, or even to the torrid zone, that there is no race of men, nor any part of the globe, in which it may not occur" (Lawrence 1819, 287). And Rudolph Wagner, who succeeded Blumenbach as professor, claimed in 1831 that "albinos are to be found among all nations" (Wagner 1831, II, 216). In the years when it was not yet certain that the Malpighian layer of the epidermis of Europeans contains a pigment (i.e., melanin), the English naturalist Robert Mortimer Glover wrote that

> the occurrence of Albinos in all races should induce us to believe the presence of a pigment universal. So that what is said of the colouring matter in the Negro may be extended to all varieties of colour, reasoning by analogy.
> (Glover 1840, 377)

The albinism of Europeans thus became evidence to argue that also their skin contained a coloring substance similar to that of blacks.

With recognition of the presence of albinism also among Europeans, the perception of albinos, Leucoæthiopes, Dondos and Kakerlakken changed among intellectuals. Albinos were no longer considered troglodytes, as Linnaeus had classified them, or a distinct race of half-humans as Voltaire had described them, or the monsters of an ancient tradition, although the notion of the latter long persisted among large numbers of Europeans. Rather, they were seen as patients, according to a definition dear to Blumenbach and adopted in the early nineteenth century by a large group of scholars. The historically important point, however, is that a pathology that could occur among all peoples of the earth became with Blumenbach an argument in favor of the thesis that all mankind has a common origin, and that a disorder of pigmentation is a sign of brotherhood.

Note

1 For the history of investigations on albinism, see the historical sections of the monograph by Pearson, Nettleship, and Usher (1911–1913), Björnberg (1960), Froggatt (1959, 1960, 1962), Kutzer (1990), Mazzolini (2006), and Curran (2009).

Bibliography

Primary sources

Anonymous. 1790. *A Description of the two Albinos of Europe* (One twenty-one, the Other twenty-four years of Age). Extracted from M. Sassure's [sic] Journey to the Alps, in the year 1785, and now published in the Encyclopædia Britannica. Liverpool: printed by Thomas Johnson.

Argensola, Bartolomé Leonardo de. 1609. *Conquista de las Islas Malucas al rey Felipe III.* Madrid: por Alonso Martin.

Blumenbach, Johann Friedrich. 1775. *De generis humani varietate nativa illustris facultatis medicae consensu pro gradu doctoris medicinae disputavit d.* XVI. Sept. MDCCLXXV [. . .]. Goettingae: typis Frid. Andr. Rosenbuschii.

———. 1781. *De generis humani varietate liber: cum figuris aeri incisis*. Editio altera longe auctior et emendatior. Goettingae: apud viduam Abr. Vandenhoek.

———. 1786. "De oculis Leucaethiopum et iridis motu commentatio recitata d. IX Octobr. MDCCLXXXIV." *Commentationes Societatis Regiae Scientiarum Gottingensis ad a. MDCCLXXXIV et LXXXV* 7: 29–61 + 1 table.

———. 1788. "Von den Augen der Kakkerlaken." *Magazin für die Naturgeschichte des Menschen* 1 (1): 71–88.

———. 1795. *De generis humani varietate nativa.* Editio tertia [. . .]. Gottingae: apud Vandenhoek et Ruprecht.

Bourrit, Marc-Théodore. 1776. *Description des aspects du Mont-Blanc [. . .]*. Lausanne: chez la Société Thypographique.

———. 1785. *Nouvelle description des glacieres de Savoye [. . .]*. Genève: chez Paul Barde.

Buzzi, Francesco. 1784. "Dissertazione storico-anatomica sopra una varietà particolare d'Uomini bianchi Eliofobi." *Opuscoli Scelti sulle Scienze e sulle Arti* 7: 81–96.
———. 1798. "Bemerkungen über die Kakerlaken." *Italienische Medicinisch-chirurgische Bibliothek* 4 (1): 17–27.
Girtanner, Christoph. 1796. *Ueber das Kantische Prinzip für die Naturgeschichte*. Göttingen: bey Vandenhoek und Ruprecht.
Glover, Robert Mortimer. 1840. "On the Functions of the Colouring Matter of the Skin in the Dark Races of Mankind." *The Edinburgh New Philosophical Journal* 29: 376–383.
Lawrence, William. 1819. *Lectures on Physiology, Zoology, and the Natural History of Man*, Delivered at the Royal College of Surgeons [. . .]. London: printed for J. Callow.
Liebsch, Wilhelm. 1806–1808. *Grundriß der Anthropologie physiologisch und nach einem neuen Plane [. . .]*. 2 vols. Göttingen: Vandenhoek und Ruprecht.
Linnaeus, Carolus. 1758–1759. *Systema naturæ* [. . .]. Editio decima reformata [. . .]. 2 vols. Holmiæ: impensis L. Salvii.
Maupertuis, Pierre-Louis Moreau de. 1744. *Dissertation physique à l'occasion du nègre blanc*. Leyde: n.p.
———. 1745. *Venus physique*. n.p.: n.p.
Saussure, Horace Benedict de. 1779–1796. *Voyages dans les Alpes, précédés d'un essai sur l'histoire naturelle des environs de Genève*. 4 vols. Vol. 1 Neuchatel: chez Samuel Fauche; Vol. 2 Genève: chez Barde, Manget compagnie; Vols. 3 and 4 Neuchatel: chez Louis Fauche-Borel.
Voltaire, François-Marie Arouet de. 1748. "Relation touchant un Maure Blanc, amené d'Afrique à Paris en 1744." In *Œuvres* de M[r.] De Voltaire. Vol. 2, 204–207. Dresden: George Conrad Walther.
———. 1765. *La Philosophie de l'histoire: Par feu l'Abbé Bazin*. Geneva: aux dépens de l'Auteur.
———. 1768. *Les singularités de la nature*. Basle [Genève]: n.p.
Wagner, Rudolph. 1831. *Handbuch der populären Anthropologie für Vorlesungen und zum Selbstunterricht*. 2 vols. Kempten: Druck, Verlag von Tob. Dannheimer.

Secondary sources

Abanime, P. Emeka. 1975. "Voltaire as an Anthropologist: The Case of the Albino." *Studies on Voltaire and the Eighteenth Century* 143: 85–104.
Björnberg, Östen. 1960. "Total Albinos Among the Cuna Indians." *Journal of the History of Medicine and Allied Sciences* 15: 265–267.
Curran, Andrew. 2009. "Rethinking Race History: The Role of the Albino in the French Enlightenment Life Sciences." *History and Theory* 48: 151–179.
Dougherty, Frank William Peter, and Norbert Klatt, eds. 2006–2015. *The Correspondence of Johann Friedrich Blumenbach*. 6 vols. Göttingen: Klatt.
Froggatt, P. 1959. "An Outline, with Bibliography, of Human Piebaldism and White Forelock." *The Irish Journal of Medical Science* 6 (2): 86–94.
———. 1960. "The Legend of a White Native Race." *Medical History* 4: 228–235.
———. 1962. "The Albinism of Timur, Zal, and Edward the Confessor." *Medical History* 6: 328–342.
Kutzer, Michael. 1990. "Kakerlaken: Rasse oder Kranke? Die Diskussion des Albinismus in der Anthropologie der zweiten Hälfte des 18. Jahrhunderts." In *Die Natur des Menschen. Probleme der Physischen Anthropologie und Rassenkunde (1750–1850)*, edited

by Gunter Mann and Franz Dumont, 189–220. (Soemmerring-Forschungen, 6). Stuttgart: Gustav Fischer Verlag.

Mazzolini, Renato G. 2006. "Albinos, Leucoæthiopes, Dondos, Kakerlakken: sulla storia dell'albinismo dal 1609 al 1812." In *La natura e il corpo. Studi in memoria di Attilio Zanca*, edited by Giuseppe Olmi and Giuseppe Papagno, 161–204. Florence: Leo S. Olschki.

Pearson, Karl, Edward Nettleship, and Charles Howard Usher. 1911–1913. *A Monograph on Albinism in Man*. (Drapers' Company Research Memoirs, Biometric Series VI–VIII). London: Dulau.

Part III

Racism, anti-racism, and Eurocentricity

8 Blumenbach's race science in the light of Christian supersessionism

Terence Keel

Determining the intellectual ancestry of a theorist often generates as many questions as it produces answers. Despite what we know about Johann Friedrich Blumenbach we lack complete certainty about the intellectual traditions that shaped his beliefs about race and the life sciences. For example, only in the last few decades have scholars begun to accept that Immanuel Kant had less influence on the great patriarch of the Göttingen School than we originally imagined (Richards 2000; Zammito 2012). If the anti-metaphysical system of the great Königsberg philosopher did not provide the groundwork for Blumenbach's vision of nature and human biodiversity, who and what traditions might we list as proper inspirations? Moreover, acknowledging some intellectual distance between Blumenbach and Kant – the latter of which is the presumed spokesperson of modern secular scientific epistemology – opens space for rethinking the relationship between the emergence of modern racial "science" (which Blumenbach helped to create) and the long history of Christian ideas about human origins that predate the field of physical anthropology.

These are the questions that animate this chapter. What I argue is that Blumenbach's conceptions of human diversity and the life sciences are an outgrowth of Christian intellectual history, and thus not merely a product of Newtonian science or simply a derivative of Kantian epistemology. Drawing upon new scholarship in the field of religious studies I explore unrecognized Christian forms of reasoning at play in Blumenbach's vision of the Caucasian and in his account of race. By using the term "Christian intellectual history" I am speaking of much more than faith, belief, scripture, and theology. Instead, I am referring to patterns of thought and habits of mind about Christian peoplehood, non-Christian others, and creationism that shaped a long tradition of theorizing about ancestry and the origins of human life. I argue that these mental formations were part of the intellectual scaffolding that made up Blumenbach's vision of race and human beginnings. Blumenbach's theory of human descent from an original Caucasian ancestor is therefore not a product of pure secular reason – a freestanding science of human becoming that emerges out of some intellectual void to then completely supersede a religious view of racial origins. Instead, Blumenbach's *De generis humani varietate nativa* is a continuation of Christian intellectual history.

In what follows I begin with an analysis of early Christian accounts of peoplehood and the theology of supersessionism. Here I focus on the work of Justin Martyr and examine the racial implications of early Christian beliefs about the superiority of Christian identity over and against that of Jews and other non-Christians. From this I then discuss new scholarship within the field of religious studies documenting the long tradition of Christian ethno-racial reasoning. This literature offers a framework for understanding how rational practices derived from Christian intellectual history bind the "ancients" and the "moderns" in terms of conceptualizing race. Drawing upon these insights I make an argument for situating Blumenbach's *De generis humani varietate nativa* at the center of an alternative "big picture" of racial science that resolves questions about the sources for Blumenbach's racial ethnology. Key to this new "big picture" is taking account of the intellectual setting in which Blumenbach published his ethnology. The nationalism burgeoning in Germany at the end of the eighteenth century brought with it a new conception of law, morality, and citizenship (Sheehan 1989, 67–71; Hess 2002, 5). These sociopolitical transformations were anchored in a uniquely anti-Jewish racial logic that has its root in the supersessionist theology of the early church. These Christian reasoning patterns appear in Kant's understanding of the ideal ethical community, Christian Wilhelm Dohm's proposal for the civic integration of the Jew, and Blumenbach's ethnology – all of which were written roughly within ten years of each other. By making this association I leave aside the question of whether these three thinkers directly influenced one another. Through this reassessment of Blumenbach's racial thought, I look instead to shed light on a tradition of Christian thought that not only provided inspiration for his ethnology but also influenced discussions about morality and citizenship in Germany near the end of the eighteenth century.

Race and the Christian imagination

During the second century the Christian apologist Justin Martyr imagined an exchange between himself and a Jew named Trypho. In his *Dialogue With Trypho* Justin articulates an early version of what came to be known as the theology of Christian supersessionism. This was the theory that the truth of Christianity supplants the law and knowledge given to the ancient Israelites. In his imagined debate Justin tries to convince Trypho that Christianity marks a new covenant superseding God's previous commitment to Israel. Justin writes, "For the true spirit of Israel [. . .] are we who have been led to God through his crucified Christ" (Martyr 2007, 200). Jesus is the messiah anticipated by the Old Testament prophets and represents for Justin the divine order (*logos*) consistent with the timeless rational structure of scripture. Justin, like many of the early church fathers, conceived of Christianity as a unique form of peoplehood where the specific social locations of its members were replaced by a superior social union (the church) – thus, the meaning of the apostle Paul's letter to Christian communities in Galatia where he writes in Galatians 3:28, "There is neither Jew nor Greek, neither slave nor free, male and female: for you are all one in Christ Jesus." To be Christian involved shedding one's inherited identity or ethnic differences and ascending into a boundless cosmological order of bodies. Not restricted by time, ethnicity, social position, or the beliefs of the Israelites, Christians are people without corporeal ancestors. In Justin's imagined dialogue, which was truly an act of self-creation, Gentiles displace the Jews and in return Christians come to sit at both the center and beginning of a truth that accounts for the complete meaning and end of human history. They are central because history is unintelligible without them. They sit at the beginning because their emergence as a social body marks a new truth (revelation) that bears on all living souls. Through its supersessionist theology, Christianity assumes the position of a universal account of the origins and ends of human history.

Justin's imagined exchange would hardly be intelligible to us were it not for the fact that within this dialogue lay the early forms of reasoning about ancestry, race, and Christianity that continue to occupy our modern ideas about human biodiversity. Recent scholarship on the formation of Christian thought has revealed the centrality of ethno-racial reasoning across the history of Christianity, leaving an intellectual legacy that shaped the development of modern systems of belief, which inform contemporary conceptions of race in European and American science (Keel 2013). For example, Denise Kimber Buell has noted that the early followers of Jesus understood their community in ways that were consistent with the ethno-racial logic of their ancient contemporaries. She argues that

> Christian texts from the late first through early third centuries do not instruct readers to understand themselves as simply members of a new "religion," a voluntary cult that entails rejection of ancestral customs (for gentiles) or a radical reinterpretation of them (for Jews). Instead, many Christian texts explicitly guide readers to understand their entrance into these emerging communities as a transformation from one descent group, tribe, people, or citizenship to a new and better one.
>
> (Buell 2009, 111–112)

Buell claims that early Christians possessed an understanding of peoplehood that functioned conceptually like an ethno-racial group and yet at the same time understood themselves to be superior to other forms of social membership by virtue of their claims to have knowledge about the destiny of all of humanity. Indeed, this was the knowledge of salvation that implicated all people.

Claims of Christian truth being superior to other ethno-religious knowledge had specific consequences for the relationship between Christianity and Judaism. J. Kameron Carter has argued that the effort of Christians to answer this question produced patterns of reasoning about race that would become deeply embedded not merely in Catholic and Protestant theology but in modern Euro-American epistemology (Carter 2008). According to Carter, "modernity's racial imagination has its genesis in the theological problem of Christianity's quest to sever itself from its Jewish roots" (Carter 2008, 4). The fallout of this severance, Carter argues, is that Jews were cast as a racial group and this racialization would later drive a wedge between the Euro-American Occident and the assumed Orientalism of Jews.

Christian thinking about Jews had a profound effect on Europe's own self-understanding as well as on its perception of native populations in the New World. Jonathan Boyarin, in his study of the religious and racial diversity in Spain before and after the colonial encounter with the New World, argues that "the troubling instability of Jewish difference shaped both Christian Europeans' self-image and their reactions to those they encountered in the course of exploration and conquest beyond what became Europe's borders" (Boyarin 2009, 1). Boyarin notes that the persistence of Jewish otherness throughout the medieval period consolidated what it meant to be Christian and European.

In this consolidation whiteness would materialize into an ideal social marker that allowed Europeans to further distinguish themselves from Jews and natives. The Christian intellectual historian Willie Jennings argues that this use of whiteness to distinguish between Christian and non-Christian reveals how "Christianity in the Western world lives and moves within a diseased social imagination" (Jennings 2010, 6). Jennings explains that in the wake of the early modern colonial encounter "whiteness emerges, not simply as a marker of the European but as the rarely spoken but always understood organizing conceptual frame" (Jennings 2010, 25). Located now at both the center and beginning of human history Jennings argues that Christian identity became fully immersed within "European (white) identity and fully outside the identities of Jews and Muslims" (Jennings 2010, 33). This construction of European Christian whiteness modernizes the supersessionist theology of the early church that displaced Israel as God's chosen people. In its new form supersessionism would place white European identity into a "boundary-less reality," thus occupying the limitless position of the body of Christ imagined by the early church fathers. Simultaneously unrestrained and central to human history European Christian whiteness touches and unites all peoples through an "ecclesial logic" that finds all bodies organized under a single ontology and conception of peoplehood rationalized through the idea of common human descent (Jennings 2010, 33).

Race in the life sciences: An alternative big picture

The recent shift in the historiography on race among scholars of religion has opened the possibility of rethinking the links between the early church and modern views of human biodiversity, which Blumenbach helped establish. Yet this connection presents a challenge for historians of science who since the 1990s have moved away from crafting a "big picture" of knowledge that draws connections across multiple historical epochs (Cunningham and Williams 1993).

In their assessment of the modern origins of science Andrew Cunningham and Perry Williams note that the first generation of historians within the field constructed a grand sweeping view of science that covered the whole of human history within a single progressive narrative. The "scientific revolution" was a culminating point within this narrative, marking philosophical and moral shifts that set the course for our current vision of science as a universal human activity. This "big picture" was common among science-supporting intellectuals during and after the World War period and was designed to establish both the importance of modern science and the need for a subfield within the profession of history. Cunningham and Williams, however, note the changing historiography on the scientific revolution that has decentered this initial "big picture" and with it the conception of science rooted in a "transcendent timeless logic and embodying absolute moral values of freedom, rationality, and progress" (Cunningham and Williams 1993, 418). As an alternative "big picture" they argue for an account of the origins of modern science that is bounded in time, space, and culture and would ultimately be concerned with identifying "the first appearance, the first practice, of something which is distinct and specific to our own region of time and space, rooted in the particular circumstances of our culture" (Cunningham and Williams 1993, 418). This formulation of scientific knowledge is thus one that emphasizes ruptures and cleavages across historical eras – most especially between ancient and modern projects of enquiry.

This shifting assessment of the modern origins of science explains two discernable features of the scholarship written about Blumenbach's racial ethnology. The first is that historians of science and anthropologists have tended to look for the origins of race science that closely mirror contemporary formulations. In this pursuit there has been a tendency to construct Blumenbach as a secular figure detached completely from Christian thought, thereby crafting a picture that conforms to our present ideas about the disaggregation of race, science, and religion.[1] Consequently we can discern a second effect, which is that scholars of religion have largely reproduced this secular interpretation of Blumenbach and the origins of post-Enlightenment race science in their understanding of the relationship between Christianity and race theory.[2] What is noteworthy here is that the unchallenged secular reading of Blumenbach has also kept in play the very old narrative of religious decline that appeared in 1959 with John C. Greene's seminal text *The Death of Adam*, written during the first generation of scholarship on the history of science (Greene 1959). In this work, and those that follow its blueprint, religion is understood as having a waning impact on the production of scientific ideas about race, beginning with Carl Linnaeus and Blumenbach during the Late Enlightenment and culminating with Charles Darwin's linkage of humans and primates. The longevity of this decline thesis explains the inability of religious studies scholars to extend their claims about the Christian roots of modern racial thought into any substantive discussion about racial science beyond the nineteenth century.[3] The alternative big picture proposed by Cunningham and Williams thus keeps us committed to the religious decline narrative that accompanied the grand picture of science during the time of Herbert Butterfield and Alexandre Koyré (Cunningham and Williams 1993, 410). What we have then is a recent reevaluation of the modern origins of science that has not fundamentally changed how historians think about the "big picture" of how the race concept is used across the life sciences.

Yet the scholarship on Christian thought and race has made it clear that it is time to reevaluate the modern origins of race science. I contend an important step toward such a reframing involves broadening our understanding of "religion" to include intellectual history – again one constituted by racial reasoning strategies and habits of apprehending the social and natural world – that cannot be reduced to belief, scripture, and theology. Religion in this sense describes a system of meaning and thought rooted in intellectual traditions that predate modern definitions of religiosity. Rather than take for granted that modern (racial) science marks a rupture from the religious traditions of the past, we should consider instead how one of the characteristics of our modern scientific perceptions of race is its indebtedness to a Christian intellectual history that scientists and the scholars who study them have actively denied and repressed. We must denaturalize this denial and repression, along with the very concept of race this refusal enables, and instead see it as an expression of a unique feature of European Christian practices of thought and reason that have harbored hostility toward the religious (Jewish) roots of Euro-Christianity's own intellectual horizon. The religious decline thesis that continues to shape the historiography of modern racial science and that constructs Blumenbach as a secular figure is merely a modern articulation of the logic of Christian supersessionism. It should be no surprise then that according to the present-day life sciences "we moderns" hold a conception of race and human origins that is unprecedented, lacks intellectual ancestors, and is thought to be superior to previous knowledge formations.

Placing the Jew in the German imagination

Blumenbach arrived at his supposedly modern secular conception of the human during a time when German states reconfigured their civic responsibility to Jews. Near the end of the eighteenth century the perceived ethnocentrism, backwardness, and moral degeneracy of Jews prompted German intellectuals to imagine a form of citizenship capable of integrating them into the social body (Hess 2002, 5). This new body politic was premised on the notion of the human as a natural being, or ontologically comparable, which was entitled to universal rights regardless of the social location of the subject in question (Sheehan 1989, 71). As the historian Jonathan Hess has argued, what emerged was a German vision of citizenry that was profoundly supersessionist and therefore Christian in its claim to having created a set of entitlements that brought various subjects into a polity that transcended cultural and social particularities, most especially Jewish life (Hess 2002, 11).

In this climate of legislative and bureaucratic modernization Kant and Dohm, a politically ambitious civil servant of the Prussian state inspired by the ideals of the Enlightenment, would articulate a vision of the ideal moral subject and the modern German *Staatsbürger* that was organized by the concept of a Christian polity and anti-Jewish supersessionist beliefs. This was not an entirely new orientation but an extension of a long-standing tradition of ethno-racial reasoning dating back to the early church. Revisiting these formulations is key for understanding how Blumenbach helped translate a Christian conception of peoplehood and the logic of supersessionism into an enduring characteristic of our modern racial science.

Let us consider the latent supersessionist Christian assumptions found within Kant's writings on religion, which were penned around the time Blumenbach began working on his theory of human biodiversity.

In 1781 Kant published the *Critique of Pure Reason*, which laid the groundwork for his writings on religion. The cornerstone of the *Critique* was Kant's belief that human thought is governed by laws of reason that structure and organize how we perceive the world around us. For Kant there was a key distinction between the world as it appears to us, which he called *phenomena*, and the life and existence of things outside and beyond our perceptions of them, or what he called *noumena*. Our minds are designed to understand the world through the universal structures of reason that figure our perceptions. Kant claimed that "we can have cognition of no object as a thing in itself, but only insofar as it is an object of sensible intuition, i.e. as an appearance" (Kant 1998 [1781], Bxxvi–Bxxvi). Given these perceived limits of reason Kant would call into question traditional religious claims about God and revelation, as well as metaphysical speculation about the inherent order of nature, all of which were premised on the notion that the human mind could transcend the restrictions of our experience as organisms bound by time, space, and the structures of the mind. Kant also had doubts about the legitimacy of the biological sciences and the aspirations of its practitioners to disclose not merely natural phenomena as they present themselves to us but the true inner workings of the natural world that leap over the limits of empirical observations (Richards 2002, 236).

Kant would later build upon the universal principles of epistemology mapped out in the *Critique* to reflect on the purpose of religion for the modern world and ultimately the history of morality. These thoughts culminated in his 1793 publication *Religion Within the Boundaries of Mere Reason*. As the very title suggests, Kant hoped to explain how traditional religious concepts like God, sin, or the organization of the church could be used to orient and guide the needs of an ethical life while keeping in mind the fundamental limits of human knowing. In *Religion*, Kant also looked to provide an account of the universal history of morality, which he conceived as the history of reason's self-awareness and self-orientation toward the good (Yovel 1980, 9).

Judaism, however, occupies a conspicuous place within Kant's understanding of original sin and in his account of the history of morality detailed in *Religion*. Kant's *Critique* gave him the ability to distinguish between events, ideas, and experiences as they must appear to us (*phenomena*) versus things as they are in themselves (*noumena*). This distinction provided Kant with a framework in *Religion* to explicitly displace the truth claims and lived experiences of Jews and the ancient Israelites. The starting point for this supersession was Kant's interpretation of original sin to mean more than an initial state of depravity all humans inherit historically from Adam and the patriarchs of ancient Israel (Kant 1998 [1793], 6:40). Instead, Kant reimagines original sin as an unavoidable consequence of humanity's innate freedom (Kant 1998 [1793], 6:41).

Kant claimed that the source of human freedom was not derived from society (*phenomena*) but from a will that is inherently good, unencumbered, and imputed to our species from a benevolent God that stands outside of history (*noumena*). For Kant, the human will was not bound by time, culture, or customs, which is to say not conditioned by acts or events that take place in history. Instead, Kant conceived of the human will as a self-legislating force (Kant 1998 [1793], 6:39–36:40). Thus, Kant concluded that original sin traditionally perceived as the beginning of evil's manifestation in history is not a permanent state of imperfection that humans inherit from our ancestors. Original sin was an ever-present possibility, a by-product of our freedom, which has no first cause beyond our own will (Kant 1998 [1793], 6:43–46:44). For this reason Kant could say,

> Whatever the nature, however, of the origin of moral evil in the human being, of all the ways of representing its spread and propagation through the members of our species and in all generations, the most inappropriate is surely to imagine it as having come to us by way of inheritance from our first parents.
> (Kant 1998 [1793], 6:40)

According to Kant's reasoning, our capacity for evil resides in our inherent freedom, not in an imagined inheritance shared with Adam and Eve.

There were clear theological and racial implications of this transformation of original sin into a consequence of a free will, as opposed to an inherited state of deficiency. Original sin and Adam's fall sat at the center of the Christian notion of common human ancestry and the redemption of Christ since the time of the early church.[4] The fallout of Adam's sin was imagined by Christian theologians to quite literally be passed down ancestrally to all races, including of course the patriarchs of ancient Israel. The early church attempted to offset original sin by seeing the salvific death of Christ as part of a new covenant with God that supplanted the bond with Israel. This new covenant was universally open to all, creating the basis whereby Christ could be imagined as the ultimate truth. Later, Protestant theologians would argue that only through God's grace could humans obtain salvation. Nonetheless, both the early church and its modern counterpart maintained a view of original sin that interpreted humanity's depravity as a bio-spiritual phenomenon that united all races through redemption in Christ.

Yet in Kant's view, original sin was not a genealogical or historical dilemma; the fall was an existential problem. Rational modern Christians were not the offspring of a bankrupt bio-spiritual inheritance. Their free will, which afforded them autonomy from history and the sociopolitical structures that shaped contemporary life, made sure of it. Indeed, freedom for Kant was the default human condition. Sin did not deform free will or place the modern German moral subject into a bio-spiritual bind with the ancient Israelites or any human predecessor. Citing the Roman poet Ovid, Kant would say, "Race and ancestors, and those things which we did not make ourselves, I scarcely consider as our own" (Kant 1998 [1793], 6:40). In Kant's mind, no longer would German Protestants be required to carry the moral debts of their ancestors.

Kant's ability to recast original sin from a problem of bio-spiritual inheritance to a consequence of free will carried implications for where Judaism sat within his vision of modern morality. This (dis)placement of the ancient Israelites reveals the profoundly anti-Jewish and uniquely Christian commitments that shaped Kant's account of human moral behavior – indeed an account that looked to obtain the same level of universal applicability as Dohm's proposal and Blumenbach's monogenist ethnology. Kant believed Christianity was the ideal model for a rational moral community. Yet it could play this role because he believed the early church took an unprecedented leap in human history (Kant 1998 [1793], 6:93–96:95, 6:101–106:102). Kant claimed that the only history relevant to the drama of human becoming was one that began after humans became aware of the self-legislating power of their free will and an innate predisposition to the moral good. Christianity was the starting point for this universal history – an origin that eclipsed Judaism as well as the religious and cultural traditions of non-Europeans (Kant 1998 [1793], 6:124). Kant argued that "Christianity marked the total abandonment of the Judaism in which it originated, grounded on an entirely new principle, effected a total revolution in doctrines of faith" (Kant 1998 [1793], 6:127). Prompting this revolution, according to Kant, was Jesus's appropriation of "Greek wisdom," which had the effect of "enlightening [Judaism] through concepts of virtue and in spite of the oppressive burden of its dogmatic faith." Through this enlightenment Kant claimed that "Christianity suddenly though not unprepared arose" (Kant 1998 [1793], 6:128). In Kant's system Jesus was no longer a Jew; he was a Greek philosopher of sorts, who was the first to model a moral life shaped by the self-legislating powers of a free will (Hess 2002, 154). This was a model to be emulated by enlightened Germans who hoped to live a moral life within the limits of reason.

Kant's claim that Christianity marked a radical rupture from Judaism is consistent with the forms of racial reasoning that have othered the Jew throughout Western intellectual history. As Jonathan Hess has argued, Kant's

> view of Christianity as a rational religion that produced itself out of itself is symptomatic of a much larger problem within Christianity, an antipathy toward the historical past grounded in Christianity's inability to give an adequate account of its own Jewish origins.
>
> (Hess 2002, 154)

Moreover, Kant diminished the significance of Judaism for modern German intellectual history by claiming that the beliefs of the ancient Israelites have no bearing on the contemporary moral subject.

In the time between Kant's *Critique* and the Christian-centric views he developed in *Religion*, German statesmen were occupied by debates over the emancipation and political integration of Jews into an emergent modern state (Hess 2000, 57). At the center of the controversy was a famous proposal written in 1781 by Dohm, titled *Über die bürgerliche Verbesserung der Juden* (On the civic improvement of the Jews). Dohm argued that the moral, political, and physical state of degeneracy that marked European Jews was the end result of their oppression and neglect at the hands of Christians rulers who discriminated against them because of their religious and cultural differences (Hess 2002, 3). To redress their disenfranchisement, Dohm put forth a comprehensive political solution that demanded the state intervene on their behalf, grant Jews civil protections under the law, and require that they serve in the military, and he proposed means to transition them into the agricultural economy, thereby moving Jews out of the practice of trade and money lending (Hess 2002, 3). Dohm had previously collaborated with his colleague and friend Moses Mendelssohn to intervene on behalf of Alsatian Jews under French territory, who were facing a current of anti-Jewish sentiment. Dohm's *Über die bürgerliche Verbesserung der Juden* was translated into French in 1782 and would sit at the center of several French treatises that eased some anti-Jewish restrictions and ultimately set the stage for the much larger debates on Jewish emancipation that ensued in the aftermath of the French Revolution.

At stake in Dohm's proposal was the viability of a modern secular conception of citizenship and the capacity of the state to transform its citizenry. If the Jews, who were perceived as morally, culturally, and physically degenerate, could be changed into productive citizens in Germany and France, this would demonstrate the success and superiority of a modern, secular state founded on the universal values of the Enlightenment (Hess 2002, 3).

Given their small numbers and marginalization across German states it would seem unusual that Jews would occupy such a large space within the burgeoning German nationalist imagination (Sheehan 1989, 67–71; Hess 2002, 5). This is a question raised by Hess, who insightfully observed, "The project of Jewish emancipation provided the ultimate test, in practice, of the rational ideals of the Enlightenment, the perfect arena for speculating about translating the lofty premises of Enlightenment universalism into concrete practice" (Hess 2002, 6). The possibility of the Jew as a citizen revealed the prospects and ideological limits of modern German notions of universal rights. However, an even deeper understanding of this utilitarian use of Jewish identity can be had if we keep in mind the long investment of Christian intellectual history in demarcating the boundary between Christian and Jew through practices of racial reasoning. Again, as Buell has argued,

> Early Christian discourses of conversion share with modern discourses about race an abundance of metaphors for evolution-change, where Christian belonging is understood as a perfection, distillation, fulfillment of individual human and collective human potential [. . .]. Early Christian supersessionist arguments, which interpret Jesus' significance and Christian belonging as the fulfillment of God's promises to Israel, as contained in scriptures, portray Christianity as the realization, maturation, and in some cases restoration of Israel and Judaism.
>
> (Buell 2009, 115)

If we take Buell's analysis into the modern context, it is imperative that we see the conceptual displacement of the Jew in German moral and political thought during the second half of the eighteenth century as an extension of the racial reasoning practices and habits of mind drawn from Christian intellectual history. The visions of the Jew that Kant and Dohm maintained in relation to their vision of universal rights and ideal moral community were not novel formations but stemmed from a long-standing tradition dating back to the early church. To view these formations as uniquely modern or unprecedented is a symptom of a Christian consciousness that harbors an antipathy toward the historical past and is incapable of acknowledging the religious inheritance that shapes its own intellectual horizon.

We will see how Christian forms of reasoning shaped Blumenbach's ideas about the Caucasian and his account of human biodiversity, which he crafted and revised over this period of tremendous intellectual productivity across the fields of science, philosophy, and theology near the end of the century. If the nationalism burgeoning in Germany at the end of the eighteenth century brought with it a new conception of the law, morality, and citizenship, all of which consolidated white German Protestant identity, Blumenbach provided a scientific account of race that translated Christian ideas of peoplehood into a monogenist ontology of race that would further the supersession of the Jew by white Europeans.

My reading of Blumenbach adds to John Zammito's and Robert Richards's reappraisal of Blumenbach's relationship to Kant. If Blumenbach's vision of the life sciences was set prior to Kant's contribution to the field, as Zammito argues, it is reasonable to assume that a deeper intellectual history was at work in Blumenbach's theory of race (Zammito 2012). As I am arguing, this was a history tied to a Christian tradition of racial and ethnic othering. Turning to this deeper intellectual history also complicates Richards's observation that Blumenbach and Kant maintained different ideas about the life sciences. My analysis here concurs with Richards's assessment: Blumenbach surely wanted to explain the cause of life's organization by imputing teleology in nature, whereas Kant believed that the life sciences could not actually prove nature was designed toward ends without transgressing the limits of reason (Richards 2000). However, both men shared a Protestant intellectual heritage that inclined them to think in universal terms, to assume common human ancestry, and to harbor antipathy toward non-Christian (and specifically Jewish) traditions. For Blumenbach, as we will soon see, this heritage manifests itself in his racial science. For Kant, this heritage presented itself in his vision of human morality, which surely was teleological. Kant's view of morality also assumed the racial logic of Christian supersession that displaced the knowledge of the ancient Israelites. On this score, we will see that Kant and Blumenbach were closely aligned. Taken together Dohm's vision of modern citizenship and Kant's moral vision gave expression to a Protestant worldview that placed the modern white German Christian at both the beginning and the center of human history. Blumenbach's ethnology would prove to be the scientific counterpart to this Eurocentric and Protestant vision of social life.

Caucasian is a Christian concept

In 1781, the very same year that Kant published the *Critique of Pure Reason* and Dohm launched a trans-European debate over Jewish emancipation, Blumenbach published the second edition of his dissertation thesis, *On the Natural Variety of Mankind*. In this version Blumenbach had yet to give proper names to the ancestral human types. He did, however, introduce the explanatory mechanism that would later allow him to do so. This was the concept of the formative drive (*Bildungstrieb*), drawn from a work also published in 1781 titled *Über den Bildungstrieb und das Zeugungsgeschäfte*, which made refinements to an essay written just the year before (Blumenbach 1781; Blumenbach 1780). In both the essay and book Blumenbach attempted to account for how previously unorganized organic matter came together to create lasting varieties of species. He also looked to explain what appeared to be nature's ability to repair itself. Blumenbach surmised there had to be a force inherent to nature capable of this task. Thus, he developed the notion *Bildungstrieb*, a concept that would have a major influence on post-Enlightenment perceptions of nature and offer Blumenbach a theoretical tool to explain human racial descent.

Blumenbach became a leading voice in the critique of preformation, devising what was then understood as a radical form of organic vitalism to explain the gradual development of embryos from unorganized matter (Richards 2002, 216). His studies on the apparent regenerative capabilities of living organisms led him to conclude that inherent to nature was a formative drive responsible for the reproduction, maintenance, and restoration of the parts of living forms (Richards 2002, 219). This *Bildungstrieb* was found within the genital fluid of living organisms and gave life to their offspring, where "its first business" was to "put on the form destined and determined" for the species in question (Blumenbach 1865 [1795], 194). After providing each organism with its species-specific form and constituent parts Blumenbach claimed that this formative force nurtured and preserved the organism. According to Blumenbach, "if by chance [an organism] should be mutilated, [it] lies in its power to restore it by reproduction" (Blumenbach 1865 [1795], 194). For Blumenbach the *Bildungstrieb* was a teleological force inherent to nature that created species out of formless organic material.

Blumenbach argued that nature's formative drive was ostensibly capable of turning "aside from its determined direction and plan" (Blumenbach 1865 [1795], 195). Climate, diet, mode of life, hybridity, and hereditary diseases were all factors that could push an organism to degenerate from a primeval type and develop novel varieties (Blumenbach 1865 [1795], 194–205). The idea that living things were capable of deviating from an original form was a notion Blumenbach borrowed from Georges-Louis Leclerc de Buffon and his work on the degeneration of animals published in 1766 (Richards 2002, 221). What differentiated their two theories was Blumenbach's introduction of the concept of formative drive to explain how original forms were maintained over time. He believed that extended periods of external stimuli on an organism "[have] great influence in sensibly diverting the [formative force] from its accustomed path." This deflection was "the most bountiful source of degeneration" which yielded nature's splendid diversity, including, of course, human racial varieties (Blumenbach 1865 [1795], 196).

Blumenbach's mature ethnology appears in Section IV of the 1795 edition of *On the Natural Variety of Mankind*. Here the Caucasian provides the means for situating white European identity within a "boundary-less reality," where it is linked to all races and remains central to human history. The colorism of Blumenbach's thinking is key to organizing humans under a single body – that of the white Caucasian – and thus within a modern ontology of race. He explains:

> Besides it is the white in colour, which we may fairly assume to have been the primitive colour of mankind, since, as we have shown above, it is very easy for that to degenerate into brown, but very much more difficult for dark to become white, when the secretion and precipitation of this carbonaceous pigment has once deeply struck root.
>
> (Blumenbach 1865 [1795], 269)

The "races of man" inherit their form out of a pluripotent Caucasian body that is present within these new races yet never fully reduced to them. The African is not Caucasian, neither is the Mongolian, even though they both "degenerate" from this original population and thus fall within a unified conception of peoplehood – much like Kant's ideal moral subject and Dohm's modern citizenry. All three men assume an equivalency between varying bodies but it is Blumenbach who represents this commensurability in ethnological terms. According to Blumenbach the primeval source of human life transcends the geographically situated and phenotypically varied instances of human life. The truth that draws together these assorted racial forms is found in an omnipresent force (the *Bildungstrieb*) that first appeared in the white Caucasian and that subsequently binds humans into a shared peoplehood (Blumenbach 1865 [1795], 269).

We have to see here that Blumenbach was predisposed to think in terms of a singular human ontology (a common peoplehood) as a result of a Christian intellectual inheritance shared with Kant and Dohm. But this equivalency between disparate bodies must itself be created – it is not simply out in nature waiting to be discovered. The idea that humans belong to a single species, that the races degenerate – to use Blumenbach's language – from an original ancestor and thus can be organized under a single story of development, is neither a necessary formulation nor one that should be assumed universally present in all peoples at all times – even though the concept of peoplehood presupposed by Blumenbach and his Christian forbearers would have us believe this to be the case. To situate Blumenbach's racial ethnology within a very specific Christian intellectual tradition forces us to recognize that the Eurocentric idea of a common peoplehood (or shared human ancestry) is not any more innate to this creature called the *human* than the notion that all people have reason, that science is a universal human activity, or that the capacity for religiosity is found within all cultures.[5] These very recent creations emerged out of a Christian European epistemic worldview forged within the context of colonialism and are indigenous to the West, even though such ideas – like the body of Christ – presume what Jennings calls "an ecclesial logic applicable to the evaluation of all peoples" (Jennings 2010, 33).

If the specificity of this Christian Eurocentric vision of common peoplehood were not clear, one simply needs to consider Blumenbach's thoughts on the Jew.

Of the many races noted in *Natural Variety*, Jews appear literally in the margins of Blumenbach's ethnology. Jews are integrated into the human species as white, a point that must be surmised from Blumenbach's mentioning in a footnote in the 1781 edition that the white variety includes "that part of Asia which lies toward us, this side of the Obi, the Caspian sea, mount Taurus and the Ganges, also northern Africa" (quoted in an editor's footnote to Blumenbach 1865 [1775], 99). In the final 1795 edition the "white variety" would take on the name "Caucasian." Yet, Blumenbach makes special note to argue in this final edition that Jews constitute a peculiar variety of Caucasian. Unlike the other white races, they appear incapable of transformation into different types and lack the aesthetic comeliness of their German counterparts. Blumenbach writes in the 1795 edition that

> The ancient Germans gave formerly instances of the unadulterated countenance of nations unaffected by any union with any other nation, and to-day the genuine Zingari, inhabitants of Transylvania do the same; and above all the nation of the Jews, who, under every climate, remain the same as far as the fundamental configuration of face goes, remarkable for a racial character almost universal, which can be distinguished at the first glance even by those little skilled in physiognomy, although it is difficult to limit and express by words.
>
> (Blumenbach 1865 [1795], 233–234)

Blumenbach elaborates on this racial character in the 1795 edition:

> The great artist Benj[amin] West, President of the Royal Academy of Arts, with whom I conversed about the racial face of the Jews, thought that it above all others had something particularly goat-like about it, which he was of opinion lay not so much in the hooked nose as in the transit and conflux of the septum which separates the nostrils from the middle of the upper lip.
>
> (Blumenbach 1865 [1795], 234)

Jews possess stable racial traits recognizable to even the most unscientific observer. We can infer from Blumenbach's writings that although Jews are white Caucasians, they lack the capacity to degenerate into other racial types because of a stubborn phenotype that consistently sets them apart from other groups.

Again we see a structure of Christian racial reasoning at play in Blumenbach's thinking. Jews have been displaced from the beginning of European history at a biological level. At the threshold of history now stands the white Caucasian – the patriarch of all races. Blumenbach integrates Jews into the drama of human becoming but unlike the white Caucasian they lack the pluripotency to yield new forms. The Jew is both displaced and provisionally integrated into Blumenbach's scientific account of human diversity.

Moreover, this supersession of the European over the Jew can be interrogated further when we consider Blumenbach's ideas on the primacy of the Caucasian form. If human varieties were created through a process of degeneration from a primeval type, from what variety did the white Caucasian degenerate? When searching for an answer to this question one begins to see very large omissions within Blumenbach's ethnology – absences that reveal the place of Christian racial reasoning.

In Blumenbach's theory of formative force, nature created living organisms out of previously unorganized organic matter. This was true even in the development of life during what was called the pre-Adamite world (Blumenbach 1865 [1806], 287). Like most of his contemporaries Blumenbach assumed the earth to be hundreds of thousands of years old but remained a traditionalist when it came to his understanding of the recent creation of human life (Greene 1959, 235–238). We know this because the need to account for the length of time in the transition from one race to another never occurred to Blumenbach as a problem to be resolved. It must be recognized then that Blumenbach's racial ontology was suspended within a non-secular, which is to say Christian, temporal framework.

Assuming a sacred chronology is consistent with the creationist logic that structures both the formation of the earth and human life out of formless organic matter, Blumenbach wrote,

> After therefore that organic creation in the Preadamite primitive epoch of our planet had fulfilled its purpose, it was destroyed by a general catastrophe of its surface or shell, which probably lay in ruins some time, until it was put together again, enlivened with a fresh vegetation, and vivified with a new animal creation. In order that it might provide such a harvest, the Creator took care to allow general powers of nature to bring forth the new organic kingdoms, similar to those, which had fulfilled that object in the primitive world.
> (Blumenbach 1865 [1806], 287)

Following the first global catastrophe, Blumenbach argued that nature's formative force retained some sort of living memory of the first organisms and drew upon these forms to create new varieties. Blumenbach claimed that

> the formative power of nature in these remodellings partly reproduces again creatures of a similar type to those of the old world, which however in by far the greatest number of instances have put on forms more applicable to others in the new order of things, so that in the new creatures the laws of the formative force have been somewhat modified.
> (Blumenbach 1865 [1806], 287)

New species, in this theory, were derived from antecedent creations whose forms functioned as a template for the creative powers of nature's formative force. All of the plants and animals thriving at the time of Adam's creation were variations from the forms that were found in the pre-Adamite world.

We arrive now at the ontological assumptions that rest behind Blumenbach's vision of the white Caucasians. Humans were obviously not present during the first iterations of life on earth. Consequently, there was no template from the pre-Adamite world for nature to draw upon to give to the original human. We know from Blumenbach's writings in *On the Natural Variety of Mankind* (1795) that he assumed the first human was a naturally domesticated species (Blumenbach 1865 [1806], 293–294). "His Creator has therefore fortified him with the power of reason and invention, in order that he may accommodate himself" to the "variety of climate, soil and other circumstances" that shape the human form (Blumenbach 1865 [1795], 183). Natural domestication was the birthright of the human being. Humans, Blumenbach asserted, possessed within their lineage no antecedent primitive form. What this means for the white Caucasian ought now to be clear. This archetypical human emerged from nothing. Richards has argued that Blumenbach's *Bildungstrieb* was responsible for giving creatures their form and "could not be an effect of organization, a property emerging out of organization"; it was instead "a cause to explain organization" (Richards 2000, 25). Thus, we must assume either that Caucasians are the product of a creation event directly from the *Bildungstrieb* or that they formed themselves out of themselves. In both instances, however, the white Caucasian stands on an ontological plane that is qualitatively distinct from the other races; they are the only population who was the result of self-creation and not degeneration.

Here the Caucasian and the Christian present themselves as interchangeable concepts. Both occupy the conspicuous role of bearing witness to a rational order (*logos*) that gives form and meaning to human history. Both displace the Jew by being situated at the beginning and the center of the human drama. For Blumenbach, this displacement creates a void out of which the Caucasian emerges *ex nihilo* to then give rise to the human species. The Caucasian is thus central to human history for all human becoming is unintelligible without exposing the natural mechanism (the *Bildungstrieb*) that gave rise to this original group. At the same time without the Caucasian there would be no template for the human form. Blumenbach's Caucasian – like Kant's enlightened moral subject, and Dohm's modern citizen – constitutes an imagined group determined by neither time, geography, or social location. Like the body of Christ imagined by Justin in his *Dialogue With Trypho*, Blumenbach's original Caucasians are people without corporeal ancestors.

Racial science, or Christian intellectual history by other means

New developments within the historiography of religious studies have provided insights for reimagining the links between Christian thought, race, and modern science. This literature reveals the commitment of Christian intellectual history to a long tradition of ethno-racial reasoning, which in turn prompts the need for an alternative "big picture" of the emergence and formation of the race concept in science. When Blumenbach's *De generis humani varietate nativa* is placed within this larger intellectual history his racial ethnology bears the influence of a Christian conception of peoplehood and a creationist epistemology that unite the human form under a single ontological vision. This history also links Blumenbach's race "science" to conceptions of the moral subject and modern citizenry that were replicas of a Christian polity. What emerges then in late eighteenth-century Germany is a new body politic organized around a set of Christian European epistemic commitments that ultimately supplant the Jew with the white European. Within this intellectual setting Blumenbach presents the pluripotent Caucasian as a self-fashioning people, unbounded by history, geography, and social location. Indeed, the Caucasian was born out of an intellectual and corporeal void and gave life to other races. As a unifying concept to explain common human ancestry the very notion of the Caucasian sustained Christian habits of mind. These rational practices have predisposed us to tell the history of racial science as one of rupture and discontinuity from the premodern world. Yet in the final analysis, Blumenbach's ethnology and his imagined racial ontology were surely extensions of a religious intellectual history often denied in the stories we tell about the origin of modern racial science.

Notes

1 See, for example, Horsman (1981), Gould (1996), and Marks (2009).
2 I realize that it is anachronistic to use the term "race science" to describe the study of human differences during the Enlightenment. However, with this usage I am intentionally situating the study of race in the West within a long historical tradition, thus avoiding hard demarcations that too often prevent us from making connections across multiple historical periods. Moreover, it is common among historians and anthropologists to situate the origins of "race science" in the Enlightenment despite the fact that the term "science" had yet to be used to describe the study of human origins. See, for example, Kidd (2006) and Livingstone (2011).
3 For an excellent discussion of race, religion, and science during the nineteenth century see Johnson (2004).
4 See Sanlon (2014) and Kolb (2014).
5 See Masuzawa (2012), Nongbri (2013), and Harrison (2015).

Bibliography

Blumenbach, Johann Friedrich. 1780. "Über den Bildungstrieb (Nisus formativus) und seinen Einfluss auf die Generation und Reproduction." *Göttingisches Magazin der Wissenschaften und Litteratur* 1 (5): 247–266.

———. 1781. *Über den Bildungstrieb und das Zeugungsgeschäfte*, 1st edn. Göttingen: Johann Christian Dieterich.

———. 1865 [1795]. *The Anthropological Treatises of Johann Friedrich Blumenbach*. Trans. Thomas Bendyshe. London: Longman, Green.

———. 1865 [1806]. *Contributions to Natural History*. Trans. Thomas Bendyshe. London: Longman, Green, Longman, Roberts.

Boyarin, Jonathan. 2009. *The Unconverted Self: Jews, Indians, and the Identity of Christian Europe*. Chicago, IL: Chicago University Press.

Buell, Denise Kimber. 2009. "Early Christian Universalism and Modern Racism." In *The Origins of Racism in the West*, edited by Miriam Eliav-Feldon, Benjamin Isaac, and Joseph Ziegler, 109–131. Cambridge: Cambridge University Press.

Carter, J. Cameron. 2008. *Race: A Theological Account*. Oxford: Oxford University Press.

Cunningham, Andrew, and Williams, Perry. 1993. "De-Centering the 'Big Picture': 'The Origins of Modern Science' and the Modern Origins of Science." *The British Journal for the History of Science* 26 (4) (December): 407–432.

Gould, Stephen Jay. 1996. *The Mismeasure of Man*. New York: W.W. Norton.

Greene, John C. 1959. *The Death of Adam: Evolution and its Impact on Western Thought*. Ames: Iowa State University Press.

Harrison, Peter. 2015. *The Territories of Science and Religion*. Chicago, IL: University of Chicago Press.

Hess, Jonathan. 2000. "Johann David Michaelis and the Colonial Imaginary: Orientalism and the Emergence of Racial Antisemitism in Eighteenth-Century Germany." *Jewish Social Studies* 6 (2): 56–101.

———. 2002. *Germans, Jews and the Claims of Modernity*. New Haven, CT: Yale University Press.

Horsman, Reginald. 1981. *Race and Manifest Destiny: The Origins of American Anglo-Saxonism*. Cambridge, MA: Harvard University Press.

Jennings, Willie James. 2010. *The Christian Imagination: Theology and the Origins of Race*. New Haven, CT: Yale University Press.

Johnson, Sylvester. 2004. *The Myth of Ham in Nineteenth-Century American Christianity: Race, Heathens, and the People of God*. New York: Palgrave Macmillan.

Kant, Immanuel. 1998 [1781]. *Critique of Pure Reason*. Trans. Paul Guyer and Allen W. Wood. Cambridge: Cambridge University Press.

———. 1998 [1793]. *Religion Within the Boundaries of Mere Reason*. Trans. and edited by Allen Wood and George Di Giovanni. Cambridge: Cambridge University Press.

Keel, Terence. 2013. "Religion, Polygenism, and the Early Science of Human Origins." *History of the Human Sciences* 26 (2) (April): 3–32.

Kidd, Colin. 2006. *The Forging of the Races: Race and Scripture in the Protestant Atlantic World, 1600–2000*. Cambridge: Cambridge University Press.

Kolb, Robert. 2014. "The Lutheran Doctrine of Original Sin." In *Adam, the Fall, and Original Sin: Theological, Biblical, and Scientific Perspectives*, edited by Hans Madueme and Michael Reeves, 109–128. Grand Rapids, MI: Baker Press.

Livingstone, David. 2011. *Adam's Ancestors: Race, Religion, and the Politics of Human Origins*. Baltimore: Johns Hopkins University Press.

Marks, Jonathan. 2009. *Human Biodiversity: Genes, Race, and History*. New Brunswick: Transaction Press.

Martyr, Justin. 2007. *Dialogue with Trypho in, The Ante-Nicene Fathers, Volume 1: The Apostolic Fathers with Justin Martyr and Irenaeus*, edited by Alexander Roberts, James Donaldson, and A. Cleveland Coxe. New York: Cosimo Classics.

Masuzawa, Tomoko. 2012. *The Invention of World Religions, or, How European Universalism Was Preserved in the Language of Pluralism*. Chicago, IL: University of Chicago Press.

Nongbri, Brent. 2013. *Before Religion: A History of a Modern Concept*. New Haven: Yale University Press.

Pinkard, Terry. 2002. *German Philosophy, 1760–1860: The Legacy of Idealism*. Cambridge: Cambridge University Press.

Richards, Robert J. 2000. "Kant and Blumenbach on the Bildungstrieb: A Historical Misunderstanding." *Studies in History and Philosophy of Biological and Biomedical Sciences* 31 (1): 11–32.

———. 2002. *The Romantic Conception of Life: Science and Philosophy in the Age of Goethe*. Chicago, IL: University of Chicago Press.

Sanlon, Peter. 2014. "Original Sin in Patristic Theology." In *Adam, the Fall, and Original Sin: Theological, Biblical, and Scientific Perspectives*, edited by Hans Madueme and Michael Reeves, 85–108. Grand Rapids, MI: Baker Press.

Sheehan, James. 1989. *German History: 1770–1866*. Oxford: Oxford University Press.

Yovel, Yirmiyahu. 1980. *Kant and the Philosophy of History*. Princeton, NJ: Princeton University Press.

Zammito, John H. 2012. "The Lenoir Thesis Revisited: Blumenbach and Kant." *Studies in History and Philosophy of Biological and Biomedical Sciences* 43: 120–132.

9 The beautiful skulls of Schiller and the Georgian girl

Quantitative and aesthetic scaling of the races, 1770–1850

Robert J. Richards

Isak Dinesen, in one of her gothic tales about art and memory, spins a story of a nobleman's startling recognition of a prostitute he once loved and abandoned. He saw her likeness in the beauty of a young woman's skull used by an artist friend.

> After we had discussed his pictures, and art in general, he said that he would show me the prettiest thing that he had in his studio. It was a skull from which he was drawing. He was keen to explain its rare beauty to me. "It is really," he said, "the skull of a young woman [. . .]." The white polished bone shone in the light of the lamp, so pure. And safe. In those few seconds I was taken back to my room [. . .] with the silk fringes and the heavy curtains, on a rainy night of fifteen years before.
>
> (Dinesen 1991, 106–107)[1]

The skulls pictured in Figure 9.1 have also been thought rare beauties and evocative of something more. On the left is the skull of a nameless, young Caucasian female from the Georgian region. Johann Friedrich Blumenbach, the great anatomist and naturalist, celebrated this skull, prizing it because of "the admirable beauty of its formation" (*bewundernswerthen Schönheit seiner Bildung*). He made the skull an aesthetic standard, and like the skull in Dinesen's tale, it too recalled a significant history (Blumenbach 1802, no. 51). She was a young woman captured during the Russo-Turkish war (1787–1792) and died in prison; her dissected skull had been sent to Blumenbach in 1793 (Dougherty and Klatt 2006–2015, IV, 256–257). On the right is the skull of Friedrich Schiller, the famous German poet, as represented by Carl Gustav Carus, premier anatomist and artist of the early nineteenth century. Though Immanuel Kant had a large, powerful skull, Carus did not think it beautiful (Carus 1845, Tafel 1). He regarded the beauty of Schiller's skull as an index of harmonious intellectual and artistic accomplishment.

These skulls and others were used to scale the human races during the late eighteenth and early nineteenth centuries. The period had seen a number of such attempts, which applied a variety of metrics: the relation of the width to the length of the skull; the dimensions of its bony plates; the so-called facial angle; the internal cranial capacity of the skull, essentially a proxy for the size of the brain. Some of these studies, as well as their successors – for instance, Francis Galton's

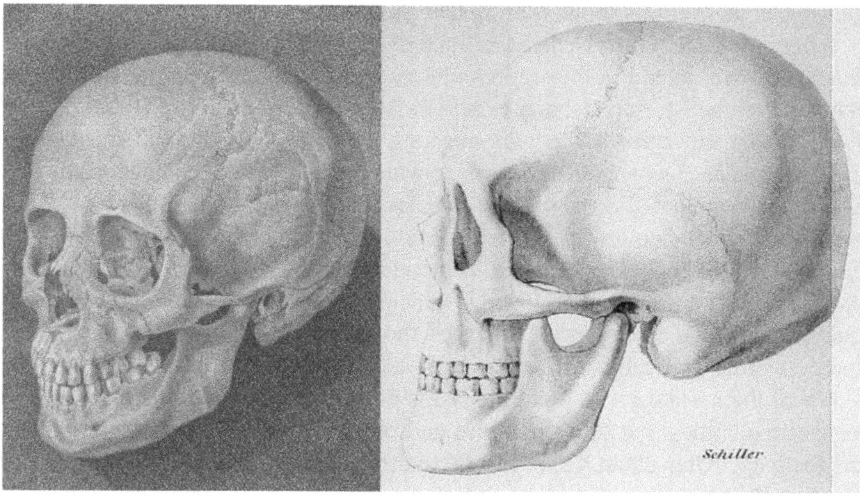

Figure 9.1 On the left, the skull of the Georgian female (from Blumenbach 1802); on the right, the skull of Friedrich Schiller (from Carus 1841).

superimposed photographs of social types, Nazi portrayals of racial types, William Sheldon's classification of body types (ectomorph, mesomorph, endomorph) – assumed that the scientific study of external, physical features would reveal psychological, intellectual, and moral characteristics. But the mathematics of skulls was not the only measure used. Surprisingly another criterion, as exemplified by the skulls of the Georgian female and Schiller, was aesthetic: the proportions, the symmetry, and the *je-ne-sais-quoi* of beauty.

Those researchers who studied skulls formed two distinct groups at the turn of the eighteenth century. One group argued that the physical features of skulls – including their aesthetic qualities – represented different human types and were permanent, revealing the mental characteristics of the several races; such physical distinctions permitted the scaling of the races into higher and lower in regard to intelligence, talent, and moral disposition. The other group also made careful measurements of the physical features of skulls – again, including judgments of beauty – but the members of this group contended that such features were impermanent and variable. They maintained that no hierarchical differences of intellect or morals could be detected among the races. What explains the different conclusions reached by the two groups? The same criteria and comparable methods of measurement were used by both, but their judgments made about the races were startlingly different. Can anything systematic be said about what led to these contrary results? I especially wish to focus on the criterion of beauty: what is a beautiful skull? Why should a beautiful skull tell you anything about the qualities of a person or race? These are the principal questions I will pursue in this essay; and they are not merely historical curiosities, since aesthetic traits today continue to be used as markers of race and are embraced in evolutionary and cultural theories. Beauty, as these skulls show, is more than skin deep.

After some preliminary considerations of the social context and racial classifications, I will attend to four representative craniologists: Friedrich Tiedemann, Samuel George Morton, Johann Friedrich Blumenbach, and Carl Gustav Carus. Both Tiedemann and Morton took exacting measurements of skulls, especially cranial capacity. Tiedemann's pioneering work on skull measurement became part of the evidence used in the slavery debates that occurred in England and Germany; he argued that the races differed little in terms of cranial size. Morton, who amassed a large collection of skulls, contended that the races demonstrated different capacities and could be arranged in a hierarchy. Blumenbach thought aesthetic qualities of skulls distinguished the races, though racial features were nonetheless quite variable and subject to environmental alteration. Carus also believed the skulls of the various races – and the skulls of individuals – differed by reason of aesthetic qualities, but those qualities and other metric features were relatively unchanging. After about 1850, the literature crackled with skulls, but most of these studies were variations on and extensions of the examinations conducted by these four naturalists (David and Thurman 1867; Engel 1851; Huschke 1854; Meigs 1857; Zeune 1846).

The social and conceptual context

Travels of adventure and trade

Several social and conceptual events helped focus interest on skull measurement during the late eighteenth and early nineteenth centuries. Travel and trade had increased significantly during the last half of the eighteenth century, bringing Europeans into contact with other peoples. Captain James Cook, for example, made three famous trips to the islands of the South Pacific, New Zealand, and Australia (1768–1771, 1772–1775, 1776–1779), and brought back unusual plants and animals, as well as tales of exotic, aboriginal peoples. The trading companies formed in the seventeenth and eighteenth centuries – especially the British East India Company and the Dutch East and West India Companies – brought Europeans into contact with a variety of different populations. The Dutch East India Company (1602–1799), for instance, carried around 975,000 merchants, traders, and workers to Indonesia, India, Ceylon, Japan, China, Vietnam, and the islands in the South China Sea; many returned to Europe with tales of the inhabitants of those foreign shores (de Vries and van der Woude 1997, 75). These tales, of course, stimulated curiosity about the range and character of the different human groups and their relation to the man-like apes discovered in Africa and the Indies.

Slavery made poignant the question of the level of humanity found in Africa, in the Americas, and in other regions of the newly explored globe. During the eighteenth century, the number of slaves carried by European ships (Dutch, British, Portuguese, and French) reached about six million individuals (Lovejoy 1982, 473–501). Both slavers and abolitionists, at least those of a sensitive nature, had elevated interests in the more theoretical question of the intellectual and moral status of Africans. In the newly established United States of America, as the Native Americans were pushed further west, the same question of justification for wretched treatment arose.

The interest in skulls

The two most important reasons for the focus on skulls in scaling the human races are quite simple: skulls encase that part regarded as the essence of the human, the brain; skulls thus evoke who we are or were – think of Dinesen's story of the young woman's skull or Hamlet's address to the skull of Yorick, a fellow of infinite jest. The skull always carries the ineradicable whiff of mortality, awakening that deeply seated fear we always carry; it thus evokes fascination and apprehension. The other reason for interest in skulls, equally important for natural science at the turn of the eighteenth century: skulls can be measured.

The great advance in science since the beginning of the Enlightenment has come through precise measurement, the mathematizing of the world picture. The historian of science Charles Gillispie has argued that disciplines became objective and thus truly scientific only when they became quantitative (Gillispie 1960). While Gillispie's criterion would leave a great deal out of the history of science that most individuals would regard as scientific, his view was quite compatible with that of Kant, whose theoretical considerations weighed heavily with German naturalists. Kant famously contended, "in every particular doctrine of nature only so much proper science can be met as there is mathematics therein" (Kant 1956, V, 14 [A IX]). Skulls could be mathematized – and they traveled well.

Classification of the races

Linnaeus's classification

In the mid-eighteenth century, Carl von Linné, the great Swedish naturalist, put graphic order into the hidden order of nature. In the first edition (1735) of his *Systema Naturae*, he divided the natural world into three kingdoms: mineral, plant, and animal. He introduced what became the standard nomenclature by which to identify plants and animals – that is, the taxonomic categories of class, order, genus, species, and variety. He placed humans in the class of *Quadrupedia*, the order of *Anthropomorpha* (also including simians – apes and monkeys – and sloths), and in the genus *Homo*, with four species: *H. Europaeus* (white), *H. Americanus* (red), *H. Asiaticus* (dark), and *H. Africanus* (black) (Linnaeus 1740, 44). By the highly revised tenth edition of his work (1758–1759), Linnaeus markedly changed his classification scheme. He replaced the class name *Quadrupedia* with *Mammalia* and changed *Anthropomorpha* to *Primates*. He divided the order of primates into four genera: *Homo*, *Simia*, *Lemur*, and *Vespertilio* (bats). The genus *Homo* now had two species, H. sapiens and H. troglodytes (caveman or, as he was also termed, "man of the night") – the latter included the subspecies Orang

Outang and Kakurlacko (perhaps a gibbon). The species *H. sapiens* had six varieties: *H. Americanus*, *H. Europaeus*, *H. Asiaticus*, *H. Africanus*, *H. Ferus* (wild man – chimpanzee?), and *H. Monstrosus* (macrocephalics, etc.). The first four human racial varieties, the ones that survived Linnaeus's classification scheme, carried several descriptors, including a too-neat application of the four Hippocratic temperaments: the Americans were red, choleric, straight-standing, and governed by custom; the Europeans were white, sanguine, muscular, and governed by law; the Asians were dark, melancholic, stiff, and governed by opinion; and the Africans were black, phlegmatic, languid, and governed by caprice (Linnaeus 1758–1759, I, 20–24).

Blumenbach's scheme

Blumenbach, in his doctoral dissertation – *De generis humani varietate nativa* (1775; *On the Natural Variety of Humankind*) – initially adopted the fourfold division of the human races established by Linnaeus (Blumenbach 1775, 41–42), and he insisted upon a basic principle: the unity of humankind (Blumenbach 1775, 40–41). Using Georges-Louis Leclerc de Buffon's criterion of hybrid fertility as an indicator of membership in a particular species, Blumenbach represented humans as forming four varieties of a single species. He also constructed their geographical boundaries somewhat differently from Linnaeus – for instance, he included among the European group individuals living just west of the Ganges River (thus including Indians and, further north, the Caucasians); the Mongolian group he made to include Native Americans living in the northwest areas of the American continent. He later added, in his *Handbuch der Naturgeschichte* (1779–1780; Handbook of natural history), one more variety or race: the south Asian and Polynesian (later called the Malay). During the late eighteenth and early nineteenth centuries, Blumenbach's fivefold division of the races became standard.

All men, according to Blumenbach, descended from Adam and Eve, and thus formed a unity; but because of environmental forces, they developed into five varieties or races. The closest to the original progeny were the Caucasians, whom he named after the Caucasus Mountains, today on the border of Georgia and Russia. They were "the original and greatest race," of white skin color, which must have been the original color of men. He supposed that the skin color of the original group had been white since it would be easier to transition from white to other skin colors than the reverse (Blumenbach 1795, 303). In appearance, the Caucasians were "according to our judgment of symmetry, the most beautiful and the best formed of men" (Blumenbach 1795, 289). The other races showed dominant characteristics: the Asiatics, with yellow-brown coloring, flat face, and small eyes; the Africans, of black color, wooly hair, squat nose, and full lips; the Americans, of copper-red color; and the South Asians and Polynesians (i.e., the Indonesians and Filipinos) with dark hair and light brown skin (Blumenbach 1779–1780, I, 63–64). The differences among men, according to Blumenbach, shaded into one another, so that the pale white skin at one end of the spectrum in the German population transitioned to the red skin of the Americans at the other end, and so through the other populations; the same gradations could be found in stature and other features. The various groups displayed predominant traits by reason of particular environmental forces but expressed features just as variegated as the distinctive nutrition and climate of those environments and the diverging customs of the peoples. He thought Linnaeus's monstrous men (e.g., albinos) were mostly men suffering from some disease, and the troglodytes were confections of tales about matings of men and orangutans (Blumenbach 1779, I, 63–64).

In the first edition (1775) of his treatise *De generis humani varietate nativa,* Blumenbach, following Buffon, assumed that climate, nutrition, and habit had altered originally created stocks of animals and men, producing the varieties of species then populating the earth. But in the second edition of his treatise (1781b, 1–2), he introduced an additional, explanatory factor, the *Bildungstrieb* (formative drive). This was an independent vital agency, which he postulated initially to explain the epigenetic formation of the fetus, as opposed to the supposition that the fetus was already preformed, a kind of miniature adult that had only to unroll (*evolvere*) (Blumenbach 1781a). This force also accounted for the regular growth, maintenance, and repair of the individual, functions that could not be given a simple, mechanistic interpretation. He regarded this power as comparable to Newton's gravitational force – a power postulated to explain phenomena, even though the ultimate cause remained obscure. Blumenbach thought the *Bildungstrieb* could thus explain general species structures, but additionally supposed that it could be deflected from its regular operations by different climates, diets, and habits; the force, in concert with the environment, should thus be regarded as "the mother of varieties properly so called" (Blumenbach 1795, 88). The concept of the *Bildungstrieb* became a staple in German biology at the end of the eighteenth century, being adopted by the likes of Johann Wolfgang von Goethe, Johann Gottfried Herder, Alexander von Humboldt, and even Kant (Richards 2002, 216–37).

Through the latter part of the eighteenth century, the classification of the races and the nature of racial traits grew in moral urgency along with the escalation of the slave trade. In Britain and the new American nation, various abolition movements brought the moral question constantly before the public.

The quantitative analysis of skulls

Tiedemann and slavery

The German lands were not engaged in the slave trade, and many prominent Germans (e.g., Alexander von Humboldt, Friedrich Blumenbach, and Georg Forster) expressed indignation at its evils. Friedrich Tiedemann (Figure 9.2), a liberally inclined anatomy professor at Heidelberg, deployed not a philosophical argument against slavery but a scientific argument: he measured skulls.

Tiedemann was born in 1781 into a professorial family; his father, Dietrich Tiedemann, was a professor of philosophy in Cassel and an opponent of Kant. At the time, Cassel had a small Negro population, some arriving in the

Figure 9.2 Friedrich Tiedemann (1781–1861). Lithograph (from Schott 1836).

mid-seventeenth century and others a century later from the United States, families of individuals who had fought with the Hessian allies of the British during the American Revolutionary War (Jones 2013). Tiedemann would quite likely have been acquainted with individuals of this group. He studied medicine at Bamberg, Würzburg, and Marburg, receiving his degree from the latter in 1804. During a three-year itinerant period, he came to study with Blumenbach at Göttingen, the philosopher Friedrich Wilhelm Joseph Schelling at Würzburg, and the zoologist Georges Cuvier at Paris. Along the way, he heard lectures from Franz Joseph Gall, who argued for a close connection between configurations of the skull and mental traits – the foundations for his doctrine of phrenology. The anatomist Samuel Thomas Soemmerring recommended this extraordinarily talented naturalist for a position at Landshut (in southeast Bavaria), where he stayed for ten years, focusing his attention on comparative anatomy. In 1816, he received a call from Heidelberg, where he happily spent the rest of his professional life, some thirty-three years.

While at Landshut, Tiedemann conducted studies of brain development, using two approaches that allowed him to move beyond description to explanation: comparative anatomy, on the one hand, and studies of the development of the human fetal brain on the other. He regarded this latter as "the thread of Ariadne for this Labyrinth" (Tiedemann 1816, 2). Following that thread led him to a cautious statement of a natural law of human development: the human fetal brain in its growth recapitulates the stages of brain structures represented by the series of animals from lowest to highest (Tiedemann 1816, 148). So, for example, though the thickness of the spinal cord in relation to the cerebrum of the human adult is quite small relative to that of other animals, in the early fetus it is relatively large but gradually diminishes in size comparable to the narrowing of the cord when one passes from fish, to amphibians, to birds, and finally to mammals (Tiedemann 1816, 91–92). Tiedemann likely would have been familiar with recapitulation theory at Würzburg, especially in his interactions with Schelling and the naturalist Lorenz Oken.[2] Tiedemann's own attachment to this conception directed him to that "daring adventure of reason" of which Kant spoke – namely, an evolutionary hypothesis about the origins of life on earth.[3] The developmental and comparative focus of German zoologists like Tiedemann in the early nineteenth century made the acceptance of Darwin's theory later in the century much smoother than in other European countries, including England.

Tiedemann insisted that human physiology was an experiential science (*Erfahrungs-Wissenschaft*), a science in which one gathered facts through systematic observation and performed experiments to discover their causes. "In observation we listen to nature, as it were; in experimentation we ask nature for advice." The attentive ear was important since many false moves arose from being in thrall to a "favorite theory" (*Lieblings-Theorie*). Any speculations had to be tested by experience (Tiedemann 1830, I, 9–11). Tiedemann brought this exacting experimental attitude to the study of human skulls. He wished to test whether there were significant differences among the races that might indicate differences in intellect, talent, and moral capacity. His first publication on the comparative anatomy of skulls came, quite surprisingly, in English: a paper read before the Royal Society of London, "On the Brain of the Negro," and printed in the *Transactions* of the Society in 1836 (Tiedemann 1836). The paper had been originally sent to the journal in 1835. Shortly thereafter, in 1837, he published a small book in German on the same subject but with a larger set of data and an extended analysis (Tiedemann 1837).

Pertinent and enduring conclusions came forth from Tiedemann's extensive study – ultimately of some 430 skulls (Figure 9.3). Among those conclusions

1) Daſs die Schädelhöhle zwischen 59 Unzen, bei einem Amerikaner, und 13 Unzen, bei einem Mongolen, spielte.

2) Daſs die Geräumigkeit der Schädelhöhle von 430 Männern aller Menschen-Rassen bei der gröſsten Anzahl zwischen 42 und 32 Unzen betrug, und zwar

unter	70	Männern	der	Aethiopischen Rasse	bei		64
—	186	—	—	Kaukasischen	—	—	144
—	45	—	—	Mongolischen	—	—	29
—	31	—	—	Amerikanischen	—	—	20
—	98	—	—	Malaïschen	—	—	63

3) Daſs eine Schädelhöhle, mehr als 42 Unzen haltend vorkam

bei	5	Männern	der	Aethiopischen Rasse	
—	42	—	—	Kaukasischen	—
—	10	—	—	Mongolischen	—
—	7	—	—	Amerikanischen	—
—	21	—	—	Malaïschen	—

Figure 9.3 Summary table from Tiedemann's *Hirn des Negers* (1837), showing (1) that the range of seed weights, used as an index of skull volumes for 430 skulls, was 59 ounces for a Native American to 13 ounces for an Asian; (2) that the largest number of skulls had a range between 42 and 32 ounces (e.g., the seventy males of the Ethiopian races having sixty-four individuals within that range, etc.); and (3) that of the races, those exceeding 42 ounces were Negros at five individuals, Caucasians at forty-two, Asians at ten, Native Americans at seven, and Malays at twenty-one.

were: the brain, relative to body size, is largest during infancy and reaches its mature size in humans at about the seventh to the eighth years of age (we would now say tenth to twelfth years); brain size is generally proportional to body size, though with great variation; females, having smaller bodies, tend to have slightly smaller brains than males – though in relation to body size, slightly larger brains. Tiedemann dissented vigorously from the racial evaluations of Peter Camper, Cuvier, and especially Soemmerring, each of whom held that the Negro brain was more like that of the orangutan than that of the European. Tiedemann found no morphological differences between the Negro brains and the European brains that he dissected. Brain size, however, was less easy to determine and might be different. Size was the crucial factor since there was, as he thought, "a very close connexion between the absolute size of the brain and the intellectual powers and functions of the mind" (Tiedemann 1836, 502). Soemmerring placed the African skull at the diminished end of the series running from the European to the Asiatic to the African (Sömmerring 1785, 19). His study, though, was quite circumscribed. His investigations of the Negro body relied on one female and a couple of males that he dissected at Cassel. Tiedemann's sample was vastly larger and the measurements more than eyeballing. His conclusion: "our investigation demonstrates undeniably that those anatomists and naturalists have been caught in error who have attributed to the Negro a cranial cavity of less volume and a smaller brain than the Europeans and the peoples of other human races" (Tiedemann 1837, 47).

Tiedemann used a fairly simple method of determining cranial capacity. He first weighed skulls of different races and genders, as well as the skulls of apes; he then filled the cranial cavities with millet seed and weighed the skulls again. By subtracting the results, he got the size of brains as weighed in millet seeds. He found that most of the individual skulls of each of the five races – Ethiopian, Caucasian, Mongolian, American, and Malay – could be found in a range of 42–32 troy ounces. He also noted the number of skulls that exceeded 42 ounces (see Figure 9.3). Tiedemann concluded rather carefully, "The cranial cavity and the brain of all the human races show a similar middling size, within certain fluctuating limits." Thus, the majority of the sampled individuals of each race lay within the 42- to 32-ounce range. The next sentence, however, seemingly qualifies his results: "The most that can be shown is that among the peoples of the Caucasian and Malay races, according to the facts I've laid out, some men more often achieve a considerable size than with the peoples of the other races" (Tiedemann 1837, 47). His phrase "most that can be shown" is borne out by the figures: 5/70ths of the Negro skulls (about 7%) but 42/186ths of the Caucasian skulls (about 23%) and 21/98ths of the Malay skulls (about 21%) were over 42 ounces.[4]

Carus, who believed there were permanent differences separating the races, critically noticed that Tiedemann's own figures showed that the number of Negro skulls in the sample that exceeded 40 ounces amounted to 6 of 54, or about 11%, while the number of Caucasian skulls over 40 ounces was 64 of 141, or about 45% (Carus 1841, 12). Read that way, the portion of big-skulled Caucasians stands out, a jutting conclusion of which Tiedemann seems to have been aware. In anticipation, Tiedemann provided some mitigating considerations. He suggested, for instance, that the bodies of Negros were generally of smaller stature than those of Caucasians; thus, relative to body size the two groups of skulls simply did not differ that much – though Tiedemann admitted to not having exact figures. He also observed that slaves, whose skulls generally constituted the bulk of the Negro numbers, were usually taken from coastal areas of Africa, where the blacks lived a more degraded life than in the highlands. He also mentioned the case studies made by his old teacher Blumenbach of Negros who had achieved considerable intellectual renown in theology, philosophy, poetry, and science.

Samuel George Morton, the American measurements

The best-known quantitative effort to scale race in the nineteenth century is that of Samuel George Morton (Figure 9.4), a Philadelphia physician and naturalist. During his own time, numerous researchers – Louis Agassiz, Paul Broca, and Carl Gustav Carus – as well as Southern slave owners admired his work and used his skull measurements as evidence of racial hierarchy. Even Alexander von Humboldt, no friend of slavery, sent a letter of congratulations upon receipt of Morton's principal study of skulls, *Crania Americana* (1839); the German adventurer prized the "profundity of its anatomical views" rendered without "poetical

Figure 9.4 Samuel George Morton (1799–1851) (from Meigs 1851, frontispiece).

reveries."[5] In our time, Stephen Jay Gould (1996, 82–101) found more poetry than science infusing Morton's craniology; he reanalyzed the measures and methods, judging them tainted by prejudicial assumption and poor technique. Then again, these very charges have recently been redirected at Gould himself.

Morton was born at Philadelphia in 1799 of Irish descent and Quaker upbringing. While taking private lessons in medicine, he attended lectures in anatomy and physiology at the University of Pennsylvania, from which he received an MD in spring of 1820. He undertook further medical training at Edinburgh beginning later that year, arriving in the city a few years before the young Charles Darwin began his own medical studies (1825). The next year he went on to Paris, lingering long enough to let slip away his Quaker observance and to suffer the enticements of the bright, flickering lights. He made his way to northern Italy and then in fall 1822 back to Edinburgh, fulfilling the requirements for a diploma in medicine in August 1823. He returned to Philadelphia in June 1824, where he punctuated a desultory medical practice with the stimulus of research, initially in pathology. He also quickened a recumbent interest in geology, paleontology, and comparative anatomy, all of which he had pursued as a member of the Philadelphia Academy of Natural Sciences, initially becoming a member just after receiving his MD. He worked his way up the administrative ladder of the Academy, eventually becoming its president in 1849. He was appointed professor of anatomy in 1839 at the Pennsylvania College of Medicine, the same year he saw published his extraordinary quantitative study of American Indian skulls, his *Crania Americana* (Figure 9.5).[6]

For *Crania Americana*, Morton not only measured the skulls taken from various Indian tribes but also compared them with skulls of other races, those groups discriminated by Blumenbach. The study also included considerable ethnographic materials culled from reports of travelers, missionaries, and naturalists. Morton's fame rests principally on this book, replete as it is with precise measurements of some 253 skulls and dramatic lithographs of those immortal parts. He supplemented this study with two other compendia of skull measurements, his *Crania Aegyptiaca* (1844), for which he measured the skulls of Egyptian mummies, and his *Catalogue of Skulls of Man and the Inferior Animals* (1849), which corrected some errors of the *Crania Americana* and both expanded the number of human skulls and added those of animals for comparison.

In *Crania Americana*, Morton and his assistant calibrated the exterior dimensions (longitudinal diameter, parietal diameter, facial angle, etc.) and internal cranial capacity of adult skulls from the several races, represented by: 147 North and South American Indians, fifty-two Caucasians (Germans, Celts, Semites, South Asians), ten Mongolians (Turks, Chinese, and Tartars), eighteen Malays (Indonesians and Polynesians), and twenty-nine Ethiopians (Africans). He further

RACES.	No. of skulls.	Mean internal capacity in cubic inches.	Largest in the series.	Smallest in the series.
Caucasian.	52	87.	109.	75.
Mongolian.	10	83.	93.	69.
Malay.	18	81.	89.	64.
American.	147	80.	100.	60.
Ethiopian.	29	78.	94.	65.

Figure 9.5 Morton's summary table of cranial capacities for the five races, with the numbers of individuals, the average capacity of the skulls, the largest capacity of an individual, and the smallest (from Morton's *Crania Americana* 1839).

divided the five races into component "families," some twenty-two different families in all. Prefacing the report on the actual measurements, Morton provided extensive anthropological descriptions of the races and their constituent families, descriptions collected from travel literature and popular assumption. To measure the skulls, Morton and his assistant used white pepper seed (having a hard and uniform consistency) to fill the cranial cavities of the skulls, and then poured the resulting volume of seed into a standard measuring tube to gauge comparative cranial capacities in cubic inches. In the later study of 1849, he used lead shot, size BB, instead of seeds, to obtain more reliable measures; and he expanded the number of skulls to 623. The measures in the 1839 study yielded a hierarchy of mean cranial volumes: the Caucasians showing the highest capacity, followed by Mongolians (Asians), Malays, Americans, and Ethiopians (see Figure 9.5). The 1849 reevaluation emended the hierarchy, making the African group penultimate and the American group last. These measures, in Morton's estimation, provided scientific evidence confirming the large number of ethnographic reports he amassed, reports, however, that were often no more than casual observations of travelers. He epitomized his analysis of the races this way: Caucasians have attained "the highest intellectual endowments"; the Mongolians are "ingenious, imitative, and highly susceptible of cultivation"; the Malay are "ingenious, and [possess] all the habits of a migratory, predaceous, and maritime people"; the Americans are "averse to cultivation, and slow in acquiring knowledge, restless, revengeful, and fond of war, and wholly destitute of maritime adventure"; and the Ethiopian are "joyous, flexible, and indolent," showing diverse intellectual character, of which the "extreme is the lowest grade of humanity" (Morton 1839, 5–7). These descriptions exemplify not so much careful conclusions based on extensive observation as on inertial traditions of description, similar to those easy Hippocratic epitomes made by Linnaeus. Such descriptions usually failed to account for differences in education, a factor Blumenbach made diagnostic of human potential (see ahead).

In *Crania Americana*, Morton attempted to answer a perplexing question of the time: are the members of the various Indian nations of one species with several races or do they constitute several different species altogether? He decided that the Native Americans formed one species with two great families, the Toltecans (the "demi-civilizations" of Peru and Mexico) and the "Barbarous Tribes" of North America, Brazil, Patagonia, and the tip of South America. This initial question and its answer make sense, however, only if you are of a disposition to regard the various human groups worldwide as specifically different from each other. And Morton was of that disposition, though cautious. He had provisionally adopted Blumenbach's division of the races, but never explicitly affirmed that all of the races constituted different species. He did conclude, however, that the Americans were specifically different from the other races. Morton, nonetheless, treated Blumenbach's groups as if they were separate species, even suggesting that their component subgroups – which he called families – had species-like sets of innate features. Their traits were not shaped by the environment after descent from a primordial couple; rather, as he urged, the Creator must have initially rendered each race fit for its specific environment. The purity of these racial lines could be sullied only by hybridization. Morton's argument for the independent origins of the human races precipitated a storm of religious objection from biblical literalists, which he weathered dangling from the elastic threads of Episcopalian theology (Meigs 1851, 34–36).[7]

Though many prominent naturalists of the period accepted Buffon's criterion of interbreeding fertility as the marker of species unity, the issue remained in contention. Darwin would later use morphological resemblance as the standard and concluded that the judgment was arbitrary as to whether human beings comprised one species with several varieties or several different species (Darwin 1871, I, 235). Yet in light of the scientific acceptance of Buffon's standard and the theological heat of the issue, Morton recognized he had to justify his conclusion, which he attempted to do by citing several studies of hybrid fertility in crosses of different animal species and by providing evidence that certain human hybrids had diminished fertility (Morton 1847). So, as he judged, distinct species could hybridize, but not well. He gathered evidence for diminished fertility from incidences of "half-caste" offspring between Caucasians and Native Australians, which by his estimate was very low, about 200 mulattos in a native population of 15,000. However, Morton did not contrast this roughly 1.5% rate against any reasonably expected rate – it just seemed small (Morton 1851). The criterion for species designation he adopted was simply that of consistent morphology over long periods of time: "when races can be proved to possess certain primordial distinctions, which have been transmitted unbroken, they should be regarded as true species" (Morton 1847, 263). The ancient and recent American skulls of his collection, he argued, provided examples of such stable transmission over time. Yet Morton would have required a special vision to perceive, say, the ideal Peruvian skull lying beneath the variability in size, the induced malformations, and the asymmetries of disease and injury, all of which were found in his collection (see Figure 9.6). Herein, I believe, lies a significant difference between Morton and Tiedemann: shining through the variability of individual skulls, Morton perceived the type, while Tiedemann saw only individuals.

Morton thought of his *Crania Americana* as a treatise in phrenology, and solicited an essay from his friend George Combe, a follower of Franz Joseph Gall, on the general philosophy of phrenology. When Combe wrote the essay, which was appended to Morton's book, he had only a few of Morton's plates to consult and none of the measurements. Little matter. Basically, only two general features of the doctrine seemed to concern Morton – namely, that internal psychological dispositions were manifest in external physical structure and that intelligence was proportional to skull size. The Peruvian and Incan skulls did make Morton more cautious about the size-to-intelligence relationship, however. He observed that "it would be natural to suppose that a people [the ancient Peruvians] with heads so small and badly formed would occupy the lowest place in the scale of human intelligence" (Morton 1839, 99). He yet recognized that their architectural accomplishments and the other monuments of a great civilization gave evidence of high intellectual achievement. He thought the same story could be told of the Incas, whose mean skull capacity was 73 cubic inches, lower than that of other peoples,

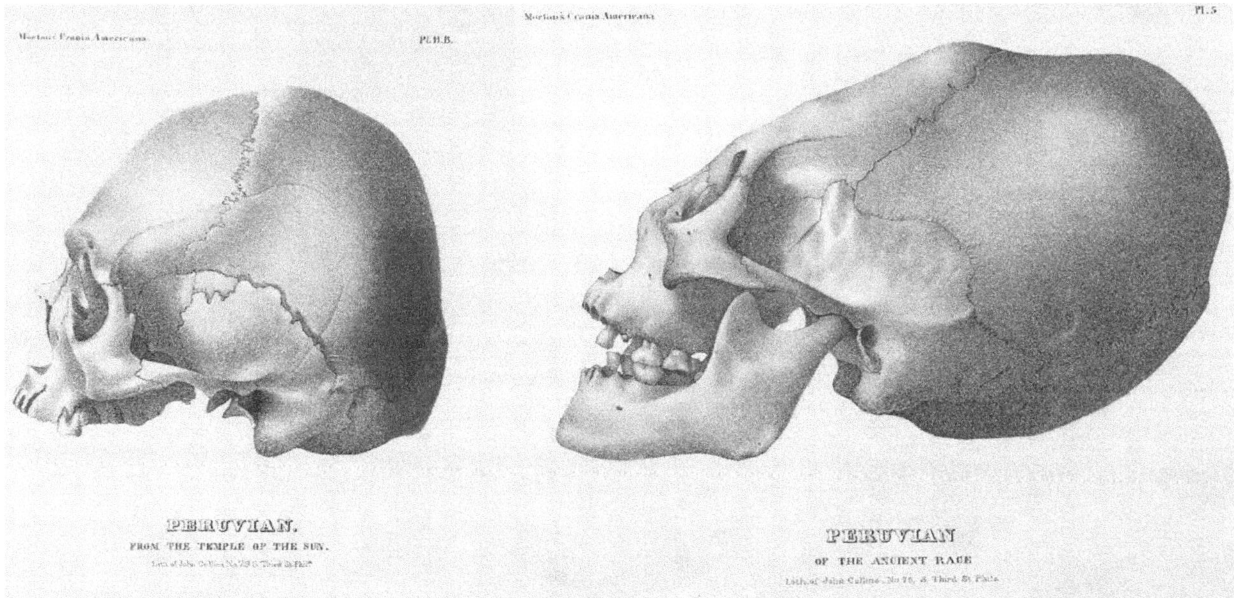

Figure 9.6 Two Peruvian skulls, both likely manipulated and shaped when the individuals were infants (from Morton 1839).

though the remnants of their civilization showed they were superior to most other peoples (Morton 1839, 132). In his review of *Crania Americana*, Combe, writing anonymously, lauded Morton's accomplishment but did take issue with the assumption that the ancient Peruvians had small heads but a great civilization. He rather thought Morton might have been misled by skulls that had been artfully shaped, since an important part of the brain might well have been pushed into an area not properly measured (Combe 1839–1840, 363–64).

Like Tiedemann, Morton did recognize the great variability of skulls. The largest Incan skull in his collection ran to 89.5 cubic inches and the smallest at 60, with an average of 73. The same scope of variability was evidenced throughout his total collection of over 800 human skulls. Moreover, he knew that the skulls of infants could be molded into different shapes, an art performed by many of the American groups (Figure 9.6). But unlike Tiedemann, he seems not to have recognized the basic principle of allometry – the smaller or bigger the skeleton, the smaller or bigger the skull. For the most part, he had only skulls and not complete skeletons, so judgments of body size were precluded as well as adjustments for size differences between males and females. Of course, distinctions of sex, as well as variability of environment and social status – and what these entail – will alter body size tremendously and thus head size.

Morton prided himself on the meticulous care he took with his skull measurements, and in his *Catalogue of Skulls* (1849), he corrected many errors that had made their way into *Crania Americana*. Gould believed Morton's pride was misplaced. In his *Mismeasure of Man* (1981; 1996), Gould reanalyzed Morton's tabulations and discovered what he thought to be unconscious errors that had warped the measures, the calculations, and the conclusions. Morton's figures, Gould asserted, were "a patchwork of fudging and finagling in the clear interest of controlling a priori convictions" (Gould 1996, 86). He discriminated two general kinds of errors – errors of seed measurement and errors of statistical aggregation. Gould assumed that for Caucasian skulls, Morton may have, perhaps unconsciously, packed the seeds into the skulls more tightly, thus giving larger values for cubic capacity – rather unlikely, at least as an error on Morton's part, since his assistant did the measuring. He also argued that Morton simply took the average of the total number of Peruvian and North American skulls, rather than taking averages of the different tribes and then taking the mean of those averages. (Why this latter method was to be preferred is quite unclear. Gould's method assumed that each of the tribes was equally represented in the total Native American population, yet neither Gould nor Morton knew what proportion of the whole populations each of the tribes represented.) Gould then corrected Morton's figures and determined that the five human races differed little from one another in average cranial capacity. Morton himself recognized the measurement errors in *Crania Americana*, which is why he started using lead shot instead of seeds when he later remeasured skulls in his 1849 study. He also thought he had left too much of the actual work of measurement to a careless assistant, whom he subsequently fired. For the new study of 1849, he did all of the measurements himself and added a much larger sampling of skulls. For instance, he doubled the size of Negro group, bringing the number of their skulls to fifty-eight, and then filled the skulls with lead shot. The new measures elevated the estimate of Negro cranial capacity markedly, from 78 cubic inches to 83 cubic inches for the highest subgroups. Not exactly what one would suspect from a racist finagler, even if this final measure of the highest Negro subgroup stayed below the Caucasian average of 92 for the highest subgroup.[8]

Very quickly after the publication of Morton's 1849 study, Sir William Hamilton, polymathic naturalist and philosopher, vehemently rebutted Morton's conclusions concerning racial differences in skull sizes. He observed that Morton did not (and could not) distinguish male from female skulls: "Now, as the female encephalos is, on an average, some four ounces troy less than the male, it is impossible to compare national skulls with national skulls, in respect of their capacity." (Nor, one might add, could Morton distinguish adolescent from adult skulls.) Hamilton himself, in 1831, had measured skull capacity using sand, and determined that "the Negro encephalos is not less than the European, and greatly larger than the Hindoo, the Ceylonese, and sundry other Asiatic brains." He specifically mentioned that he agreed with Tiedemann's conclusion concerning the Negro brain (Hamilton 1850, 330 and 332).

Tiedemann, Morton, and Gould could not provide perfect measures of imperfect objects. They could not control the multitude of factors affecting biological specimens of uncertain provenance. Each trusted in numbers, which the scientific ethos requires. They were less cautious, however, about which numbers to trust. Both Tiedemann and Morton passed their gaze over hundreds of skulls, which varied greatly in size and shape. Morton detected types within the mass of individuals; Tiedemann saw in the proclaimed types only the individuals. These problems of critical judgment and scientific discernment stand highlighted when the standard becomes not quantitative measures of skulls but aesthetic measures of their beauty.

The aesthetic evaluation of skulls

Blumenbach's Georgian female

Johann Friedrich Blumenbach (Figure 9.7) was the most influential and widely read naturalist at the turn of the nineteenth century. His dissertation, *De generis humani varietate nativa*, was published in 1775 (reprinted in 1776), the year he received his medical degree. The book went through two further editions (1781b, 1795) and many translations, into English, French, Dutch, and German. The same is the case for his many other publications. His interlocutors included Kant, Humboldt, Herder, and Goethe, in addition to numerous anatomists and physiologists of the period. His classification of the races became standard, used by those friendly to his persuasion that race was fluid and a matter of environmental conditions (e.g., Tiedemann) and those who disputed his view, naturalists who regarded race as fixed and typically impervious to environmental alteration (e.g., Morton). Blumenbach's collection of skulls, numbering some 240 at his death in 1840, was

Figure 9.7 Johann Friedrich Blumenbach (1752–1840). Engraving by Ludwig Emil Grimm, 1823.

famed and continued to expand at Göttingen even after he died, the tradition of collecting being continued by colleagues. Those skulls revealed something of a hierarchy, based not on cranial capacity but on beauty.

Blumenbach received his medical degree from Göttingen University in 1775. He was obviously a favored student, for the next year, 1776, he was appointed *extraordinarius* professor of medicine and curator of the university's natural history museum. With the publication of his dissertation, he was recognized immediately as an important scholar, and in 1778 he became *ordinarius* (full professor) in the medical faculty. His real interest, however, was natural history, especially comparative anatomy, with a focus on human beings and their skulls. As his network of correspondents grew, so did his skull collection. He asked these interlocutors, especially those in distant lands, if they would mind sending him a skull or two. He received such gifts from the likes of Humboldt and Goethe, the latter sending a cast of Raphael's skull. Joseph Banks, of the Royal Society of London, provided him skulls from Cook's voyages. Even the Bavarian king Ludwig I made him the present of an Etruscan skull. His greatest supplier was probably Georg Thomas Ash, military surgeon to the Russian czar and graduate of Göttingen. Ash typically responded with enthusiasm: "No effort will be spared to acquire for you the requested skulls from the Asiatic peoples. It will make me very happy if I succeed in enlarging your excellent collection." Though in a letter the same day to one of Blumenbach's colleagues, Ash confessed that "considerable patience will be required until that request can be fulfilled" (Dougherty and Klatt 2006–2015, II, 312–313).

Blumenbach recognized the great variability of skulls and the way environmental impact could further alter them. For instance, he noted that Germans tended to have block heads, because their mothers had the habit of keeping infants on their backs with their heads usually flat against a firm surface. He was aware, as well, that ancient peoples often manipulated infant skulls, as if they were wet clay, to produce a pleasing shape; and he assumed with the Hippocratics that these alterations could be inherited by subsequent generations (Blumenbach 1795, 214–221). Yet through the varieties of possible alterations, Blumenbach, like other anatomists of race, believed he could still detect racial types, but not in terms of cranial capacity – rather in their aesthetic features. In the third edition of *De generis humani varietate* (1795), he included a spectrum of aesthetically arranged skulls of the five races he had discriminated (this volume, Figure 5.1).

Blumenbach arranged his series of skulls with the central figure that of the Georgian girl, whose skull he thought "exquisitely symmetrical, rather globular, with a forehead moderately expanded, the zygomatic bones a bit narrow but not protruding" (Blumenbach 1795, 206). This "most beautiful cranium" he situated between two extremes: the Mongolian skull, which was "like a block, with zygomatic bones extending prominently," at one end, and the Ethiopian at the other end, with "narrow head compressed laterally, and forehead bumpy [tuberosa] and arched [fornicata]" (Blumenbach 1795, 207–208). Blumenbach confessed that given his experience with the varieties of skulls, he could find no quantitative measure to distinguish the races – certainly not the facial angle devised by Camper. Yet there were fairly constant differences distinguishing the races. He thought this could best be perceived by looking vertically down on the skulls of the Mongolian, Georgian, and Ethiopian (this volume, Figure 5.2).

In Blumenbach's judgment, the Georgian skull was "highly symmetrical and most beautiful [maxime symmetricum et venustissimum], while on either side were skull bones quite opposite and different from it." The skull of the Georgian female had

> the sides of the orbits, as well as the zygomatic bones more elegantly narrowed; they and the mandible itself are concealed under the periphery of the moderately expanded forehead; the former [Ethiopian], by contrast, has the maxillary bones compressed and protruding, and the latter [Mongolian] the zygomatic bones are placed on the same horizontal plane as the small

bones of the nose and the glabella [bridge of the nose], and they extend enormously and are prominent.

(Blumenbach 1795, 205)

Why did the skull of the Georgian female strike Blumenbach's fancy, and what other values did his aesthetic judgments imply for the five races that he discriminated? Klatt has observed that Blumenbach likened the Georgian skull to the ideal of female beauty in ancient art (Klatt 2008, 90). The elegant symmetry and the cool marble-like whiteness of the female's skull seem to have evoked from Blumenbach such comparisons with classic statuary. He did on occasion mention that his judgment was based on *our* standards of beauty, but at other times he described the beauty of the Georgian female's skull in more absolute terms (Blumenbach 1795, 289). The backstory of a captured young woman and her mysterious death undoubtedly made the aesthetic experience even more piquant (see Rupke's introduction to this volume and the beginning of this essay). Add to that the reputation of the Georgian women for comely beauty (Blumenbach 1795, 303), befitting a race connected by legend to the area where Noah's ship beached after the flood, and Blumenbach's judgment is rendered more comprehensible. As in Dinesen's story, the skull of the Georgian female became the repository for a history of singular personal meaning to the great naturalist. But what did Blumenbach's aesthetic judgments imply for the other races? Strangely, very little.

A year after the publication of the third edition (1795) of his *De generis humani varietate*, Blumenbach began issuing a series of pamphlets with illustrations of natural historical objects (Blumenbach 1796). In the first series, he provided faces of known individuals – copper etched portraits – who were to represent the different races, thus giving flesh to each of the skulls he had described the previous year. Each individual pictured was a member of one of the five races, and each had either been raised in Europe or spent significant time there. The visages were certainly less primitive than those usually depicted by other authors. But Blumenbach also intended, by the brief accompanying biographies, to suggest that the various races had individuals of conspicuous talents who exercised those talents in European pursuits.

The Negro Jacob Johan Eliza Capitein, for instance, had been taken as a child from Ghana, raised by a Dutchman, and given an early education in classical languages and mathematics (Blumenbach 1796, no. 5). He attended the University of Leiden and became a theologian and preacher in the Dutch Reformed Church. His sermons and Latin poetry revealed to Blumenbach the innate capacities of the African race. In another essay, occasioned by a trip to Switzerland in 1787, he mentioned several other cases of Africans living in Europe whose features varied as to skin color and other traits and who demonstrated obvious talents (Blumenbach 1787). He was especially admiring of a young Congolese woman whom he met while visiting a chateau in Yverdun, at the southern end of Lake Neuchâtel. The naturalist's eye immediately fell on "features that had they been in white skin would certainly have been regarded quite agreeable" (Blumenbach 1787, 3). Moreover, she was learned in obstetrics and had become noted in the region for her abilities in midwifery. Blumenbach insisted that his interactions with a variety of Negros and an investigation of their abilities made him realize that "Negros, in respect of their natural mental capacity and abilities, certainly do not appear inferior to the other human races" (Blumenbach 1787, 4). What the Negroes generally lacked, as did other races, were opportunities of civilized living and education. Hence, in Blumenbach's mind, the great evil of slavery.

Carus and the aesthetics of Schiller's skull

Carl Gustav Carus (Figure 9.8) was a physician, an anatomist, an artist – a friend of the Romantic painter Caspar David Friedrich, with whom he would travel on painting excursions – and a protégé of Goethe, who greatly admired his magnificent treatise on comparative anatomy, *Von den Ur-Theilen des Knochen- und Schalengerüstes* (1828).

As a young physician, Carus helped direct the field hospitals during the Battle of Leipzig (October 16–19, 1813), the bloodiest of the Napoleonic campaigns. After his retreat from Moscow, the emperor recruited a second Grand Army to secure his earlier German acquisitions. The largest and deadliest of his efforts occurred in and around the city of Leipzig. French forces totaled almost a quarter of a million men, and the coalition facing them – composed of German, Austrian, Russian, and Swedish troops – numbered about 350,000. After four days of slaughter, the French fell back into a costly retreat, leaving the battlefields littered with 90,000 casualties from both sides. Carus himself almost died from the typhus that blazed through the forests of wounded. In his autobiography, he reflected on this experience:

> I understood for the first time [. . .] how little a human life seemed to count in the account-books of the world. A rich country was drained of the blood of its young men. Thousands of families must send off what had been cultivated for long years with love and care and full of hope – so that they would be tossed aside without a thought. [. . .] Whole generations were cut down by the merciless angel of destruction and there was no one there who seemed to have noticed. [. . .] Certainly it is not possible to have attained the elevated concept of the wonderful structure of man and of the value of the character of the human spirit and not feel a deep shudder when one – one cannot express it otherwise – becomes aware of the contempt had for humanity in its masses.
>
> (Carus 1865–1866, I, 122–123)

After Napoleon's forces fled Germany, a measure of peace returned to the land, and Carus again took up interests cultivated in medical school – namely, research in anatomy and physiology – ultimately composing during his lifetime some eight or so major monographs and numerous lesser studies. His field of interest expanded to psychology and natural philosophy, the latter reflecting his reading of the works of Schelling and his friendship with Oken. In 1862, his scientific eminence brought him the presidency of the Deutsche Akademie der Naturforscher (Leopoldina), a position he held until his death in 1869. And through his long, rich

Figure 9.8 Carl Gustav Carus (1789–1869) (arm resting on his famous book, *Von den Ur-Theilen des Knochen- und und Schalengerüstes*, 1828). Oil by Julius Hübner (by permission of the Frankfurter Goethe-Museum).

life, he pursued drawing and painting, especially landscapes in a Romantic style. He expressed his artistic talent also in his science, in the anatomical drawings illustrating his numerous monographs. The depictions that crowded the plates of his great work *Von den Ur-Theilen* won the praise of Goethe.

In the decade after the completion of his *Von den Ur-Theilen*, Carus's energies did not flag. He was occupied with travel, considerations of aesthetics – especially Goethe's *Faust* (Carus 1835) – a second edition of his manual of comparative anatomy (Carus 1834), and a comprehensive study of human physiology. This latter effort yielded his *System der Physiologie*, three large volumes (1838–1840) that explore the different animal systems – the vascular, muscle, pulmonary, nerve, skeletal – and that pay considerable attention to the origins of human beings, their races, and their psychic life.

In Volume 3 of the *System der Physiologie*, which he completed in June 1840, Carus had begun a study of skulls that would be expanded the next year into his *Grundzüge einer neuen und wissenschaftlich begründeten Cranioscopie* (Foundation of a new and scientifically grounded cranioscopy; finished on in February 1841). Several events seem to have initially stimulated him to work out a theory of cranioscopy. In 1833, the third edition of Combe's *System of Phrenology*, an explication of Franz Joseph Gall's theory, appeared in a German translation (Combe 1833). Carus specifically mentioned the book in his *System der Physiologie* (Carus 1838–1840, III, 350). He thought the phrenologists had made a good start, but exhibited a distinct lack of significant anatomical and physiological knowledge of the human brain and skull. Then in spring of 1840, Carus obtained four illustrations taken from George Morton's *Crania Americana*, though not the book itself. Initially he was intrigued by the images of Peruvian, Mexican, and Carib Indian skulls that had been distorted by artificial means (Carus 1838–1840, III, 351). A few months later, in August 1840, he read a detailed account of Morton's work in a German medical journal (Anonymous 1840); the long, three-part article included Morton's table comparing average cranial capacities of the different races, measurements Carus would cite in his subsequent work. In passing, the article also mentioned Tiedemann's skull measurements, and criticized his conclusion that the average African skull fell within the range of the average Caucasian skull (Anonymous 1840, 212). The images and the literature pushed Carus a little further along a path he had already begun.

In his *System der Physiologie* and in his *Cranioscopie*, Carus developed a theory of skull measurements directly tied to brain formation. He contended that his theory was grounded in the most recent science and that it was far superior to anything suggested by Gall or Combe. He first distinguished three brain areas, which were quite evident in lower animals and in the early human fetus: the hindbrain, or cerebellum; the midbrain, or corpora quadrigemina; and the forebrain, or cerebral hemispheres (Carus 1838–1840, III, 341). Ablation experiments and postmortem pathology examinations indicated the functions of each: the hindbrain governed willful behavior, desires, and sexual impulses; the midbrain gave expression to feeling, especially self-feeling (Gemeingefühl), and, in humans, self-awareness (Gemüth); and the forebrain received perceptions, constructed images, and was the locus of intellect in humans (Figure 9.9, on the left). As one passed from more primitive creatures to more advanced – or from the early stages of the human fetal brain to that of the adult – marked changes in brain morphology could be observed. First, the three brain areas gradually became more tightly bound together through

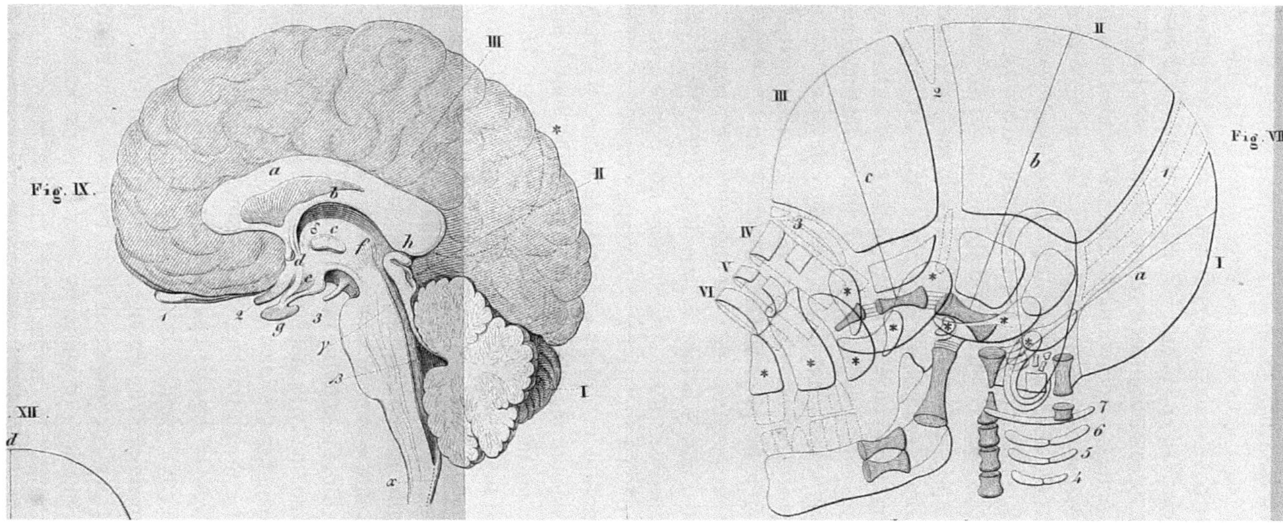

Figure 9.9 On the left, human brain (Fig. IX), with three areas indicated: I, the hindbrain or cerebellum (will, behavior); II, the midbrain or corpora quadrigemina, at the top of the brain stem (feeling); III, forebrain or cerebral hemispheres, which cover the other structures of the adult brain (perception and intellect). On the right (Fig. VIII), the human skull with the three principal plates labeled I, II, III (from Carus 1841).

a multitude of nerve connections, such that functions initially characteristic of one area would be distributed throughout the whole brain; and second, the cerebral hemispheres, in the human adult, had grown to cover the midbrain and most of the hindbrain (Figure 9.9, on the left).

Shielding these three brain areas were the three plates of the skull, those transformed vertebrae whose development through lower species to higher Carus had traced in his *Von den Ur-Theilen* (Figure 9.9, on the right). He maintained that the dimensions of those three plates – their length, breadth, and height – might be diagnostic of racial capacities and individual abilities. Here then was the basis for a truly scientific cranioscopy – or so Carus argued. Yet, if that were the main thrust of Carus's science, it would seem no better than that of Gall, perhaps even less refined – for in the adult human, for example, if the cerebral hemispheres covered most of the other two areas, how could the mid-skull plate and the hind-skull plate be indicative of any features of those parts of the brain they no longer covered? Moreover, why would one suspect in the first place that those three transformed vertebral skull plates could tell you anything about the brain underlying them and be diagnostic of psychological abilities? None of this would make any sense in the absence of the Romantic metaphysics that does provide answers to these questions. We may no longer be receptive to Carus's particular metaphysical views, but they were not foreign to his place and time.

Carus's metaphysical assumptions derived ultimately from Spinoza, but more proximately from Goethe and Schelling (Carus 1865–1866, III, 134–135). The one substance in existence was *Deus sive Natura* – the divine spirit and nature were two expressions of that underlying substance. Organisms embodied this dual character and more fully expressed it over time, so that all of nature moved from more primitive stages to more developed stages. The human individual as well as the human species underwent continuous development. The individual moved, both bodily and psychologically, through stages of fetal life, childhood, and adulthood. The person's inner life began at the unconscious, barely feeling stages of embryogenesis, moved through the childhood stages of the dawning of consciousness, and finally achieved the mature stage of rational life. The human species itself went through comparable developmental periods, from the misty obscurity of prehuman life, through the more primitive races of mankind, to the more advanced races, and finally to the most elevated individuals, those geniuses who came closest to realizing the ideal of humanity (more of this ahead). Like Schelling, Carus held that the abstract idea of humanity inclusively contained the ideas of the various levels of organic development; and like Goethe, he understood this idea to be creative, yielding over time the various physical manifestations of organisms, from simplest up through the races of man (Carus 1838–1840, I, 349). In a given individual, development of the body would be mirrored by the development of that divine idea of humanity, now in its particular instantiation as the human soul. At the very beginning of life, the fetal brain, its nascent skull covering, and its concomitant psychic idea (the soul) were, in Carus's theory, bound to one another, such that the skull plates would be impressed with that original binding.[9] So, for example, even though in the adult, the midbrain lay below the cerebral hemispheres, the mid-skull plate would still reflect the mental dispositions with which it originally corresponded at the beginning of fetal life; the psychic energies of the various brain areas (*die Energie des Hirns*) would thus manifest themselves in the dimensions of the skull plates of the adult (Carus 1838–1840, III, 342). Carus's idealist metaphysics would be shared by the likes of Goethe, Johann Gottlieb Fichte, Schelling, and Hegel, but certainly these individuals did not set that metaphysics to do the kind of intricate physiological work that supported Carus's cranioscopy. His developmental theory of race would also rest on idealist metaphysical assumptions.

In his *System der Physiologie* and later in his *Denkschrift zum hundertjährigen Geburtsfeste Goethe's* (Memorial for Goethe's hundredth birthday, 1849), Carus distinguished four races of the one human species (Carus 1838–1840, I, 122–123): men of the day (Caucasian-Europeans), men of the night (Ethiopians), men of the eastern twilight (Mongolian-Mylan-Hindus), and men of the western twilight (Americans). He assumed the original Caucasian race appeared after the age of the great lizards, and, with Blumenbach, suggested that the original group first appeared on the high Asian plateau, around Mount Ararat. Ultimately whether humans arose from a more primitive form was, Carus admitted, lost in "a mysterious darkness" (Carus 1838–1840, I, 113). He yet thought several propositions could be established with certainty: (1) that the development (*Entwicklung*) of humanity was essentially and necessarily spiritual (*geistig*); (2) that the development had occurred through the social action of individuals manifesting different attributes (especially the duality of the sexes); (3) that its highest expression was in particular individuals (e.g., Goethe); and (4) that this development had occurred in different regions of the earth in different ways (Carus 1838–1840, I, 113–117). Like Blumenbach, Carus assumed the original Caucasian group spread to different parts of the world and adapted to different regions. Citing Herder's essay on language, Carus maintained that crucial to the development of humankind were the advent of language and the interactions of individuals within a society. The different races represented a progressive scale, with the people of the night at the lowest rank, then the people of the western twilight, then those of the eastern, and finally with the most developed being the people of the day. In this developmental scheme, the people of the night were still at the fetal stage, though with progressive potential, while people of the day exhibited the most advanced form of humanity (Carus 1838–1840, I, 114–115). Carus's developmentalism came very close to an authentic biological evolutionism, yet at the end of his life, when he was fully apprised of Darwin's theory, he would not take the final step.[10]

Carus justified his classification of the races through the kinds of measurements his cranioscopy suggested, though later he would seek additional support from Morton's measurements (Carus 1849, 19). These measurements also permitted him to determine the cognitive gifts of particular individuals, such as Kant, Napoleon, and Schiller (Figure 9.10). The empirical sampling of skulls on which his measurements were based and from which conclusions were drawn was minimal, however. In the *Cranioscopie*, Carus recorded measurements of a motley

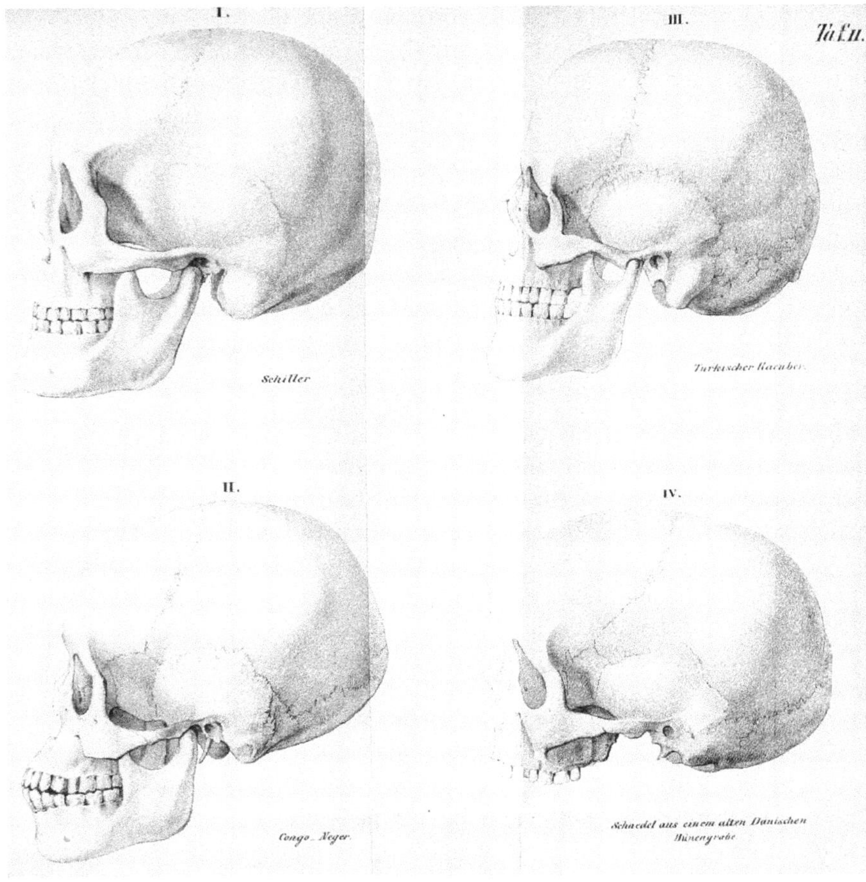

Figure 9.10 Schiller's skull (I), compared with the skulls of a Congo Negro (II), a Turkish thief (III), and a Dane, taken from an old grave (IV) (from Carus 1841).

of seventeen skulls, some of which were not even the original skulls but plaster casts (e.g., Napoleon, Charles-Maurice de Talleyrand, and Schiller), and others still harbored a living soul (e.g., Ludwig Tieck and Carus himself). The measures were done with calipers, which could give the length, width, and height (this last from the ear-opening to the highest part of the plate). Here is a sampling of the measures (Table 9.1).

What do these measures purportedly mean? Carus reckoned that the low measures for the frontal plate of the Negro slave meant a low intelligence; his mid-plate likewise showed deficient sensitivity and self-reflection. The hind-plate, though, indicated a strong will and sexual impulses – even having a greater height than Schiller's. Napoleon's measures revealed an extremely strong intelligence, great sensitivity, and an iron will. Schiller's skull, in contrast to the others, showed

Table 9.1 Table of skull measurements (extract from Carus 1841). The measures are given in inches and twelfths of an inch. Napoleon's death mask was missing the rear portion, hence the absence of scores. Carus probably could not accurately measure the length of his own skull plates, not knowing where the mid-plate and hind-plates began and ended – hence the blanks for lengths of the various plates.

	Schiller	*Negro slave*	*Napoleon*	*Carus*
Frontal plate				
Height	5"	4"6'''	5"8'''	5"
Width	4"8'''	3"10'''	5"5'''	4"9'''
Length	4"8'''	4"2'''		
Mid-plate				
Height	5"4'''	4"7'''	5"11'''	5"7'''
Width	5"10'''	4"6'''	5"8'''	5"8'''
Length	4"8'''	4"2'''		
Hind-plate				
Height	3"7'''	4"	4"2'''	
Width	4"	3"2'''	4"1'''	
Length	3"7'''	3"5'''		

that "the regions of intelligence, feeling, will, and desire are quite harmoniously developed [*sehr harmonisch entwickelt*]. The first two regions are significant and the relationships are quite felicitous throughout" (Carus 1844, 48). What is it that makes a poet's skull? For Carus, a head, developed generally in "beautifully harmonious structures [*schönen harmonischen Formen*], with a well-formed front head, a modest rear head, but a decidedly powerful middle head – these features indicate a poetic human being" (Carus 1841, 57). None of the other skulls that Carus examined exhibited the graceful harmony of Schiller's, not Napoleon's and not Kant's. The more a skull would deviate by reason of a one-sided development in either height, breadth, or length of the skull plates, "the more generally the form would represent a lower, unbeautiful [*unschöne*] and, in its psychic significance, an unfavorable form" (Carus 1841, 59). In all of Carus's six works devoted to cranioscopy, Schiller's skull served as the standard by which to evaluate all of the other skulls, just as the Georgian girl became the aesthetic standard for Blumenbach.

Carus based his cranioscopy on exacting anatomical descriptions and on powerful generalizations from that anatomical work. These latter contributions entered the mainstream of biology during the mid-part of the nineteenth century, especially through Richard Owen's conception of homology. Carus's developmentalism stopped just short of a full-blown evolutionary conception, and represents a stage in German scientific life that prepared the way for a rapid acceptance of Darwinism. Carus's empirical measures of skulls won the admiration of many for their precision, but the effort to derive portraits of intelligence and talent from such measurements seems to us little better than the efforts of the phrenologists – perhaps even less availing, since Carus's interpretations of skull plates had to be justified by a metaphysics in overdrive.

Note on Schiller's skull

Carus presented Schiller's skull as a singular standard for the harmonious development of poetic sensibility. There are two problems with his choice. First, as mentioned, he didn't measure Schiller's skull directly, but rather a plaster cast of the skull. But the second difficulty completely undermines his effort. There is strong evidence that the skull thought to be Schiller's is not really his. At his death in 1805, Schiller was buried in a mass grave for distinguished individuals in Weimar. Twenty-one years later, Karl Schwabe, the Weimar Bügermeister, decided to retrieve Schiller's remains, which by then had become mixed with the bones of many others. He pulled out twenty-three skulls and judged the largest "must be Schiller's skull" (Schöne 2002, 14). Goethe revered this skull, making it into a small shrine to his friend. It was this skull that provided the plaster cast that Carus used for his measurements (Carus 1845, Tafel 1). In later years, doubts arose concerning the skull, and another skull was recognized as more likely Schiller's. In 2008, DNA from both skulls was extracted and compared with DNA from the remains of known relatives of Schiller. Neither skull was a match (Smee 2008).

Conclusion: Exacting measurement and ineffable beauty of skulls

Global travel during the late eighteenth through the mid-nineteenth centuries revealed the great variety of mankind and brought to the fore the question of the very nature of the human. During this same period, slavery as a political and moral question grew in volatility and nearly destroyed a young nation. Science, especially the introduction of exact measurement into new areas of inquiry – into anatomy, psychology, and anthropology – should have been able to provide, or so it was thought, objective methods for coming to conclusions about such social issues. Techniques of measurement thus came to be applied to that most durable and iconic feature of the human – namely, the human skull.

Four scientists of acknowledged ability undertook the measurement of man, or at least his skull – Tiedemann, Morton, Blumenbach, and Carus. All four assumed that external physical characteristics might reveal internal mental traits and talents or affiliation to an ethnic group. Each, however, brought to his effort different assumptions and different techniques of measurement. The former two focused on quantitative determinations of cranial capacity – indicative of mental ability – and the latter two were concerned with aesthetic evaluations, which might suggest innate talents or racial origins.

All four of the naturalists recognized that there was significant variability both across races and within races. Tiedemann and Blumenbach, the former measuring cranial capacity and the latter aesthetic features, found no significant differences in innate qualities of the races, while Morton and Carus thought the races formed a hierarchy ranging from inferior to superior. Tiedemann recognized that women had smaller skulls than men, but actually larger skulls relative to body size; Carus judged the anterior and posterior plates smaller in women, which implied smaller intelligence and weaker will, though a mid-plate that was relatively larger, suggestive of greater sensitivity. Despite the variability found within and across races, Morton and Carus detected beneath such differences stable and unchanging types, while Tiedemann and Blumenbach seemed to perceive only individuals.

From our perspective, Morton and Carus appear simply to have endorsed cultural stereotypes, yet we should not demand them to be wiser than their times would allow. One likely explanation for their assumption that general types lay beneath variable structures was their training. Both were illustrators, and Carus, of course, an extremely accomplished artist. As Ernst Hans Gombrich has shown, from the late medieval period through the first half of the nineteenth century (and even today), drawing manuals instructed the novice to practice schemata – that is, patterns for drawing *the* bird, *the* tree, *the* human figure, as opposed to drawing a particular bird, a particular tree, a particular human being. As the student advanced, he or she could begin adding individual details to the schematic drawing, turning the universal into the particular (Gombrich 1984, 146–178). So the artist and illustrator, at least as part of their training, would have reflexively perceived the schema – or as Carus called it, the archetype – beneath the particularity. Blumenbach and likely Tiedemann may have been more disposed to see only individuals instead of types because they had personal acquaintance with Negros, and so stereotyping would be more difficult than it would be in the absence of such interaction. Of course, personal acquaintance is not an infallible protection against prejudice, as slaveholders in the American South make evident. This array of social causes and professional inclinations does seem to explain differences existing among the four naturalists, but, of course, something must be attributed to individual disposition and personal psychology.

In this essay, I have paid special attention to the aesthetic judgment of skulls, since such evaluation seems not only unusual but also subjective, not a scientific measure. Both Blumenbach and Carus were readers – and followers – of Kant, who argued, in the third *Critique*, that the judgment of beauty made a universal claim on others. It was subjective, according to Kant, but nonetheless universal since judgments of beauty depended on an aesthetic feeling arising from an ineffable relationship between reason and imagination, traits common to all humans. Since the judgment of beauty was grounded in a feeling that lacked a conscious, rational component, Blumenbach and Carus, as Kantians, could do hardly better than point to the object, while uttering terms like "symmetry," "harmony," and "graceful arrangement of parts," which were little more than synonyms for "beauty." For the Kantian, direct experience was crucial, not argument, in making a judgment of beauty.

At the end of the nineteenth century and through the early part of the twentieth, intelligence tests began to be constructed and personality tests devised. These had the potential for revealing inner human traits more directly than evaluations based on external measurements. External, physical features were still used, but the Nazi experience quashed most such efforts. So the period discussed in this essay represents a particular moment in the evaluation of human beings. Aesthetic evaluation of skulls is even more distant from our present expectations. But I have seen the Georgian female's skull in the Blumenbach collection at the Georg-August-Universität in Göttingen. It is quite beautiful.

Notes

1 I am grateful to Marina Bell for putting me on to Dinesen's short story.

2 Lorenz Oken advanced the hypothesis of recapitulation in an early work. See Oken (1805, 164–167). I discuss the hypothesis in Richards (2002, 493–494).

3 Tiedemann (1808–1814, I, 64–65): "Just as each individual begins with the simplest formation and during its metamorphosis becomes more evolved [*entwickelt*] and developed, so the entire animal organism [i.e., animal kingdom] seems to have begun its evolution [*Entwicklung*] with the simplest animal forms that is with the animals of the lowest classes." While studying in Paris with Cuvier, Tiedemann would have become quite familiar with Lamarck's version of evolution. See also Tiedemann (1830, I, 102–104). I have discussed Tiedemann's theory of recapitulation and that of many others, including Darwin, in Richards (1992).

4 Stephen Jay Gould wrote an admiring essay on Tiedemann (Gould 1999). Gould said that Tiedemann "offered no summary statistics for groups – no ranges, no averages"; this gave Gould something to do – he provided the averages. Gould, however, relied on the English version of Tiedemann's study; in the German version, as in Figure 9.3, Tiedemann certainly did give summary statistics and ranges, if not averages. Gould thus missed the last set of summary tabulations showing the greater proportion of large skulls for the Caucasian and Malay races. By depending only on the English version Gould was led to speculate: "Did Tiedemann calculate these means and not publish them because he sensed the confusion that would then be generated – a procedure that I would have to label as indefensible, however understandable? Or did he never calculate them because he got what he wanted from the more obvious data on ranges and then never proceeded further – the more usual situation of failure to recognize potential interpretations as a consequence of unconscious bias? I rather suspect the second scenario" (Gould 1999, 69). Gould simply missed Tiedemann's obvious worry about his numbers.

5 Charles Meigs, a long-time friend and colleague of Morton, included Humboldt's letter as an appendix to Meigs (1851). See also Kelly (1912, II, 192–197) and Stanton (1960, 24–44). Fabian (2010) gives a detailed account of Morton's efforts to collect skulls from friends, traders, travelers, and grave-robbers.

6 Morton's *Crania Americana* was simultaneously published in London. The price of $20 was prohibitively expensive and Morton had to use an inheritance to cover his costs in production and printing. As a result of lack of sales, he sent complimentary copies to many individuals and learned societies in America and Europe (Fabian 2010, 87–91).

7 The unity of mankind was not only a theological issue; it also engaged naturalists on either side of the divide between monogenists and polygenists. James Cowl Prichard led the partisans of human unity and Louis Agassiz represented those who believed humans to consist of several distinct species. Agassiz strongly supported Morton in the fray (Lurie 1954).

8 In his *Catalogue of Skulls* (1849), Morton provided the averages of the families within the five races. So the Negro race, in his tables, comprised four families: Native African, American-born Negros, Hottentot, and Australian. The first two had the highest cranial capacities of 83 and 82 respectively. The Caucasian race had eight families, with the Teutonic family having an average of 92. Jason Lewis and colleagues at Stanford remeasured a sampling of Morton's skulls and found his final measures using lead shot

to be decently accurate. They rejected Gould's claims about the Morton's analyses of group means and subgroup means. See Lewis et al. (2011) for the particulars. Weisberg and Paul (2016) have entered the fray, and support Gould's conclusions. They point out that Gould did not dispute the accuracy of Morton's shot calculations; they focus on the fact that Morton's errors in the seed measurement were pronounced in regard to the African skulls. They contend if Morton's errors were not due to unconscious bias, the errors should have been systematically the same. Since they were not systematically the same, "Gould's claim that this is prima facia evidence of unconscious bias in Crania Americana remains intact" (Weisberg and Paul 2016, 3). This does not follow at all. Morton's racial attitudes are clear from his anthropological discussions. But if he were unconsciously manipulating the seed calculations to meet those prejudices, why did he fire his assistant and redo all the calculations with more reliable lead shot? After all, his prejudices would have been satisfied with the original seed calculations. There are many other possible reasons for the non-systematic errors in the seed calculations than unconscious prejudice. First, it was Morton's assistant, not Morton, who did the actual measurements. Second, if the assistant were making careless errors, there is no reason to assume he would be carelessly systematic.

9 Carus (1841, 8 fn 1): "At the first disposition [Anlage] of the brain, the first, second, and third brain areas and the first, second, and third skull plates completely correspond, so that with the progressive formation of the brain, that is, the greater development of the forebrain area, [. . .] the original relationship of the skull plates in relation to the three brain areas remains the same."

10 Carus published several essays showing not only differences in morphology between man and gorilla but also differences in spirit, such that "man is raised to something qualitatively other than the animal" (Carus 1863a, 30). See also Carus (1863b; 1865).

Bibliography

Anonymous. 1840. "Vergleichende Betrachtung der Schädel der verschiedenen ureingebornen Nationen von Nord- und Südamerica." *Neue Notizen aus dem Gebiete der Natur- und Heilkunde* 15 (13–15): 193–202, 211–216, 227–232.

Blumenbach, Johann Friedrich. 1775. *De generis humani varietate nativa.* Göttingen: Rosenbuschius.

———. 1779–1780. *Handbuch der Naturgeschichte.* 2 vols. Göttingen: Johann Dieterich.

———. 1780. "Über den Bildungstrieb (Nisus formativus) und seinen Einfluß auf die Generation und Reproduktion." *Göttingisches Magazin der Wissenschaften und Litteratur* 1 (5): 247–266.

———. 1781a. *Über den Bildungstrieb und das Zeugungsgeschäfte.* Göttingen: Dieterich.

———. 1781b. *De generis humani varietate nativa,* 2nd edn. Göttingen: Vandenhoek et Ruprecht.

———. 1787. "Einige naturhistorische Bemerkungen bey Gelegenheit einer Schweizerreise, von den Negern." *Magazin für das Neueste aus der Physik und Naturgeschichte* 4 (3): 1–12.

———. 1795. *De generis humani varietate nativa,* 3rd edn. Göttingen: Vandenhoeck.

———. 1796. *Abbildungen naturhistorischer Gegenstände.* Heft 1. Göttingen: Johann Christian Dieterich.

———. 1802. *Abbildungen naturhistorischer Gegenstände.* Heft 6. Göttingen: Heinrich Dieterich.

Carus, Carl Gustav. 1834. *Lehrbuch der vergleichenden Zootomie.* 2 vols., 2nd edn. Leipzig: Gerhard Fleischer.

———. 1835. *Briefe über Goethe's Faust.* Leipzig: Gerhard Fleischer.

———. 1838–1840. *System der Physiologie*. 3 vols. Leipzig: Gerhard Fleischer.
———. 1841. *Grundzüge einer neuen und wissenschaftlich begründeten Cranioscopie*. Stuttgart: Balz'sche Buchhandlung.
———. 1843. *Atlas der Cranioscopie*. Heft 1. Leipzig: August Weichardt.
———. 1844. *Vom gegenwärtigen Stande der wissenschaftlich begründeten Cranioscopie*. Nürnberg: Theodor Cramer.
———. 1845. *Atlas der Cranioscopie*. Heft 2. Leipzig: August Weichardt.
———. 1849. *Denkschrift zum hundertjährigen Geburtsfeste Goethe's*. Leipzig: Brockhaus.
———. 1863a. "Die Gorilla-Hand." *Leopoldina* 4: 28–30.
———. 1863b. "Weiteres über den Gorilla und gegen die Hypothese Darwin's." *Leopoldina* 4: 59–61.
———. 1865. "Ueber Begriff und Vorgang des Entstehens." *Leopoldina* 5: 139–152.
———. 1865–1866. *Lebenserinnerungen und Denkwürdigkeiten*. 4 vols. Leipzig: Brockhaus.
Combe, George. 1833. *George Combe's System der Phrenologie*. Trans. Stephan Eduard Hirschfeld. Braunschweig: Friedrich Vieweg und Sohn.
———. 1839–1840. [Review of Crania Americana by Samuel George Morton]. *The American Journal of Science and Arts* 38: 341–375.
Darwin, Charles. 1871. *The Descent of Man and Selection in Relation to Sex*. 2 vols. London: Murray.
David, Joseph, and John Thurman. 1867. *Crania Britannica: Delineations and Descriptions of the Skulls of the Aboriginal and Early Inhabitants of the British Islands*. London: Taylor and Francis.
De Vries, Jan, and A. van der Woude. 1997. *The First Modern Economy: Success, Failure, and Perseverance of the Dutch Economy, 1500–1815*. Cambridge: Cambridge University Press.
Dinesen, Isak. [1934] 1991. "The Old Chevalier." In *Seven Gothic Tales*, edited by Isak Dinesen, 81–108. New York: Vintage International.
Dougherty, Frank William Peter, and Norbert Klatt, eds. 2006–2015. *The Correspondence of Johann Friedrich Blumenbach*. 6 vols. Göttingen: Klatt.
Engel, Joseph. 1851. *Untersuchungen über Schädelformen*. Prag: Calve'schen Buchandlung.
Fabian, Ann. 2010. *The Skull Collectors*. Chicago, IL: University of Chicago Press.
Gillispie, Charles Coulston. 1960. *The Edge of Objectivity: An Essay in the History of Scientific Ideas*. Princeton, NJ: Princeton University Press.
Gombrich, Ernst Hans. 1984. *Art and Illusion: A Study in the Psychology of Pictorial Representation*, 2nd edn. Princeton, NJ: Princeton University Press.
Gould, Stephen Jay. 1981. *The Mismeasure of Man*. New York: Norton.
———. 1996. *The Mismeasure of Man*, 2nd edn. New York: Norton.
———. 1999. "The Great Physiologist of Heidelberg." *Natural History* 108 (July/August): 26–29, 62–70.
Hamilton, William. 1850. "Remarks on Dr. Morton's Tables on the Size of the Brain." *Edinburgh New Philosophical Journal* 48: 330–333.
Huschke, Emil. 1854. *Schädel, Hirn and Seele des Menschen und der Thiere*. Jena: Mauke.
Jones, Jeannette. 2013. " 'On the Brain of the Negro': Race, Abolitionism, and Friedrich Tiedemann's Scientific Discourse on the African Diaspora." In *Germany and the Black Diaspora: Points of Contact, 1250–1914*, edited by Mischa Honeck, Anne Kuhlmann-Smirnov, and Martin Klimke, 134–152. New York: Berghahn Books.
Kant, Immanuel. 1956. "Metaphysische Anfangsgründe der Naturwissenschaft." In *Immanuel Kant: Werke in Sechs Bänden*, edited by W. Weischedel. Wiesbaden: Insel.

Kelly, Howard. 1912. *A Cyclopedia of American Medical Biographies*. 2 vols. Philadelphia: Saunders.

Klatt, Norbert. 2008. "Klytia und die 'schöne Georgianerin' – Eine Anmerkung zu Blumenbachs Rassentypologie." In *Kleine Beiträge zur Blumenbach-Forschung*, edited by Norbert Klatt. Vol. 1, 70–101. Göttingen: Norbert Klatt Verlag.

Lewis, Jason, et al. 2011. "The Mismeasure of Science: Stephen Jay Gould Versus Samuel George Morton on Skulls and Bias." *PLOS Biology* 9 (6) (June): 1–6.

Linnaeus, Carolus. 1740. *Systema Naturae, sive Regna Tria Naturae* (Latin and German), edited by Johann Langen. Halle: n.p.

———. 1758–1759. *Systema Naturae, per Regna Tria Naturae*, 10th edn. 2 vols. Stockholm: Laurentius Salvius.

Lovejoy, Paul. 1982. "The Volume of the Atlantic Slave Trade: A Synthesis." *Journal of African History* 23: 473–501.

Lurie, Edward. 1954. "Louis Agassiz and the Races of Man." *Isis* 45: 227–242.

Meigs, Charles. 1851. *A Memoir of Samuel George Morton, M.D.* Philadelphia: T.K. and P.G. Collins.

Meigs, J. Aitken. 1857. "The Cranial Characteristics of the Races of Men." In *Indigenous Races of the Earth*, edited by Alfred Maury et al., 203–352. Philadelphia: Lippincott.

Morton, Samuel George. 1839. *Crania Americana or: A Comparative View of the Skulls of Various Aboriginal Nations of North and South America*. Philadelphia: Dobson.

———. 1844. *Crania Aegyptiaca; or, Observations on Egyptian Ethnography, Derived from Anatomy, History and the Monuments*. Philadelphia: Penington.

———. 1847. "Hybridity in Animals, Considered in Reference to the Question of the Unity of the Human Species." *Edinburgh New Philosophical Journal* 43: 262–288.

———. 1849. *Catalogue of Skulls of Man and the Inferior Animals*, 3rd edn. Philadelphia: Merrihew & Thompson.

———. 1851. "On the Infrequency of Mixed Offspring Between European and Australian Races." *Proceedings of the Academy of Natural Sciences of Philadelphia* 5: 173–175.

Oken, Lorenz. 1805. *Die Zeugung*. Bamberg: Goebhardt.

Richards, Robert J. 1992. *The Meaning of Evolution: The Morphological Construction and Ideological Reconstruction of Darwin's Theory*. Chicago, IL: University of Chicago Press.

———. 2002. *The Romantic Conception of Life: Science and Philosophy in the Age of Goethe*. Chicago, IL: University of Chicago Press.

Schöne, Albrecht. 2002. *Schillers Schädel*. Munich: C. H. Beck.

Schott, Johann Adam Christoph. 1836. *Die Controverse über die Nerven des Nabelstrangs*. Frankfurt am Main: Wilmans.

Smee, Jess. 2008. "DNA Tests End Debate Over Schiller Skulls: Neither Is His." *The Guardian*, 5 May. URL: www.theguardian.com/stage/2008/may/05/theatre.germany.

Sömmerring, Samuel Thomas. 1785. *Ueber die körperliche Verschiedenheit des Negers vom Europäer*. Frankfurt am Main: Varrentrapp Sohn.

Stanton, William. 1960. *The Leopard's Spots: Scientific Attitudes toward Race in American 1815–1859*. Chicago, IL: University of Chicago Press.

Tiedemann, Friedrich. 1808–1814. *Zoologie. Zu seinen Vorlesungen entworfen*. 3 vols. Landshut: Weber.

———. 1816. *Anatomie und Bildungsgeschichte des Gehirns im Foetus des Menschen nebst einer vergleichenden Darstellung des Hirnbaues in den Thieren*. Nürnberg: Steinischen Buchhandlung.

———. 1830. *Physiologie des Menschen*. 3 vols. (projected, one produced). Darmstadt: Leske.

———. 1836. "On the Brain of the Negro, Compared with That of the European and the Orang-Outang." *Philosophical Transactions of the Royal Society of London* 126: 497–527.

———. 1837. *Das Hirn des Negers mit dem des Europäers und Orang-Outangs verglichen.* Heidelberg: Winter.

Weisberg, Michael, and Diane Paul. 2016. "Morton, Gould, and Bias: A Comment on "The Mismeasure of Science." *PLoS Biology* 14 (4): e1002444. URL: https://doi.org/10.1371/journal.pbio.1002444.

Wood, George Bacon. 1853. *A Biographical Memoir of Samuel George Morton, M.D.* Philadelphia: T. K and P. G. Collins.

Zeune, August. 1846. *Über Schädelbildung zur festern Begründung der Menschenrassen.* Berlin: Vereins-Buchhandlung.

10 Ethnographic exploration in the Blumenbachian tradition

Peter Hanns Reill

Prince Maximilian Wied-Neuwied is not a household name. Even in well-educated circles, mention of him more often than not draws a complete blank. Why, then, should I be devoting an essay to him in this volume? One answer is that he was an important German learned explorer shaped by the Enlightenment who undertook two arduous "scientific" trips to South and North America and wrote fascinating accounts of these journeys. A second is that his narratives provide us with subtle ethnographic descriptions of the native inhabitants he encountered in eastern Brazil and the upper Missouri Valley in North America that are still invaluable for scholars and those interested in these peoples. The third, and most important for this volume, is that his ethnographic investigations and descriptions were shaped by his studies in Göttingen under Johann Friedrich Blumenbach, whose program he sought to implement on the ground in these relatively unexplored lands. In this sense, Maximilian's works embody the type of ethnographic knowledge Blumenbach called for to understand peoples of other lands and cultures; it was Blumenbachian ethnographic science-in-action.

Blumenbach's ethnography

Johann Friedrich Blumenbach was one of the most important and influential naturalists of the Late Enlightenment. He spent his whole career teaching and writing at the University of Göttingen, the most progressive university in Germany at the time. There he trained three generations of naturalists, physiologists, comparative anatomists, and countless students interested in the life sciences and their relationship to the cultural sciences. His standard texts went through many editions and were translated into all the major European languages. His correspondence network was large; his contacts reached across Europe and the New World, a reach intensified by his editing of the influential journal *Medicinische Bibliothek* (1783–1795). For German thinkers like Immanuel Kant, Johann Wolfgang von Goethe, Johann Gottfried Herder, and Alexander and Wilhelm von Humboldt, to mention but a few, Blumenbach's anthropological vision was captivating, as it was also for many British, American, and continental European thinkers.

For many years, Blumenbach was a forgotten figure in the historiography of the Enlightenment. But there has been a more recent revival of interest in Blumenbach that makes his anthropologic and ethnographic views very controversial. For many scholars, especially Anglo-American commentators, he has been described as one of the fathers of modern scientific racism. This view is underpinned by reference to the highly skewed, racist-inspired mid-nineteenth-century English translation of his work on human variety (Blumenbach 1865). But those who have read Blumenbach in the original know how wrong this judgment is. Blumenbach's original texts, especially the many editions and translations of his highly influential *Handbuch der Naturgeschichte*, demonstrate clearly that Blumenbach was a fierce advocate of the spiritual and mental equality of all members of the human race, even though he enunciated a heuristic model of physical racial differentiation. But, and this must be emphasized, these differentiations were not based on superficial observations of skin color. Rather, like many late eighteenth-century physiologists and anthropologists, Blumenbach charted varietal differences by comparing skeletal examples, especially those drawn from a comparative examination of human skulls. Blumenbach believed that the human race had its origins at one place – the Caucasus – and then was dispersed over many different climates, geologies, and geographic locations. The combined effects of these dispersals led to physical changes driven by the *Bildungstrieb* (formative drive) in the original type (which he named the Caucasians) in order to allow them to best thrive in these environments. Blumenbach's idea of the *Bildungstrieb* was one of his most attractive ideas for late eighteenth-century life scientists and cultural commentators. Blumenbach called the *Bildungstrieb* a "free organic power" that shaped a living "organized body."

Thus, for Blumenbach, varietal generation was a spontaneous process in which the hidden formative drive separates a specific quantity of matter from the mass and gives it a unique and individual form. Each of the varieties Blumenbach discerned was, he argued, shaped by natural forces unique to living entities, closely linked to the environment in which they thrived. Though Blumenbach sought to classify different human varieties, he did not reify these distinctions; he recognized the arbitrary nature of such classifications, which he considered to be, at best, helpful heuristics in classifying peoples. Instead, he argued that no hard and fast differentiations could ever be made within the vast variety of human forms (Blumenbach 1781, 49). None of these varieties were fixed. They were always open to transformation over time. According to Blumenbach, the "varieties were literally infinite and can never be reduced to any certain and definitive class" (Blumenbach 1795, I, 44). In short, Blumenbach denied the existence of the great chain of being, where all species, from the simplest organism to the human being, were linked through minute variations in a continuous progressive ladder of development. For higher species an argument based on the great chain of being meant that each "race" or variety was an improvement on the preceding one, usually with the European at the apex. In Blumenbach's view, each species was unique onto itself, not necessarily linked to other species, such as apes. Rather, Blumenbach posited a form of infinite gradation or continuity *within* species. An infinity of possible combinations existed within the limits assigned to any living species, such as the human race. Thus, Blumenbach negated any theory grounded upon hard and fast physical distinctions in human varieties. In Blumenbach's view, all humans, being part of the same species, were equally endowed with the same kinds of creative powers and intellectual capabilities.

Often, scholars represent Blumenbach's categories by pointing to the different skull types he associated with the five varieties he claimed to have discerned. Since these skeletal varieties were devoid of living content, such differentiations could, under different interpretive schemes, be seen as validating distinct racial types. And by the mid-nineteenth century, this validation happened very easily. But Blumenbach made his advocacy of the intellectual and spiritual equality of all peoples clearly evident in one of his works, *Abbildungen naturhistorischer Gegenstände*, in which he offered his readers fully fleshed-out images of real people to illustrate his five broad categories of human varieties: namely, the Mongolian, American, Caucasian, Malaysian, and Ethiopian (Blumenbach 1796). His five images and short characterizations made it clear that all of these people were highly intelligent, capable, and worthy of equal respect. In his classification, the Caucasian was the stem or "middle race." The other four were variants of the stem people. The two extremes were the Mongolian on one side and the Ethiopian on the other, while the remaining two, transitions between middle and extreme, were portrayed on a horizontal scale without any hierarchical ordering. The personages Blumenbach chose to portray each "race" are fascinating: the Mongolian was the artist Feodor Ivanovitsch, who had gained renown in Rome; to represent the American variety Blumenbach chose Tayadaneega or Captain Joseph Brant, a chief of the Mohawk tribe, a "sogenannte Wilde" (so-called savage) who, Blumenbach countered, had published learned articles in the *Philosophical Transactions* (Blumenbach 1796, 16). His representative of the Caucasian, the group Blumenbach claimed embodied "*our* ideas of beauty" (emphasis in original) was Jusuf Aguiah Efendi, the Turkish ambassador to London – namely, someone who was not normally considered white by European standards, either then or now (Blumenbach 1796, 19). The Malaysian was O-mai, the Tahitian who accompanied and advised Captain Cook on his voyage to Tahiti.

Blumenbach's choice for the Ethiopian variety was the most politically charged, given the fierce debates at the time of whether blacks were by nature intellectually retrograde. Blumenbach chose Jacob Johan Eliza Capitein, a black theologian known for his publications in Latin and Dutch, whom Blumenbach first discussed in an earlier work, *Beyträge zur Naturgeschichte* (Blumenbach 1790, 99). In *Beyträge*, Blumenbach also summarized examples of many blacks who distinguished themselves by their literary and scientific accomplishments. In *Abbildungen* he cited another example, Benjamin Banneker, an accomplished astronomer from Philadelphia, a self-taught man who had acquired his knowledge through his studies of James Ferguson's works and Tobias Mayer's tables. According to Blumenbach these examples demonstrated "that intellectual ability is not influenced by skin color" (Blumenbach 1796, 28). In short, Blumenbach did not link physical differentiation with spiritual, moral, or intellectual abilities. Human physiognomy did not express moral or intellectual hierarchies, a position that corresponded to Blumenbach's Göttingen mentor and colleague Georg Christoph Lichtenberg's critique of Johann Caspar Lavater's physiognomy (Lichtenberg 1972, 258).

Over and above Blumenbach's anti-racist conceptions of human variety, what has often been overlooked is his wide-ranging vision of what ethnographic research should do beyond discovering or demonstrating the existence of the physical differentiations of the peoples in the world. Blumenbach not only taught physiology and comparative anatomy at Göttingen but also was in charge of the ethnographic collections of Göttingen University's *Academisches Museum*, which he expanded exponentially. Thus, besides the physical aspects of the "science of man," Blumenbach was deeply involved in the ethnographic side of this inquiry (Vermeulen 2015, 381). His ethnographic research agenda incorporated the established Göttingen tradition of learned travel, a wide-ranging view of the intimate interconnections between humans and the environment in which they were nourished, and the cultural world they inhabited. These positions were made evident by his commitment to what I have described elsewhere as Enlightenment vitalism, which infused nature with life (Reill 2005). Vitalists considered nature's activities as founded upon reciprocal interaction, its apprehension a combination of hard observation and sympathetic, aesthetic recreation. For them, it was imperative that a natural historian go out into the field in order to acquire firsthand exposure to the region's plants, animals, and its human inhabitants in order to perceive the harmonious relationship existing between all aspects of living nature. And though Blumenbach hardly left Göttingen, he was instrumental in expanding Göttingen's rich collection of travel literature, wrote many introductions to travel accounts, reviewed scores of others in many learned journals, and encouraged his students to undertake such voyages. As a leading vitalist, Blumenbach argued that the living world of nature was constituted by complex *rapports* linking all living beings within an organized system of active forces, each part of nature mirroring and influencing the other parts. Finally, reflecting the vitalist views that nature cannot be simply comprehended by abstract reason, Blumenbach, following the aesthetical theory then being formulated in Germany, called for a form of sensate knowledge, which spoke to both humans' reason and their sensations, emotions, and passions. This program was carried out by Alexander von Humboldt, who had studied in Göttingen under Blumenbach and Lichtenberg, in his explorations of South America (Ackerknecht 1955). Maximilian zu Wied-Neuwied employed the same approach, which he learned from his Göttingen studies and especially from Blumenbach.[1] Both Humboldt and Maximilian believed that peoples cannot be understood without examining the whole range of their experiences, from their way of livelihood to their customs, religions, dress, food, and, in true Göttingen manner, their languages.[2]

Maximilian faithfully carried out this program in both of his expeditions. From 1815 to 1817, at the instigation of Blumenbach,[3] he explored the east coast of Brazil, thereafter publishing a two-volume narrative of his trip with a third volume of images (Maximilian 1820–1821). In 1832, he undertook another journey, this time to the relatively unexplored regions of the upper Missouri in North America. He also published a two-volume account of this journey, accompanied by a magnificent book of illustrations, executed by the painter Carl Bodmer, who accompanied him on the trip (Maximilian 1839–1843). The Brazil trip framed

Maximilian's overall research program; the second journey into North America expanded it and reflected a more mature appreciation of the material with which he dealt. These works present a fascinating picture of the regions and their native inhabitants, written by an astute observer deeply enmeshed in the discourse of Enlightenment vitalism.

Maximilian's critique of Euro-American colonialism

Maximilian sought to recapture nature as it existed prior to Euro-American colonization, to find and portray in immediate ways the *Urwälder*, *Urtiere*, and *Urvölker* (original plants, species, and indigenous peoples) of the New World. Of course, this mission was impossible, since the Portuguese had been in Brazil since the sixteenth century and white colonists in North America since the seventeenth; so Maximilian modified his project by seeking to explore those parts of Brazil and North America that were relatively untouched by massive Euro-American cultural aggrandizement. His search for original species, plants, animals, and especially peoples runs like a red thread through all of his writings. In his ethnographic observations, Maximilian sought to dispel both false European narratives of the peoples he encountered and what he deemed false descriptions of the lands they inhabited. He directed his scorn especially toward "arm-chair philosophers," who uncritically adopted accounts of Native Americans, portraying them as cannibals, half-humans, and peoples without a sense of religion, culturally deficient and therefore objects to be controlled, enslaved, converted, obliterated, and, when not killed, robbed of their unique cultures. This critical deconstruction of accepted Euro-American stereotypes led him to reconsider the nature of the peoples and the lands he was exploring. In so doing, Maximilian offered a searing critique of how white colonists, both in South and North America, treated the land and peoples they had conquered. In the process, he proposed an alternative vision of empire strongly anti-imperialist in substance and intent.

Maximilian's critical stance toward colonial imperialism not only reflected his ethnographic interests but also was rooted in his own experiences as a member of the ruling family of the *Grafschaft* (loosely equivalent to an earldom), and later in 1784 the *Fürstentum* (princedom) of Wied-Neuwied. For many readers, mention of Wied-Neuwied probably conjures up an image of an obscure, out-of-the-way place, part of the unwieldy Holy Roman Empire then in its death throes. Yet, for many well-educated eighteenth-century travelers, Wied-Neuwied was judged one of the most progressive states in the Holy Roman Empire, comparable to other highly esteemed states, such as Weimar and Anhalt-Dessau (Geuns 2007, 183–185). Under *Graf* and then *Fürst* (prince) Johann Friedrich Alexander, Maximilian's grandfather (who died in 1791 when Maximilian was nine), Wied-Neuwied became a prime exemplar of what Enlightenment policy could create within the flexible structures of the Holy Roman Empire.

Graf Friedrich Alexander made religious toleration a cornerstone upon which the state was built. Every religious group found a welcome in Wied-Neuwied; they were allowed to build their own places of worship and have their own burial grounds and their own schools. Thus, though originally Calvinist, Wied-Neuwied attracted Lutherans, Catholics, Moravians (*Herrnhuter*), Mennonites, and Jews as well as many small radical Protestant sects, usually referred to as the *Inspirierten*. Alexander improved the school system, founded a Masonic order, which later became part of the *Illuminati*, supported reading clubs, and instituted freedom of the press, attracting publishers and writers who published journals and books of all political stripes. He also expanded the town's economic base by founding factories and providing liberal tax-breaks for new industries. He was supported in all of these endeavors by his wife and by Maximilian's mother, both extremely gifted women committed to Enlightenment ideas and in close contact with the *Illuminati*. They supervised the children's education, supplying them with tutors who were drawn from this circle, many of them educated at Göttingen.

This commitment to Enlightenment ideals and practices seemed to come to a screeching halt when Maximilian's father, Friedrich Carl, succeeded Alexander. Carl turned his interests elsewhere, running up enormous debts in the process. He fired most of the bureaucrats, drastically cut the budgets of the court, making it almost impossible to clothe and feed his large number of children and his now estranged wife, Maximilian's mother, and made Wied-Neuwied a haven for French aristocratic *émigrés* during the French Revolution. On top of all this, his sexual peccadillos and seemingly irrational behavior led the *bürgerliche* elite of the town, supported by Maximilian's mother and members of their noble and ruling extended family, to seek to have him deposed. They brought the case before one of the two imperial courts, the *Reichskammergericht* – the suit was called the *Imbecillitäts-Prozess* – and surprisingly they won. The *Reichskammergericht* deposed Carl in 1792 and the victory was hailed throughout Germany as proof of the spread of Enlightenment and of the reform potential of the Holy Roman Empire. But the victory was short-lived. Carl appealed to the council of princes (*Reichsrat*), which as the French wars progressed had become increasingly reactionary. In his suit, he invoked what we might call the red scare, accusing his "enemies," especially Maximilian's mother, of being Jacobins. Carl was reinstated in 1795 and Maximilian's mother along with all her children fled to her parents' home in Meiningen until Carl abdicated in 1802 in favor of his eldest son, August. Maximilian was twenty. The return to independent power did not last long. When the Holy Roman Empire was abolished in 1806 after the Allies' defeats against Napoleon (Maximilian lost two brothers in the war and fought against Napoleon twice), Napoleon brought Wied-Neuwied under the control of Nassau. With Napoleon's defeat, Wied-Neuwied became a part of Prussia, though still retaining some of its local sovereign powers. In 1848, it became fully integrated into Prussia.

These experiences, especially the early ones, tied Maximilian very closely to his mother, his brothers and sister, and to the Enlightenment vision his mother, grandmother, and grandfather had espoused. They also made him very wary of what he considered undue authoritarian power, a position reinforced by his opposition to Napoleon's imperial-inspired conquests and by his subsequent military service in the wars against Napoleon. His experience in Wied-Neuwied thus raises an important question concerning how a "German" of the late eighteenth century perceived empire. In the current debates about empire, exploration, and science, where explorers have been characterized as being complicit in the colonial project, most of the focus has been on expansive and colonizing empires such as England, France, Spain, Portugal, and by extension, the United States. Little has been said about the Holy Roman Empire. It may be that many scholars today still seem to agree with Voltaire's sarcastic characterization of it as neither holy, nor Roman, nor an empire. But there is now a substantial and growing body of literature that reevaluates the Holy Roman Empire positively, seeing it as a confederation of states with an emperor as head, but constrained by imperial institutions, such as the two imperial courts of law (*Reichskammergericht* and *Reichshofrat*), the assembly of princes (*Reichsrat*), and the ten circles that oversaw the functioning of the constituent member states. The empire had also developed the most efficient and extensive postal system on the continent and because it was not "holy" (that is, committed to one religious persuasion), it allowed for a diversity of opinion and approaches unthinkable in more monolithic states, such as France, Spain, and even England.

And most important for thinkers such as Maximilian and Humboldt, the empire was not an expansive entity driven by the desire to conquer, annex, and colonize new territories. After the Peace of Westphalia (1648), its whole raison d'être was built upon compromise, negotiation, and allowing its individual members to develop in ways unique to them. Thus, a good measure of Maximilian's opposition to colonial empires derived from his experience of an empire where the theater of empire was at home, and where diversity was central to its makeup. Maximilian's critical stance toward colonial empires was reinforced and magnified by his commitment to Blumenbach's ethnographic agenda focusing on the equality of all peoples, a stance Maximilian often referred to in his travel accounts.

Maximilian's opposition to colonial imperialism manifests itself on two levels. The most obvious is the direct indictments he made of colonial rule. These indictments fill both travel reports. In the Brazilian account, for example, Maximilian blamed the ferocity of Native Americans toward the whites on the manner in which white planters had treated the Native Americans: "simply, the planters treat these people wrongly. They consider them to be animals and speak immediately of using the whip. In this manner they naturally provoke rage and cause hate and conflict" (Maximilian 1820–1821, I, 144–145). He explained planter behavior by citing the Europeans' *Gewinnsucht* (avarice) and *Goldgier* (greed), a central theme in both of his travel narratives. "In earlier times, avarice and greed suppressed all feelings of humanity amongst the European settlers. They considered brown and naked people only as animals, who were merely created for their use" (Maximilian 1820–1821, I, 162). This theme of the Indians being treated as slaves constantly emerges in his Brazilian narrative. It is encapsulated in his description of the treatment of Native Brazilian Americans on a plantation.

> One tyrannizes the poor Indians, uses them as slaves, transports them, commands them to build roads, to cut down trees, to run errands, forces them to

> fight against the enemy Tapuyas, doesn't pay them a thing or just a pittance, prejudicing them greatly against their oppressors because of the Indians' inherent love of liberty.
>
> (Maximilian 1820–1821, II, 211)

Maximilian's travel narrative of North America is even more negative. His initial entrance into Boston and his travels in the Northeast shocked him even more than his arrival in Rio de Janeiro. In Rio he encountered a Creole culture of whites, African slaves, free blacks, indentured Indians, and people of mixed blood. In the Eastern US, by contrast, he sought in vain for a hint of Native American culture; he observed that the white colonists had erased most memories of Native Americans. The black presence was there but largely segregated from the rest of the population. For Maximilian, blacks did not fare very well, even in so-called slave-free states, where they performed most of society's menial and difficult tasks. In slave-owning states blacks were treated horribly, as was the case, Maximilian argued, in all slave-holding societies. His journey from European-like places, such as Boston, New York, and Philadelphia, to ones that had original Native Americans in the upper Missouri was long and difficult. During his travels, Maximilian encountered white social types whose basic goal, he believed, was money making, who hated Native Americans, who destroyed the environment, and who were uneducated and uncultured. It took him almost a year to see "real" Americans, not, as he put it, the white "so-called Americans" (*sogenannten Americaner*). In Brazil, it had taken him just a few days.

Maximilian expressed his anger about the activities of the "so-called Americans" and their "destructive rage" (*Zerstörungswut*) in terms that characterized the colonists as "usurpers" and "displacers" (*Usurpatoren* and *Vertreiber*). He believed their goal was to exterminate the Native Americans and replace them with the "Back-Woodsman," a white American type he characterized as "half-wild," without any education, besotted by whisky, land-hungry, and interested only in making money, whether by legal or illegal means. So as he traveled through the state of Indiana he sarcastically remarked, "There is no trace of Indians in Indiana, who all have been exterminated or deported. The land is now populated and blessed by the backwoodsman!" (Maximilian 1839–1841, I, 186). Reflecting upon the fate of Native Americans in Pennsylvania Maximilian lamented, "The thought saddened me that of all the original inhabitants in this large state of Pennsylvania, there is not a trace of them left. O land of the Free!!!" (Maximilian 1839–1841, I, 76–77). Perhaps, for Maximilian, the most blatant attack upon native cultures by white colonists was that they had erased most memories of native cultures, as witnessed, for example, by the destruction of the impressive burial mounds in North America. And Maximilian believed that erasing a people's memory was the prelude to erasing the people.

Not only did Maximilian bemoan what he considered the American colonists' destructive rage against Native Americans, but also he complained bitterly that they ravaged the environment. This theme does not appear in the Brazilian travel account, partly because of Maximilian's belief that the Brazilian jungles were so lush and fertile that they could recover from the settlers' intrusions through self-regeneration. In North America, he thought, this was not the case. Thus, as Maximilian traveled through regions already conquered and cultivated by the colonists, he observed the lack of underbrush or of second growth in the forests, which he considered a proof that these forests were on their way toward obliteration (Maximilian 1839–1841, I, 45). He was amazed that the white colonists did not have an inkling of the most basic principles of forestry.

Maximilian found the widespread depletion of indigenous animals and birds even more disturbing. He made his trip up the Missouri on a steamboat and then on keelboats belonging to the American Fur Company, owned by the German immigrant entrepreneur Johann Jacob Astor. The Fur Company had penetrated deeply into Indian country in order to trade with Native Americans and brought back a staggering number of pelts and skins, from which they made an enormous profit. Maximilian believed virtually every fur-bearing animal was threatened with extinction but especially the beaver and the bison, two animals central to the cultural, social, and economic integrity of the Plains Indians.

Beavers, which provided the fur for the ubiquitous beaver hats prized by Europeans and white Americans alike, were, Maximilian claimed, enormously plentiful when the whites first arrived. Now they were decimated, the result of the white fur traders' greed. A similar fate awaited the bison. As Maximilian traveled down the Missouri on his way back to Germany, he observed, "In the past year one saw from the steamboat a region where the whole prairie was covered by herds of bison. Now the area is almost lifeless except for a few wild geese and ducks" (Maximilian 1839–1841, I. 323). The destruction of the bison, already underway and dramatically increased as the American imperium over Indian lands expanded, could clearly be recognized by the whitened bones of the slaughtered animals, which dotted the prairie.

In these critiques, Maximilian kept asking, where was the American government? Why did it not stop the excesses that were against the law? Why did it let the backwoodsmen infringe and steal federal property? Why did it look the other way when whisky, which he considered one of the greatest dangers to Native American culture, was sold illegally to the Native Americans in the Indian territories? Why did it not support an enlightened policy of forestry in the new and old lands? Why did it not control the fur trade? And most important, why did it allow white settlers to steal and expropriate Indian lands? In a footnote he chided the then American president, Andrew Jackson, for doing nothing to counter the aggressive colonists' push into Indian territory, not knowing, or feigning not to know, that Jackson was one of the central architects of the removal policy that brutally expelled Native Americans from their own lands (Maximilian 1839–1841, II, 392 fn).

From these few examples, Maximilian's direct critique of colonial activities is glaringly obvious. But there is another dimension to his work that demonstrates his comprehensive anti-imperialism in a more profound manner: his ethnographic portrayals of the Native Americans of Brazil and North America. As many writers on imperialism and ethnography have argued, ethnographic reports that homogenized native peoples into single undifferentiated categories paved the way for their dehumanization. Thus, to counter such dehumanization, native peoples would have to be seen as culturally, socially, and economically unique (Muthu 2003). A true Enlightenment anti-imperialism had to avoid the twin dangers of characterizing native peoples as either noble savages or murderous barbarians. To acknowledge these peoples' real humanity would be to portray them as part of integrated and unique cultures with all of their strengths and weaknesses, peoples different from Euro-Americans but admirable in their own individual way and masters of their own agency. Maximilian accomplished these aims in his ethnographic analyses.

Maximilian dealt with many peoples in South and North America. In his ethnographic observations, he discussed fourteen different Native American peoples in Brazil and twenty-three in North America. Some received passing review, others more detailed descriptions. But in both Maximilian's Brazilian and North American travels, he spent much of his time living among two particular peoples, stays that allowed him to probe deeply into each of their cultures. In Brazil it was the people called the Botocudo, with whom he spent three months; in North America it was the allied tribes of the Mandans and Hidatsas, with whom he lived for the winter of 1833–1834. In this essay, I will concentrate on the Botocudo and Mandans/Hidatsas.

Maximilian's Brazilian ethnography: The Botocudo

The Botocudo were one of the most feared of Brazilian native peoples (Langfur 2006). They were fierce warriors but, more significantly, they were believed to be bloodthirsty cannibals. This reputation instilled terror in those who ventured into their territory, but also made the Botocudo perfect targets for extermination. Thus, for example, Prince Regent Joao, newly arrived in Rio after moving the capital from Portugal due to Napoleon's invasion, used this claim to justify his 1808 declaration of war against the Botocudo. In his proclamation, he asserted that when the Botocudo killed whites or "tame Indians," "they opened wounds in their victims and drank their blood and consumed their 'sad remains'" (Langfur 2006, 273). The accusation of cannibalism allowed him to proclaim the battle against the Botocudo a "just war," which condoned their extermination or enslavement. Initially, Maximilian shared the general European apprehension about the Botocudo. Still, he desired to get to know these people and forged on into the jungles of eastern Brazil in his quest. At his first meeting with them, he was instantly repelled, finding them ugly and deformed because of the wooden plugs they wore in their lower lips and ears (Maximilian 1820–1821, I, 237). Once he overcame his original distaste, Maximilian then followed what became his normal pattern of characterizing native peoples. He always began with a description of their physiognomy, followed by how they wore their hair, how they painted their bodies, their dress, what decorations they wore, their weapons, what they ate, and the nature of their dwellings. All of these observations dealt with what he called their outer form. Thus, for the Botocudo, he noted their absolute nakedness, judged the men's bodies positively – they were muscular, slim, and well built. His characterization of the female physiognomy was less positive, calling them ugly, small, broad, and fleshy (Röder 1954, 42). He then went on to describe their hair and what ornaments the males and females wore. Since all Botocudo went naked, there was no clothing to describe. In fact, the only piece of cover was worn by the men: a sack made of dried leaves covering the penis. The Botocudo painted themselves, using red and black, sometimes painting one side of the body black and other parts red (Maximilian 1820–1821, I, 135, 352; II, 11, 50). Their huts were extremely simple.

From descriptions of external life forms, Maximilian turned to the social, religious, and emotional aspects of their cultures. Central to all of his ethnographic depictions, both in Brazil and North America, were gender relations, including topics such as marriage patterns, child rearing, domestic life, sexual relations, and gender-related work. With respect to gender relations, Maximilian observed that Botocudo men were polygamous, having up to five wives, whom the men, Maximilian claimed, treated well, but as their own property (Maximilian 1820–1821, I, 332–336; II, 1–70). Like other European observers, he believed that the women had to do most of the hard work, taking care of all household duties and childcare, and lugging all of the heavy goods while the men were basically concerned only with hunting and fighting. Marriages were formed without ceremony and easily dissolved; in fact, when a man was away, the wife could use the time to find another mate and, if she succeeded, there were no negative consequences. In all, native Botocudo women were, he declared, rather uninhibited in their sexual practices. Maximilian also discussed religion, ritual activities, warfare and hunting, leisure activities, and the Botocudo's psychological and emotional attitudes. He finished with an attempt to provide a dictionary of their language, for Maximilian believed, in true Göttingen fashion, that language offered the key to understanding other peoples' cultures and ways of life (Maximilian 1820–1821, II, 305–318). In Maximilian's desire to capture natives' mentality, to give indigenous people their own voice, he sought out reliable native informants who would allow him both to correctly learn and transcribe their language and understand their culture. In the case of the Botocudo, his major informant was a Botocudo youth, nicknamed Quäck, who Maximilian acquired as a servant and traveling companion. Quäck accompanied him on his whole sojourn and Maximilian brought him back to Neuwied when he returned, where Quäck lived the rest of his life, serving as Maximilian's personal servant.[4]

Three of Maximilian's observations of the Botocudo are crucial. First, he addressed the central issue of whether the Botocudo were cannibals. As mentioned earlier, the assumption that the Botocudo were cannibals allowed the Portuguese to wage a war of extermination against the Botocudo, a war still raging when Maximilian visited them. As Maximilian became more acquainted with the Botocudo, he concluded that they did not eat human flesh for food, a judgment that directly contradicted the official Portuguese position justifying the war against the Botocudo. But in a few cases, Quäck confessed to Maximilian, they would consume the flesh of a mighty enemy warrior after he was killed in a symbolic act of revenge. The meat of the warrior was cut off and boiled; all ate the cooked meat and then danced and sang. The enemy's head was placed on a pole and used as a target for aspiring warriors to shoot (Maximilian 1820–1821, II, 51).

A second theme of Maximilian's observations focuses on the Botocudo's religious sensibilities. Many earlier commentators, among them Felix de Azara, had claimed that all Native American peoples had no idea about religion, in essence a claim that justified their repression by European colonists. Maximilian was adamant in disputing this claim, asserting there was no single people on earth who did not have religious ideas (Maximilian 1820–1821, I, 147–147). Thus, when speaking about the Botocudo, he did the best he could to sketch out the beliefs that animated them. He observed that the Botocudo had many "strange" ideas of evil spirits or devils, both large and small, who inhabited the world. According to Maximilian, if one of the great devils appeared in one of their huts, all those who looked at him would die. If he arrived at a hut and there was no fire burning on the graves, he exhumed their dead. At times he would grab a stick and beat the dogs to death, and he also preyed on children when they were sent out to bring back water. Because of their fear of these devils, the Botocudo hardly ventured out at night alone, but preferred to travel in groups. Maximilian believed that the Botocudo considered the moon the most powerful of heavenly forces, for they explained most natural occurrences by referring to it. Thus, they believed that the moon caused thunder and lightning. Maximilian claimed that as in many other mythic traditions of native peoples in South and North America, the Botocudo believed that an earlier great flood covered the earth, or the part of the earth they inhabited. But after discussing these "fanciful ideas," Maximilian modified his Enlightenment critique by remarking that Botocudo religious ideas were not much more absurd than those of the "raw" Portuguese settlers in Brazil (Maximilian 1820–1821, II, 58–60).

A third key element of Maximilian's travel narrative is his characterization of the Botocudo psychology and emotional structure. He framed it around the two conceptual categories of "civilized" on the one hand and "raw" or "primitive natives" on the other, highly charged concepts that still plague ethnographic discourse. Maximilian judged the Botocudo to be among the "rawest" of peoples. But his valuation of the two concepts, civilized and raw, was highly ambivalent, not a one-sided praise of either. On the negative side, Maximilian characterized the Botocudo as slaves to their senses and passions. Not hampered by the restraints of *bürgerliche Ordnung*, instincts and sensations shaped their imagination. Passion, once awakened, became their overriding impulse before subsiding. On the positive side, they demonstrated love for a free, independent life, which was deeply implanted in them from their earliest youth (Maximilian 1820–1821, II, 17). Coupled with this love of freedom was their love of place, which Maximilian juxtaposed to chaotic city life. This tension between *Ordnung* and *Freiheit*, modern turmoil and simple *Vaterlandsliebe*, informed all of Maximilian's portrayals of Native American peoples. In his description of the Botocudo, he made these qualities evident in his juxtaposition between Botocudo battle scenes, where he commented upon the ferocity of the Botocudo warriors, and idyllic depictions of life in what he called a "Wilden-Republik," images of which he sketched in his travel notes.[5] In these images one sees naked youths of both sexes frolicking and swimming, climbing trees to gather fruit, and hunting for fish and food. He remarked that in these scenes one could observe with delight the Botocudo's lively activity (Maximilian 1820–1821, I, 364–365).

Maximilian employed a stadial description to determine what differentiated a raw from a civilized people, where the stages ranged from hunter-gatherer to more advanced and complex levels of communal life. But, despite Maximilian's use of stadial theory to locate Native Americans within the parameters of raw and civilized, he was not a proponent of the idea that it was the task of Euro-Americans to "civilize" the Native Americans, to take them under their care and make them more like Euro-Americans. Rather, Maximilian hoped that some sort of agreement could be reached in which Native Americans were allowed space and time to evolve from hunters and gatherers to land-tilling inhabitants (which Maximilian equated with being civilized) through their own efforts within their own cultural contexts. Maximilian asserted, following Blumenbach's lead, that the indigenous peoples of Brazil possessed intelligence, wit, and the ability to master many tasks (Maximilian 1820–1821, II, 15). Left to their own devices they had the agency to develop from hunter-gatherers into cultivating societies. He did believe that this shift could be helped through "enlightened" exchange and trade between indigenous peoples and white colonists, through what he described as "friendly contact" (Maximilian 1820–1821, II, 52). He saw hints of this in his Brazilian travels and even more in North America. But these instances were countered by the actions of most whites, who sought to "civilize" Native Americans. In his opinion, whenever Euro-Americans interfered with native cultures, they did so out of their own self-interest, with very little concern about the natives they sought to control and "civilize."

North American ethnography: The Mandans/Hidatsas

Maximilian's South American journey convinced him of the correctness of Blumenbach's idea that all North and South American native peoples were part of the American race or variety, which led him to search for the similarities between native North Americans and the Botocudo and to substantiate that they were part of one larger people, though very different in their individual cultures. His sojourn with the Mandans and Hidatsas reinforced this view. He thought he saw the Botocudo in the Mandans and vice versa. Maximilian stayed at Fort Clark (in present-day North Dakota, near the city of Bismarck) in the winter of 1833–1834, adjacent to their major village, Mih Tutta-Hangkusch. Both the Mandans and Hidatsas had a mixed economy. They hunted bison and other animals but also grew crops. Their villages were fixed, though they did move from one location in the summer to another in the winter.

As in all of Maximilian's ethnographic work, he followed the descriptive patterns he employed for the Botocudo. The physiognomic renditions by the artist Carl Bodmer who accompanied him are magnificent, the product of Bodmer's skill and Maximilian's scholarly input. The illustrations were meant to deepen the type of *anschauende Erkenntniss* (immediate knowledge) Maximilian sought to convey of the land and the peoples who inhabited it. Maximilian investigated many topics, but two key themes were his description of Mandan gender relations and his portrayal of the intersections between religion, ritual, social organization, and dance.

In many ways, Maximilian's discussion of Mandan gender relations reveals his underlying European assumptions about how women should be treated. He understood the relationship between Native American men and women to be extremely unequal, where the women did most of the work while the men sat around smoking and talking until they engaged in their major activities, hunting, stealing horses, and fighting battles with their enemies. In fact, unlike European women, Maximilian claimed, Native American women received very little compensation for their hard work, not even a pretty piece of clothing; it was the men who dressed beautifully. To Maximilian, the men appeared extremely vain and he observed they carried little mirrors with them to make sure they looked as best they could. But despite interpreting Mandan society as male-dominated, he also commented upon the power accorded the sisters of the male warriors. Thus, the horses a young man stole (stealing horses was a sign of great skill and was honored) or won in battle belonged to his sister (Maximilian 1839–1841, II, 130). Maximilian also remarked on another display of feminine power within the household – namely, that the mother-in-law of a young warrior would refuse to speak to him until he brought her the scalp of an enemy he killed, along with the enemy's rifle (Maximilian 1839–1841, II, 132). Only then was he accepted into the family.

According to Maximilian, marriage among the Mandans was an arranged affair. When a young man wished to marry a young woman and she agreed, the prospective groom sought the permission of the woman's father. If the father agreed, then an elaborate ritual of mutual gift giving was enacted that led to the marriage. The young man brought two to eight horses to the prospective bride's dwelling and tied them up at the entrance. The woman gave the horses to her father. Then a reciprocal action took place. The father presented an equal number of horses to his prospective son-in-law's father. The important part of this gift-giving ritual was that the same number of horses exchanged hands. There were no winners or losers in this transaction. From that moment another ritual was enacted. The young woman cooked maize and brought a bowl to the young man's house every day. After a while, the young man was allowed to enter the woman's dwelling and slept with her and the marriage was completed. The young couple either moved in with the father-in-law, who ruled the whole household, or they built their own house. If the couple later separated, the father-in-law played the central role in directing the fate of their children.

Maximilian was also fascinated by the sexual practices of the people he observed and sought to understand. Unlike many of his contemporaries, he showed little surprise about practices that were much more open than was normal in Euro-American society. He remarked that "Sprödigkeit" (prudery) was not one of the characteristics of Mandan women (Maximilian 1839–1841, II, 130). Many had two, three, or more lovers. If the woman was married and discovered to be unfaithful she was not punished, as was the case in other tribes, such as the Blackfeet, who cut off the women's noses in retribution. Rather, the husband whose wife slept with another man would demand compensation from the man who seduced her and the seducer had to comply. But, Maximilian noted, the wife was never taken back by the aggrieved husband. If the wife was the oldest daughter, the husband also had a right to her sisters. According to Maximilian, the men were even more promiscuous. One of the major activities of young men was to seduce as many women as they could; they roamed the village for targets and when successful displayed the sign of their conquests on a colored staff (Maximilian 1839–1841, II, 131).

Even more fascinating, at least from our vantage point, was the non-condemnatory manner in which Maximilian dealt with what we would call transgender males in Mandan and Native American society. He described a group of men called *Bardaches*, whom he called "Mannweiber." They were men who dressed as women and performed all of the women's duties, including sexual ones. The men officially treated them as women, often lived with them as such and, according to one of Maximilian's French-Canadian sources, many preferred them to women. Maximilian reports that the *Bardaches* were guided by a dream or by a higher calling, which made them impervious to entreaties to give up their chosen path. He recounts the story of how fathers had tried to dissuade their offspring *Bardaches* by giving them masculine robes and weapons and, when that failed, tried to beat them into submission. Nothing worked, Maximilian remarked. The general acceptance of the *Bardaches* was sealed by a mythic story in which a warrior tried to physically force a *Bardache* to give up his ways. A furious fight ensued in which the warrior shot and killed the *Bardache* with his bow. But instead of finding a corpse, one discovered a pile of stones from which the deadly arrows were protruding. This myth ensured the *Bardaches*' right to live as they chose, for the Mandans believed that the *Bardaches* were protected by a higher power. And though Maximilian, in true Enlightenment form, usually judged Native American myths as superstitions, he refrained from mocking this story (Maximilian 1839–1841, II, 132–133).

This myth is closely linked to a second of Maximilian's themes: the intersection between religion, ritual, social organization, and dance. As with the Botocudo, many contemporary writers claimed that Native North American peoples had no sense of religion, but Maximilian instead argued that religious beliefs were fundamental in structuring the social and emotional life of the groups he described. True, he considered many of the Mandans' beliefs superstitious (and here his European prejudices were clearly evident), but he nonetheless presented a very comprehensive account of these beliefs and myths, for he was convinced that they shaped the peoples' way of life (Maximilian 1839–1841, II, 149). Maximilian asserted, for example, that the Mandan world was infused with an acute awareness of the existence of spiritual forces that shaped their life and whom they could call on for help, forgiveness, and good fortune or whom they had to avoid because of the evil they could do (Maximilian 1839–1841, II, 148). For these reasons, they erected sacred places where they made offerings to powerful spirits, such as the Lord of Life, the original creator who lived in the sun, the First Man, an almost equally strong spirit, an evil spirit who also influenced the people, a powerful older woman whose sphere was the moon, and many other spirits, which they called medicines. Thus, for example, certain animals, such as the eagle, owl, bison, and beaver, were considered powerful medicines, as were certain sacred places where they went to worship. Dreams, often preceded by long fasting and painful penitence, directed them in how they would worship and to whom they should turn; they also served as prophecies. In this process of imploring and placating the myriad spirits inhabiting nature, certain medicine men played crucial roles in mediating between the humans and the spirits.

Maximilian was also fascinated by their rituals, which were organized within this corpus of belief. Each ritual was designed to serve a unique role in the people's lives and, as a rule, was celebrated by special dances, chants, and practices. The dances were very structured, with rules about who could dance, what the dress for the dancers should be, and how long the ceremonies should last. There were dances in which only men participated and ones in which only women danced. There were many such diverse dances and festivals, among them the scalp dance, the buffalo dance, the corn dance, and the penitence ritual. Finally, there were dances associated with the way in which society was structured.

Mandan society, Maximilian observed, was organized around a number of male and female associations; he listed four for the women and six for the men. These associations, most of which regulated life in the community, were based on age and also achievement, each marked by a specific dress and having their own dances and chants. Members of one association "bought" their way into a higher one from a member who was leaving. The newcomers had to offer the leaver gifts. As with Maximilian's views on gender relations, he was adamant in observing that gift giving played a central role in all the Mandans' societal interactions. Thus, aspiring males had not only to give the departing member horses and precious articles but also to make his wife or wives available to him for a month. If the aspiring candidate did not have a wife, he had to ask a relative or friend to lend him one for the period in question. The prospective member also had to undergo certain exacting initiation ceremonies, such as sacrificing a finger joint to the Lord of Life. In all of these ritual enactments, usually designed either to insure the return of the bison, the flowering of the grain, and the populations' fertility or to celebrate victories or a successful bison hunt, smoking the peace pipe marked its symbolic beginning and end.

During Maximilian's stay with the Mandans, he assiduously engaged in his language studies and ethnographic inquiries. He formed close attachments with some of the elders of the tribe, the keepers of sacred information and knowledge of the tribe's past. The most important was the elder Dipauch (Broken Arm), one of the most respected of Mandan chiefs. Maximilian and he along with the translator James Kipp would spend hours together talking about Mandan traditions. They were assisted by the warrior chief Mató-Tópe (whom Bodmer twice portrayed), the Hidatsa elder Addih-Hiddisch, and other elders of both tribes. These elders were not just "native informants" in the modern sense, as they developed close friendships with Maximilian and Bodmer. Due to this intense cultural interchange, Maximilian was able to present an extremely vivid portrayal of the Mandans' myths and beliefs and acquire a large sampling of Mandan vocabulary. Maximilian's intention was to compose a Mandan grammar, but he failed to do so because he came down with a serious, incapacitating disease, which most people thought would kill him quickly. He was fortunate that the fort's black cook recognized the disease as scurvy and knew of a native cure – namely, the eating of wild onion leaves, which slowly restored Maximilian to health (Maximilian 1839–1841, II, 317–318).

Maximilian and Bodmer had gloried in their Mandan surroundings, despite enduring bitter cold, searing heat, hardship, hunger, and – in Maximilian's case – life-threatening disease. Their return through the upper northern states of the United States was a depressing reintroduction into the bustling world of Anglo-American culture. The remnants of the native populations Maximilian encountered along the way only made him more disconsolate when he saw how much of their culture they had lost while gaining little in return. But later, as he was putting the finishing touches on his travel narrative in Neuwied, that loss was made even more visceral. He ended the account with a plaintive *Schlussnachricht* (afterword): an infectious disease had carried away virtually all of the Mandans and Hidatsas (Maximilian 1839–1841, II, 687). In a letter to Maximilian, the American artist and part-time explorer George Catlin described the horrible scene of the spread of the disease and the tragic death of Mató-Tópe, who, after watching his fellow tribesmen and his own family perish, starved himself to death (Roth 1995, 23–24).

Conclusion: Blumenbachian ethnography in action

In all of Maximilian's discussions of other peoples, he like Blumenbach never used skin color as a defining element in characterizing them, though he commented on it, often praising the beautiful skin color of the indigenous peoples he observed. When he discussed the enslaved Africans he encountered, he repeated Blumenbach's defense of their intellectual capabilities and despaired at how Euro-Americans treated them. But his ethnographic goal was to understand Blumenbach's "American type" in its natural surroundings. In this undertaking, he never demeaned the indigenous peoples he met, praising their skills, mental quickness, oral traditions, and in the North American case, their beautiful artifacts. When he criticized these peoples, the critiques were founded upon what level of stadial development Maximilian thought they had reached. But these critiques were open-ended, both allowing for the possibility of further natural development and recognizing admirable traits that more "civilized" societies had lost. Throughout his undertakings, he kept in frequent contact with Blumenbach. He sent artifacts and "American" skulls to Blumenbach from both Brazil and North America, adding to what Blumenbach called his "Golgatha."[6] Blumenbach read over the drafts of Maximilian's travel narratives, adding comparisons and corrections to his ethnographic observations, all of which Maximilian incorporated in his texts.[7] In short, Maximilian's expeditions reveal his commitment to a scientific vision shaped by Blumenbach's vitalism, Blumenbach's belief in the mental equality of all peoples, his contention that there was a distinct American variety, and his argument that there is a vital interchange between land and peoples, which helps shape their cultures. Joined to these assumptions was the imperative to go out into the field, observe these interconnections as closely as possible, and then present them in as lively and comprehensive a manner as possible.

Despite Maximilian's attempts to celebrate indigenous peoples' individual humanity in his Blumenbach-inspired, anti-imperial travel narratives a great deal of what we know of these peoples, including the Botocudo and the Plains Indians, is drawn from Wied's narratives and, for North America, from Bodmer's plates. They show us how the world once was and how, perhaps, it could have developed differently, driven by its own internal dynamics, were it not for European and American colonial imperialism.

Notes

1 Maximilian is often seen in the few interpretations of him that exist to have been a follower and therefore a student of Alexander von Humboldt. There is no doubt that Maximilian greatly admired Humboldt, especially his *Ansichten der Natur* and later his travel accounts of South America. But Maximilian met Humboldt only once or twice in Paris before he began his first voyage. In fact, he was much closer to Cuvier and to Bonpland, Humboldt's travelling companion, than to Humboldt. The letter exchanges between Maximilian and Humboldt were minimal. What united them was the vitalist vision they adopted from their Göttingen mentors, especially Blumenbach, who, I believe, played a much greater role in shaping Maximilian's research tradition than did Humboldt.

2 There is no space in such a short essay to summarize this tradition, but it had long roots leading back to Albrecht von Haller, intensified by Johann David Michaelis, who pioneered language research of the Near East and was the driving force for the so-called Niebuhr expedition, to August Ludwig Schlözer, and to the linguistic research of Johann Christian Adelung, Johann Severin Vater, and the young Wilhelm von Humboldt. These and other influences were apparent in Alexander von Humboldt's pioneering efforts in his vast research program in South America. He, too, was a Blumenbach student, as was his brother Wilhelm.

3 The central role Blumenbach played in determining Maximilian's choice to travel to Brazil is made clear in the correspondence between Maximilian and his friend and former tutor Christian Friedrich Hoffmann. Hoffmann tried to dissuade Maximilian from following Blumenbach's advice that Maximilian go to Brazil and argued that he should travel to the southern parts of the United States. Hoffmann to Maximilian, January 1, 1812. *Brasilien-Bibliothek der Robert Bosch GmbH, Stuttgart.* The Bosch library has an extensive collection of Maximilian materials and also over fifty items from the correspondence between Blumenbach and Maximilian, including two important *Beilagen.* In the first Blumenbach outlines what Maximilian should look out for on his travels (April 4, 1815). The second outlines Blumenbach's corrections and expansions of Maximillian's first draft of his travel journal (March 30, 1818). Maximilian integrated all of them into his published narrative. Maximilian wrote Blumenbach frequently and sent him many natural historical specimens, including skulls from Brazil and North America.

4 It would be fascinating to know more about Quäck and his stay in Neuwied. From the few brief discussions of him, it appears that Maximilian first showed him off to the gaping Neuwieders, demonstrating his skills with the longbow. It appears obvious that Quäck was not very happy in the cold winter climes of Germany. The few reports of his death say he became an alcoholic and fell while drunk from the first story of the castle in the snow, caught pneumonia, and died. His death occurred when Maximilian was already on his second expedition. But there must have been some close feelings between them, at least from Maximilian's side. He had three separate paintings of Quäck done and upon hearing of his death Maximilian wrote, "Sadly my poor, good Quäck has died when I was away. My brother Karl luckily painted a very fine portrait of him in oil. The memories of him remain very vivid" (quoted by Roth 1995, 19). Quäck was buried in Neuwied as a citizen of the town; his skull was sent to the University of Bonn for its collection (Roth 2012, 34).

5 The publication of many of Maximilian's sketches can be found in Röder and Trimborn (1954). They are drawn from Maximilian's notebooks, which are in the *Brasilien-Bibliothek der Robert Bosch GmbH, Stuttgart.*

6 *Brasilien-Bibliothek der Robert Bosch GmbH, Stuttgart.* Blumenbach letters to Maximilian, November 9, 1818, and November 28, 1834. The reference to Blumenbach's Golgatha is in his letter to Maximilian, April 27, 1835.

7 *Brasilien-Bibliothek der Robert Bosch GmbH, Stuttgart. Beilage* to March 30, 1818.

Bibliography

Ackerknecht, Erwin H. 1955. "George Forster, Alexander von Humboldt and Ethnology." *Isis* 46: 83–95.

Blumenbach, Johann Friedrich. 1781. *De generis humani varietate liber: cum figuris aeri incisis*. Editio altera longe auctior et emendatior. Goettingae: apud viduam Abr. Vandenhoek.

———. 1790. *Beyträge zur Naturgeschichte*. Vol. 1. Göttingen: Johann Christian Dieterich.

———. 1795. *Elements of Physiology*. Trans. Charles Caldwell. 2 vols. Philadelphia: Thomas Dobson.

———. 1796. *Abbildungen naturhistorische Gegenstände*. 1. Heft. Göttingen: Johann Christian Dieterich.

———. 1799. *Handbuch der Naturgeschichte*, 6th edn. Göttingen: Johann Christian Dieterich.

———. 1865. *The Anthropological Treatises of Johann Friedrich Blumenbach: With Memoirs of Him by Marx and Flourens*. Trans. and edited by Thomas Bendyshe. London: Longman, Green, Longman, Roberts and Green.

Brasilien-Bibliothek der Robert Bosch GmbH, Stuttgart. Nachlass Maximilian zu Wied-Neuwied.

Burgdorf, Wolfgang. 2006. *Ein Weltbild verliert seine Welt. Der Untergang des Alten Reiches und die Generation 1806*. Munich: Oldenbourg Verlag.

Evans, Robert J.W., Michael Schaich, and Peter H. Wilson, eds. 2011. *The Holy Roman Empire 1495–1806*. Oxford: Oxford University Press.

Evans, Robert J.W., and Peter H. Wilson, eds. 2012. *The Holy Roman Empire, 1495–1806: A European Perspective*. Leiden: Brill.

Geuns, Steven Jan van. 2007. *Tagebuch einer Reise mit Alexander von Humboldt durch Hessen, die Pfalz, längs des Rheins und durch Westfalen im Herbst 1789*, edited by Bernd Kölbel and Lucie Terken. Berlin: Akademie Verlag.

Hartmann, Peter-Claus, and Florian Schuller, eds. 2006. *Das Heilige Römische Reich und sein Ende 1806: Zäsur in der deutschen und europäischen Geschichte*. Regensburg: Verlag Friedrich Pustet.

Langfur, Hal. 2006. *The Forbidden Lands: Colonial Identity, Frontier Violence and the Persistence of Brazil's Eastern Indians, 1750–1830*. Stanford, CA: Stanford University Press.

Lichtenberg, Georg Christoph. 1972. "Über Physiognomik; wider die Physiognomen. Zu Beförderung der Menschenliebe und Menschenerkenntnis." In *Georg Christoph Lichtenberg. Schriften und Briefen*. Vol. 3, 256–296. Munich: Carl Hanser.

Maximilian, Prinz zu Wied-Neuwied. 1820–1821. *Reise nach Brasilien in den Jahren 1815 bis 1817*. 2 vols. Frankfurt am Mian: Heinrich Ludwig Brönner.

———. 1839–1841. *Reise in das innere Nord America in den Jahren 1832 bis 1834*. 3 vols. Coblenz: Hoelscher.

———. 1843. *Travels in the Interior of North America*. Trans. H. Evans. London: Ackermann & Co.

———. 2008–2009. *The North American Journals of Prince Maximilian of Wied*, edited by Stephen S. Witte and Marsha V. Gallager. Trans. William J. Orr, Paul Schach and Dieter Karch. 3 vols. Norman, OK: University of Oklahoma Press.

Muthu, Sankar. 2003. *Enlightenment Against Empire*. Princeton, NJ: Princeton University Press.

Reill, Peter Hanns. 2005. *Vitalizing Nature in the Enlightenment*. Berkeley: University of California Press.

Röder, Hermann Joseph, and Hermann Trimborn, eds. 1954. *Maximilian Prinz zu Wied: Unveröffentliche Bilder und Handschriften zur Völkerkunde Brasiliens*. Bonn: F. Dümmler.

Roth, Hermann Josef. 1995. *Westerwald und Amerika: Prinz Maximilian zu Wied: Jäger, Forscher, Reisender. Leben und Werk*. Montabaur: Verlag der Museen des Westerwaldkreises.

———. 2012. "Mit Buch und Büchse durch die Welt: Maximilian Prinz zu Wied." In *Übergabe der Sammlung Hans Engländer zur Geschichte der Ornithologie an die Universitäts- und Stadtbibliothek Köln am 18*. April 2012, edited by Wolfgang Schmitz, 19–56. Cologne: Universitäts Bibliothek. (Kleine Schriften der Universitäts- und Stadtbibliothek Köln 35).

Schnettger, Matthias, ed. 2002. *Imperium Romanum – irregulare corpus – Teutscher Reichs-Staat: Das Alte Reich im Verständnis der Zeitgenossen und der Historiographie*. Mainz: Philip von Zabern.

Stollberg-Rilinger, Barbara. 2008. *Des Kaisers alte Kleider: Verfassungsgeschichte und Symbolsprache des Alten Reiches*. Munich: C. H. Beck.

Whaley, Joachim. 2012. *Germany and the Holy Roman Empire. Vol II: The Peace of Westphalia to the Dissolution of the Reich, 1648–1806*. Oxford: Oxford University Press.

Wilson, Peter H. 2016. *Heart of Europe: A History of the Holy Roman Empire*. Cambridge, MA: Harvard University Press.

Vermeulen, Han F. 2015. *Before Boas: The Genesis of Ethnography and Ethnology in the German Enlightenment*. Lincoln, NE: University of Nebraska Press.

11 The rise of paleontology and the historicization of nature

Blumenbach and Deluc

John H. Zammito

When we reconstruct the conceptual horizons of the emergent earth and life sciences over the eighteenth century, as these sciences worked through new discoveries and longer-term debates, we can trace a "historicization of nature" across the whole epoch, gaining impetus from mid-century onward to culminate in the outright affirmations of "transformism" of life forms by figures like Jean-Baptiste de Lamarck around 1800.[1] In shifting from traditional natural history (German: *Naturbeschreibung*) to a new and provocatively literal *history* of nature (German: *Naturgeschichte*), Georges Leclerc de Buffon proved of pivotal importance (Lyon and Sloan 1981; Schneider 2001; Lepenies 1976; 1978). In a paper, I have demonstrated the extraordinary attentiveness to this whole matter in the work of Immanuel Kant (Zammito, forthcoming a; see also Fritscher 1992). Here, I want to explore the place of Johann Friedrich Blumenbach in this historical development.

There has been some question regarding whether Blumenbach's notion of natural history can really be assimilated to the stronger sense of a *history* of nature. While in an older paper, from 1963, Walter Baron argued that Blumenbach was central in this late eighteenth-century establishment of the "historicity of nature," Frank William Peter Dougherty twenty years thereafter suggested that we should be much more careful about the "historical model" in Blumenbach and guard against anachronistically reading back into it our own notions of "historicity" (Baron 1963; Dougherty 1986). Still more recently, Robert Bernasconi has written that "Blumenbach seems never to have embraced natural history in Kant's sense" (Bernasconi 2014, 245). I propose to defend a position closer to that of Baron.

Two impulses animate the strategy I will pursue. The first is to exploit the widely received notion of a "Göttingen School" in science developed some time ago by Timothy Lenoir (Lenoir 1978; 1980; 1981; 1989). As I have indicated elsewhere, I have many reservations about Lenoir's particular construction which features the affiliation of Blumenbach and his students with the thought of Kant (Zammito 2012). But here I note that Blumenbach's most famous students – figures like Alexander von Humboldt, Heinrich Friedrich Link, Gottfried Reinhold Treviranus, and Carl Friedrich Kielmeyer – manifested a more or less explicit "transformism." Blumenbach must have been a significant stimulus for their having done so. More specifically, I contend that it was the *turn to paleontology* in

Blumenbach's natural history over the course of the 1780s and 1790s that instigated their eventual research programs. Very revealing, here, is that the career paths of some of them led from study with Blumenbach to study with Abraham Gottlob Werner at the Freiberg School of Mines, as in the key cases of Humboldt and Ernst von Schlotheim.[2]

The second and defining strategic impulse is to connect Blumenbach's ideas on paleontology with the Genevan geologist Jean-André Deluc.[3] I suggest that, among many other naturalists in the field, it was the specific hypotheses of Deluc which stimulated the turn toward paleontology in Blumenbach's work. The triggering stimulus may well have been Blumenbach's reading and published review of Deluc's *Lettres physiques et morales sur l'histoire de la terre et de l'homme: addressées à la Reine de la Grande-Bretagne* (Deluc 1779/1780), which appeared in 1779 – or 1780, as Deluc reported the actual publication of his work. Blumenbach reviewed that work in *Göttingische Anzeigen von gelehrten Sachen* [*GAgS*] in 1780 (Blumenbach 1780b). This was the same moment in which Blumenbach was reading Eberhard August Wilhelm Zimmermann and the travel reports concerning the South Seas by the Forsters, son and father, as well as the studies of Petrus Camper on the anatomy of the orangutan (Zimmermann 1778–1783; Georg Forster 1777; Johann R. Forster 1778; Camper 1779; 1782; see also Soemmerring 1781). All these together constituted the decisive elements out of which Blumenbach forged his notion of natural history as set forth in the many editions of his *Handbuch*, starting in 1779. That work, in combination with his personal teaching, decisively shaped a whole generation of German naturalists.

A key connection, here, was with the emergent field of "biogeography" (see Ebach 2015). Field study, often in remoter regions, had been central in the collection of fossils and in recent disputes about the meaning of vertebrate remains, most prominently in Siberia and Ohio.[4] The controversies among naturalists concerning these problematic vertebrate fossils became a salient interest for Blumenbach as he concentrated his natural history on questions of zoology.[5] Zimmermann had made biogeography central to his major publication at the close of the 1770s, widening the question from the distribution of humankind into a more broadly zoological inquiry (Zimmermann 1778–1783; see Feuerstein-Herz 2004). This had substantial impact on the field and on Blumenbach specifically.[6] It would become a major concern of Blumenbach's key student, Treviranus (Treviranus 1802–1822, II [1803]; see Wingler 1958). The connection of what we would call ecological systems and biodiversity with a paleontological concern lay precisely in the increasing awareness that alterations in the geological environment must have had decisive impact on the distribution and survival of life forms. This informed projects with disparate theoretical orientations, most prominently that of Buffon in his 1778 masterwork, *Époques de la nature*, whose theory of diminishing earth heat was aimed to explain the presence of animal fossils in northern latitudes where they were no longer viable (Buffon 1778). Buffon was a fundamental force in the development of eighteenth-century natural history, crucial for Blumenbach and for Deluc as well. Blumenbach's sense for the development of geological thought was probably shaped by Buffon's evaluation of the literature, *Histoire et théorie de la terre*, in the first volume of the *Histoire naturelle* (Buffon 1749b), and then by Deluc's extensive critical survey of the literature – including not only Buffon's work of 1749 but also the new work of 1778 (Deluc 1779/1780).

To be sure, Blumenbach had been interested in "fossils" since his student days in Jena, where he had heard the lectures of Johann Ernst Immanuel Walch, who managed a famous collection of "Versteinerungen" (Dougherty and Klatt 2006–2015, I, 28; Walch 1768–1774; see Häner 2017, 139–146).[7] But "fossil" in the German context of the third quarter of the eighteenth century remained an ambiguous notion, going back to Agricola's sense of things (of all sorts) dug out of the earth.[8] The pioneer publication of the great theorist of geognosy, Werner, while its title mentioned "fossils," really treated mineralogy and crystals, with virtually no attention to specifically organic remains (Werner 1774). To be sure, fossil finds atop European mountains had become a fascination since Nicolas Steno's *De solido intra solidum naturaliter contento dissertationis prodromus* [Preliminary discourse to a dissertation on a solid body naturally contained within a solid] in 1669, but the rise of interest in "geognosy" in late eighteenth-century Germany had to do with mining and metals, and hence with the (primary) rock formations that carried them. Organic fossils, while clearly discerned and categorized with "secondary" mountain formations (or *Flötzgebirge*), remained theoretically marginal. It was Deluc, instead, who offered Blumenbach a general theory into which to situate his own issues of natural history and comparative anatomy and thus brought him to the cutting edge of the field of paleontology. Once attuned, Blumenbach recognized that paleontological evidence lay all about him, indeed, as he famously put it, in every paving stone in Göttingen.[9] But my point is not simply that Blumenbach got *interested* in paleontology. It is rather that he proved a *pioneering influence* in the field.

It is not surprising at all that in the early 1790s Deluc should have entered into important public dispute with Jean-Claude de Lamétherie or James Hutton, prominent figures in geology with whose views Deluc was in active contestation (Deluc 1790–1791). But why would Deluc in 1793 have chosen to send the remarkable series of letters summarizing his theory of geological development to Blumenbach (Deluc 1793–1796)? It is amusing to note, in the classic text of Gillispie, the following telltale phrasing; "Deluc extended his ideas further in six letters, *addressed for some reason* to Professor J. F. Blumenbach" (Gillispie 1996, 57, my italics; see Rupke 1994).[10] We know that, when Blumenbach visited Deluc in Windsor in 1791 for their first extensive conversations, he invited Deluc to provide him with a revised (more concise) formulation of the ideas he had developed in his other publications. Why did Deluc take up this invitation? My suggestion is that this was not simply to honor an acquaintance going back to 1776, or even because Deluc was interested in propagating his ideas in a German-language context. Rather, I submit, it was because Deluc recognized Blumenbach, alongside de Lamétherie and Hutton, as a significant interlocutor in the field of the history of nature. Blumenbach's development of paleontological considerations across the various editions of his *Handbuch der Naturgeschichte* after 1779, and then in the separate and crucial publication of his *Beyträge zur Naturgeschichte* (first edition, 1790), needs to be reckoned historically as an important intervention in the historicization of nature, as it was by his contemporary, Deluc (Blumenbach 1779; 1790a; 1790b).

The leading historian of this epoch in geology, Martin Rudwick, recognizes the prominence of Blumenbach in the European conversation about the historicization of nature in earth and life sciences by 1800: "Perhaps no other naturalist anywhere in Europe was in such a favorable position for turning a traditionally static 'natural history' into a dynamic *history* of nature" (Rudwick 2005, 425). For Rudwick, the publication by François-Xavier Burtin of *Réponse à la question physique, proposé par la Société de Teyler, sur les révolutions générales qu'a subies la surface de la terre, et sur l'ancienneté de notre globe* (Burtin 1789) served as a pivotal moment in the rise of "geohistory," especially in its paleontological connection.[11] Rudwick notes, "Whether or not he read Burtin's essay, Blumenbach at Göttingen was developing rather similar ideas about nature's revolutions" (Rudwick 2005, 297). In his lecture at the fiftieth anniversary of the Göttingen Royal Academy of Sciences in 1801 (Blumenbach 1801; 1803), "Blumenbach programmatically linked the museum science of fossils with the field science of geognosy in order to reconstruct the history of life in retrospective order from the present back into the deepest past" (Rudwick 2005, 426). The upshot for Rudwick is that Blumenbach prominently articulated "how a new and more rigorous 'natural history' of fossils might yield a reconstruction of the '*history* of nature', even back into the deepest past before the human species existed" (ibid., 427). That is, Blumenbach "sketched programmatically [. . .] that the earth's 'archaeology', a true geohistory, could be traced from the sequence of fossils found in its successive formations" (ibid., 647).[12] Thus, "given his growing reputation as a naturalist, Blumenbach's work was a significant example for the new emphasis being given to fossils as evidence for geohistory" (ibid., 300).

While this may have been new – "programmatic" in Rudwick's sense – for a wider European impetus toward "geohistory" after 1800, it was not new in Blumenbach's agenda, for he was – appropriately enough in an anniversary address – celebrating a project to which he had long been devoted as scholar and teacher – and museum director – in Göttingen (Blumenbach 1803). What Rudwick takes as an inaugural program statement in 1801, I suggest, was the capstone formulation of a long-standing theoretical and pedagogical focus dating back to the late 1770s and early 1780s. Blumenbach's increasing interest was clearly evidenced in revisions of his *Handbuch der Naturgeschichte* (and also quite plausibly in his *teaching* over these years), thus motivating a new generation of naturalists – above all, his own students – to become concerned with paleontology as a key to natural history and to the emergent life sciences more generally (Rupke 1983). He had already been doing this for two decades by the time he gave the anniversary lecture at the Göttingen Academy in 1801 that Rudwick features in his reconstruction.

While Rudwick recognizes the significance of Blumenbach, he also *restricts* his achievement to having suggested a program of research that he could not himself undertake.

> Blumenbach's lecture on the "archeology" of the earth was therefore most important as an inspiring programmatic sketch [. . .]. Given his prominent position among European naturalists, the geohistorical perspective [. . .] was now set to take a more prominent place in the practice of natural history. However, Blumenbach himself could hardly hope to fulfill the promise implicit in his lecture. As a busy teacher and leading member of his university, he did not have the time for extensive original research [. . .] and anyway his main efforts in research were increasingly focused on [. . .] the origin and diversity of the human species itself.
>
> (Rudwick 2005, 428)

Continuing in this deflationary vein, Rudwick observes that "taken literally, [. . .] Blumenbach's [notion] was too defeatist or at least too modest" (ibid., 427), especially in comparison with the concurrent work of Georges Cuvier, the "pivotal figure" in Rudwick's own narrative.[13] As he puts it, "it would need more than Blumenbach's sketchy speculations to establish convincingly the reality of faunal change, let alone his conception of the revolution itself as a *total* replacement of species" (ibid., 300). He explains the limitations of Blumenbach's intervention in terms of his institutional situation at the University of Göttingen, where his teaching and administrative responsibilities (notably, supervising the natural history collection) consumed the lion's share of his time. Rudwick also believes that Blumenbach's research concentrated increasingly on physical anthropology to the exclusion of the wider issues of geology and paleontology. But one could offer a different reconstruction of this situation. Without wishing to claim more for Blumenbach than is warranted, it remains to examine more carefully not only *when* he came to his key ideas but also *what* his notion of "faunal change" actually was, and what he meant by the notion of "total revolution" – which Rudwick certainly takes as his most important claim in the field.

Rudwick correctly notes that "the successive editions of [Blumenbach's] highly esteemed textbook on natural history had given progressively greater attention to fossils" (2005, 425). He highlights two shifts in Blumenbach's conceptions. The first he sets around 1790: "What he had previously treated as a subordinate branch of 'mineralogy' now became crucial evidence for the past *history* of the earth and of life, provided that fossils were treated like historical documents" (ibid., 297). According to Rudwick, in 1788 Blumenbach made a crucial discrimination between "unknown" (i.e., no longer extant) and "known" (currently observable) life forms in the fossil record, highlighting the problem of continuity and discontinuity of life forms – that is, extinction (ibid., 297). For Rudwick, the decisive change came in Blumenbach's 1799 edition of the *Handbuch*, for there he "began to modify his strongly geohistorical but still conventionally binary conception into a more differentiated history of life on earth" by inserting between his long-established categories of "unknown" and "known" life forms a third category of "uncertain" (*zweifelhaft*) (ibid., 426). Blumenbach was careful about the issue of extinction, quite aware that there were prominent naturalists, from Gottfried Wilhelm Leibniz to Jean-Guillaume Bruguière, who insisted that there were still vast regions of the ocean deep as well as the interiors of continents where ostensibly "extinct" creatures might still survive. But Blumenbach had long since committed himself to the extinction thesis; what remained uncertain – *zweifelhaft* – were only *specific* remains at particular sites, and their historical sequencing. We need to press more concretely into the emergence of Blumenbach's position, documented across the editions of his *Handbuch* and his other writings, seeking after the *sources* – both interpretive and evidential – that instigated it. There will be many revisions that need to be proposed to Rudwick's reconstruction. The place to start, I suggest, is with Jean-André Deluc.

Deluc, the "Protestant Enlightenment," and historical geology

Deluc grew up in the context of the rich scientific culture that had arisen in Geneva over the course of the eighteenth century, with its roots not only in a progressive Calvinism but also in experimental Newtonianism.[14] These Genevan savants had a European network of associations that kept them at the cutting edge of scientific and philosophical developments (Trembley 1987; Montandon 1975). Charles Bonnet is a notable instance; Deluc's colleague and rival, Horace Bénédict de Saussure, is another. The education Deluc received in Geneva made him part of this self-conscious intellectual elite of naturalists. Jean-André and younger brother Guillaume-François studied at the Genevan Academy under Gabriel Cramer and Jean Jallabert. Even after Jean-André left the Academy in 1744, he remained in close contact with Jallabert, with whom he had studied physics. He and his brother devoted themselves to experiments in electricity, which Jallabert communicated to the leading European experimental scientist of electricity, the Parisian academician Jean-Antoine Nollet (Hübner 2010, 44). Even more important for Jean-André were private lessons with the mechanistic physicist Georges Louis Le Sage (ibid., 23, 45).[15] Le Sage would prove a major figure in the debate over physics and force by the close of the century, especially with his significantly entitled work, *Lucrèce Newtonien* (Le Sage 1784). The object of Le Sage's work, as the title implies, was not only a strictly mechanistic physics but also an ideological repudiation of materialism. That is the essential vantage from which to understand Deluc.

Geneva was the center of a "Protestant Enlightenment," explicitly concerned to rebut the all-too-radical Enlightenment of the Parisian *philosophes* associated with the *Encyclopédie* of Denis Diderot and D'Alembert. The Genevan Bonnet orchestrated his association with Albrecht von Haller into this anti-*philosophe* crusade, especially against Buffon as "Epicurean" (Hoquet 2005). The program is captured in the title of an important study by Renato Mazzolini and Shirley

Roe: *Science Against the Unbelievers* (Mazzolini and Roe 1986). Famously, the *Encyclopédie d'Yverdon*, under the editorship of Haller's friend, Fortunato Bartolomeo De Felice, sought to reassert Christian values within a broadly Enlightenment – and explicitly scientific – context, against the irreligious materialism of the Parisian *philosophes* (Maccabez 1903; Donato 1993; Doig 1994; Félice 1999; Cornaz 1992). As Marita Hübner notes, "The *Encyclopédie d'Yverdon* was the most important Protestant successor [*Folgewerk*] of the *Encyclopédie* of Diderot and D'Alembert" (Hübner 2010, 84). But even more than a successor, it was intended as a revisionist rival.

Deluc belongs squarely in that vein. He identified himself explicitly with this tradition associated with Haller and Bonnet. Deluc's religious orientation has been a constant consideration in the reception of his work.[16] The most recent flurry of interest in him has underscored not just his personal commitments but also their situation in a larger current of Calvinist Enlightenment, centered predominantly in Geneva and Yverdon, but with at least one important outpost in Berlin, in the person of Jean Henri Samuel Formey, secretary of the Prussian Academy of Sciences.[17] This is the thrust of Hübner's intellectual biography: Deluc "shared his Protestant profile with a circle of scientists [*Naturforscher*], theologians, literati [. . .] primarily ensconced in Calvinist cultures [. . .] in Switzerland, England, the Netherlands, Göttingen and Berlin," who saw themselves called to campaign against "materialists and atheists" (Hübner 2010, 211).

Jean-André and his brother devoted themselves, starting in 1744, to field trips into the alpine regions, gathering fossils and building a collection that became a significant attraction for visitors to Geneva, both scientific and touristic. By 1756 they had determined to write together a theoretical work on the geology of the Alps, based on the distinction between primordial and secondary mountain formations. Given that date, the most likely inspiration of this theoretical construction was the work of the Italian scholar Anton-Lazzaro Moro (Moro 1740).[18] In the 1750s it would become the major concern of Johann Gottlieb Lehmann and a bit later also of the German geognosists of the school of Werner, as well as Peter Simon Pallas in Russia (Lehmann 1756; see the introduction by Alexander Ospovat to Werner 1971; on Pallas, see Carozzi and Carozzi 1991). In short, in Hübner's apt phrasing, "the question concerning the emergence of mountains was the Archimedean point of geology" (Hübner 2010, 116).

Deluc was caught up in a wider cultural fascination with the Alps grounded in aesthetic as much as in natural-scientific elements (Reichler 1994, and, more extensively, Reichler 2002). In significant measure for this reason, his prominent colleague, Saussure, became famous throughout Europe for his *Voyages dans les Alpes, précédés d'un essai sur l'histoire naturelle des environs de Genève* (Saussure 1779).[19] The first volume of that work appeared in 1779, an event not without a very specific impact on Deluc. There was a further concomitant of local patriotism and celebration of the rustic virtue of the Swiss mountain villagers going back to Béat Ludwig von Muralt and Haller, and certainly shared by Deluc with Jean-Jacques Rousseau. Indeed, Deluc and his family had a close personal connection to Rousseau, with whom Jean-André and his brother shared walks in the

Swiss countryside in 1754 that played a significant role in Rousseau's celebrated sentimental novel, *La Nouvelle Héloise* (1761) (Rousseau 1761; François 1924; see Hübner 2010, 71). To be sure, the Rousseau Deluc knew and favored was a Genevan Calvinist republican in the context of the 1750s and 1760s.[20] Moreover, he took Rousseau, especially in light of the famous controversy over the 1756 "Geneva" article by D'Alembert in the *Encyclopédie*, to be a defender of Christianity, and this was one of the bonds that would eventually cement his appointment to the readership with Queen Charlotte of England (Hübner 2010, 92–93). When the French Revolution erupted in 1789 and intensified in 1793, with frequent claims that Rousseau was its intellectual godfather, Deluc, grown as conservative politically as he had always been religiously, repudiated this earlier connection. That political-religious dimension played a significant role in his *Letters to Blumenbach*, published simultaneously in English and German translations between 1793 and 1796 (Deluc 1793–1796).

Between the two brothers, it was Guillaume who proved the more assiduous collector and analyst of specimens.[21] Jean-André became fascinated above all with instrumentation, especially as their early efforts to establish altitude by barometrical measurement alerted him to problems in the devices available (Ratcliff 2011; Sigrist 2011b). The result would be his first great work in natural science, *Modifications of the Atmosphere* (Deluc 1772). Delayed years in publication from its original announcement and deformed in the transition by the vagaries of Deluc's investigations, it nevertheless made him one of the celebrities of European empirical science (Hoffmann 2003). That celebrity Jean-André parlayed into membership in the Royal Society of London and a position as reader to the queen of England, starting in 1773 (Hübner 2010, 88–91, 128). Not only did the position entail extensive instruction of the queen on matters natural-scientific and otherwise, but also it led to Deluc's extensive travels with one of her ladies in waiting, Elizabeth Schwellenberg (1728–1797), ostensibly for reasons of her health. These took him not only back to Switzerland but also, and very importantly, through the German territories of the Hanoverians. One very important stop (1776 and 1777) was the city of Göttingen and its university. For the academic community at Göttingen, deeply invested in their connection with the Hanoverian royal court, Deluc's association with the queen made him especially important (Selle 1937; Neuburger 1943; Marino 1995; Gascoigne 2002). Georg Christoph Lichtenberg, avidly caught up in "Anglophilia," brought Deluc to the attention of his Göttingen colleagues, and indeed introduced him to Blumenbach during his first visit to the city.[22]

From his travels, Deluc sent Queen Charlotte extensive letters reporting all aspects of the experience. Epistolary form became the genre for his most important later publications, a genre which fused many topics, cultural as much as scientific, in a manner oriented to the wider public sphere, and to a feminine readership in particular (see Reichler 2011). Already, excerpts from his massive *Modifications* of 1772 were issued by his Francophone Dutch publisher, Pierre Gédéon Dentand, as a travelogue (1775, translated immediately into German, with supplements in 1776 and 1777), while the entire scholarly tome was translated into German (in two volumes) only in 1776–1778 (Hübner 2010, 112). The ultimate fruit of his later travels Deluc published, starting in 1778, as *Lettres physiques et morales sur les montagnes et sur l'histoire de la terre* [. . .], suspended, and then republished under a new title, *Lettres physiques et morales sur l'histoire de la terre* [. . .] in 1779 (Deluc 1779–1780). The omission of the term "mountains" in his revised title reflected Deluc's realization that his Genevan colleague Saussure had seized the European imagination on that score with his widely celebrated opening volume of *Voyages dans les Alpes*, and certainly had more to say about the specifics of primary mountain formations than he did. Instead, Deluc concentrated on his explorations of the secondary formations of the Rhine Valley and central Germany.[23] One crucial consequence was that fossils played a far more substantial part in Deluc's theorizing.

The German reception of these works of 1778 and 1779 was swift. The first version was reviewed in the bellwether journal of the German Enlightenment, *Teutscher Merkur*, in 1778. A translation appeared in that same year. The 1779 version of his *Lettres physiques et morales* was reviewed by Blumenbach in 1780 (Blumenbach 1780b), and a German translation appeared in 1781. Deluc was a highly celebrated European naturalist, and the German reading public found itself swiftly and richly provided with his works and their assessment. All this does not betoken substantial personal interaction between Deluc and Blumenbach at this time, but it makes it thoroughly unsurprising that Blumenbach would take a particular interest in Deluc's work, as reflected in the review in *GAgS*.

As the era of revolution set it, both in Geneva and of course especially in France, Deluc saw himself increasingly as the beleaguered defender of "truths" being trampled by the materialist currents of the radical Enlightenment emanating out of Paris. Rudwick reads this posture quite sympathetically: Deluc was surrounded by the "illiberally skeptical society of the cultured despisers of religion" (Rudwick 2005, 333). As regards the emerging science of geology, "for de Luc [. . .], it was the prophets of a false Enlightenment, not the religious believers trusting in Moses' veracity, who were responsible for unfounded and speculative 'fictions'" (ibid., 331).[24] That may well be how Deluc felt; whether it is an accurate assessment of the epoch is quite another matter. Hübner offers a different construction: "Deluc's *myth* of war of natural science against religion" (Hübner 2010, 200, my italics). John Heilbron's account also presents Deluc more ironically: "savior of science and society" (Heilbron 2011). In Germany, Deluc's extravagant campaign, flush with the tone of Counter-Enlightenment, drew the criticism of Immanuel Kant, among others (Kant 1791).[25] By the mid-1790s, Blumenbach, too, felt that Deluc was blurring boundaries between science and other domains in a way that was uncomfortable for him, and that he credited Kant for trying to clarify and fortify.[26]

Blumenbach's thoughts on geology and paleontology from 1779 to 1801

In the 1779 first edition, first volume of his *Handbuch der Naturgeschichte* there is a crucial section (§39) in *Part III, Of Animals Generally* that would be removed from all subsequent editions, not because Blumenbach repudiated anything in it but rather because he would expand upon it in a separate, major component of later editions (Blumenbach 1779, 43–44). That section dealt with the question of extinction. He wrote,

> Since we are acquainted with so many animals only in fossil form, and not yet in [live] nature, some famous men have concluded that probably some species [*Gattungen*] – indeed genera [*Geschlechter*] – may have died out. Against this, to be sure, one might protest that a very large part of the earth is not yet investigated, and that we cannot know what might lie hidden at the bottom of the seas, in the interior of Africa and other places where natural history has not yet made its way.
>
> (Ibid., 44)

This was cautious, to be sure, but Blumenbach clearly inclined toward the idea of extinction. He noted that there were so many fossils without contemporary living counterparts that notwithstanding the doubters "still [*doch*] we have to see from all this that our earth over time has suffered very grave catastrophes [plural!]" (ibid.).

In the second part of this first edition, which appeared in 1780, Blumenbach presented a very lopsided treatment of the other two realms of traditional natural history besides zoology: plants (only *one* section of around *twenty-five pages*, quite amazing in the age of Linnaeus) and minerals (six sections of almost ninety pages) (Blumenbach 1780a). Setting out from a section on "mineralogy in general," it culminated in a section specifically dealing with fossils [*Versteinerungen*]. The thrust of this last section on fossils was to establish clearly the distinction of organic remains from inorganic minerals, and, indeed, to discriminate among the forms of metamorphoses that organic remains underwent (Blumenbach 1780a, Part XVI, 541–559, §§243–251).[27] It is worthwhile to consider closely the treatment of this material from 1780 to 1782 and beyond.

The opening section on mineralogy in general shows no changes between the first two editions. In §221, Blumenbach presented his overview of the field in language that is crucial (and largely unmodified in subsequent editions):

> To start with, something concerning the *origin* [*Ursprung*] of minerals, namely concerning the main paths [*Hauptwegen*] by which in part in earlier times they emerged all at once [*theils vor Zeiten mit einemmal entstanden sind*] and in part [they] have emerged gradually and even now continue to emerge [*nach und nach und noch immerfort entstehen*]. To shed some light on this, we must necessarily go back to the origin of our earth itself, an investigation in which to be sure a few daring conjectures must always be allowed; although we do not want to give ourselves over to the flights of clever men who have offered comets and burned-out suns as the basis of their system of the earth [. Instead,] we offer our more modest opinion, to which we have been brought first of all through the investigation of fossils [*Versteinerungen*]

> and their observed difference from currently existing organized beings, as well as through the comparison of ancient volcanos [*ehemaligen Vulcane*] and their products with those still burning, etc., and though this remains just another hypothesis, yet it is a hypothesis that nature and what is observable can easily and effectively accommodate.
>
> (Blumenbach 1780a, 473–474, §221)[28]

Rudwick writes of Blumenbach's adherence to a "conventionally binary conception" still in 1801 (Rudwick 2005, 426). The passages from 1779 and 1780 suggest that this is a rather inaccurate reconstruction of Blumenbach's intent from the outset.

When we turn to the very next section of the 1780 text, we find further grounds to resist Rudwick's claim:

> We believe ourselves persuaded, accordingly, that our globe [*Erdkugel*] has experienced at some time at least *one* apocalypse [*wenigstens schon* einen *Jüngsten Tag einmal erlebt*], and that we have the last judgment that took place then to thank for the current state of the world.
>
> (Blumenbach 1780a, 474, §222; 1782, 477, §222)

I think the phrase "at least" in this passage needs to be registered as highly significant, though assuredly Blumenbach did concentrate on a single catastrophic event in this account.

> This great catastrophe was, in all likelihood, occasioned by a subterranean fire that presumably thrust the floor of the sea into the heights and at the same time occasioned the dry land all at once to be swamped by the sea. Thus was the entire living earth drowned [*die ganze beseelte Erde ertrunken*], and by the same token the sea animals perished in being thrust out of their element onto dry land. Hence the size and the regularity of the strata, on the highest peaks, full of the majority of fossilized mollusks, etc., which have not been discovered and are indeed unlikely to be discovered in [living] nature.
>
> (Blumenbach 1780a, 474, §222; 1782: 477, §222)

If fossil remains represented one of the two classes of empirical evidence for the ancestral earth, volcanic remains formed the other:

> In a thousand places, however, the fire broke through the crust of the earth, hence the countless extinct volcanos which in the most recent times have been recognized as such and which from Göttingen to the banks of the river Rhine alone have been counted at about fifty. Perhaps it was through this great catastrophe that granite assumed its current appearance, and consequently, like the vast majority of fossils [*Petrefacten*], along with the majority of extinct volcanos and basalt columnar formations, should be considered the ruins of the former world, that Pre-Adamitic earth, and accordingly be

> discriminated from all the other minerals which, gradually or also through similarly violent catastrophes, emerged on the subsequently cooling earth, after the Creator, in the manner recounted by Moses, gave it new life with its current creatures.
>
> (Blumenbach 1782, 478, §222, with slight changes as compared to 1780a, 475, §222)

In this passage Blumenbach both embraced the biblical account (Moses) and situated it in a context of "pre-Adamitic" times that was clearly a more recent and non-scriptural conception (Popkin 1987). But what is more central, here, is that Blumenbach arrayed two substantial classes of empirical evidence for a radically different earlier history of the earth, and that he contemplated the possibility of "similarly violent catastrophes," suggesting that the plurality of these events was already on his mind. Indeed, in §223, Blumenbach moved from the conception of the unique catastrophe to a notion of general causes of geological disruptions:

> The outbreak of subterranean fires and deluges, the two means by which, in our opinion, the former world was annihilated, are still two considerable sources also on our current planet sometimes for the destruction and other times for the transformation and emergence of minerals.
>
> (Blumenbach 1780a, 475–476, §223; 1782, 478, §223)

It should be recalled that Rudwick sees as the breakthrough achievement of Burtin in 1789 the insistence on a plurality of geological revolutions (as against Deluc) and that he charges Blumenbach with being tardy in coming to such a plurality of revolutionary episodes in earth history only around 1799. Instead, Blumenbach considered that prospect from the outset. Now, the position that Blumenbach articulated in 1780 and repeated in 1782 bears too strong a resemblance to Deluc's theory for this to be ignored. We have the confirming evidence in Blumenbach's review of Deluc's *Lettres physiques et morales* (1779), which appeared in early December 1780. That review made clear that Blumenbach was acquainted not only with the earlier 1778 version, from which he contended that the 1779 version was not particularly remote, but also with the fact that these texts were the fruit of thirty years of investigation, for the most part in collaboration with Deluc's brother (Blumenbach 1780b, 769–770).

Blumenbach discriminated two elements in the text, a rambling travelogue and a systematic scientific theory, and he concentrated entirely on the second of these, offering "merely the first outline of this edifice [of doctrines]," which he took to be a finely articulated *system*, in which Deluc had anticipated and answered many possible objections (Blumenbach 1780b, 770). Summarizing, Blumenbach observed that Deluc "divides all of world history into two great periods" – a former world [*Vorwelt*] and a "newer" earth – separated by a great catastrophe.

Our earth has its current form thanks to the deluge – but this came to be in the following manner: that the dry land, after it was gradually undermined by subterranean fire, eventually collapsed and sank below the level of the ocean of that time, which accordingly was drawn into these new depths and abandoned its former beds, which in turn, having been left dry, became the new land masses. This new, still current dry land therefore remained stable and unchanged in its original fixed stratification, and its previous mountains and valleys had almost all already existed while all this was still at the bottom of the sea. The highest peaks had stood out over the surface of the waters as islands. The rest was covered by the waters (Blumenbach 1780b, 770–771).

Since then, Blumenbach continued in his summary, the oceans and continents had remained roughly in the same configuration. Most significantly, according to Blumenbach, Deluc maintained that "since this great catastrophe [. . .] only a few thousand years could have passed" up to the historical present (Blumenbach 1780b, 771).

Deluc drew his evidence, Blumenbach noted, from the "archive" of the mountains to be found on the new continents. Deluc offered a tripartite typology of these mountain formations: primordial, aquatic, and volcanic. Blumenbach rendered these into "our" – that is, the Wernerian geognosical – terminology, with which his German readers would be familiar. What Deluc called "primordial," Blumenbach rendered *Ganggebirge*; what Deluc called "aquatic," Blumenbach termed *Flözgebirge* (Blumenbach 1780b, 772). The latter were particularly decked with strata upon strata of fossils (*Versteinerungen*). Since Deluc posited that in the former world there had been many upheavals thanks to volcanic action, this allowed for the *local* submersion and reemergence of dry land, relative to the seas, thus accounting for the fossil presence of land animals and plants in strata beneath those of sea-life, and conversely. The volcanic explosions were outbursts of a subterranean fire, which, combined with explosions resulting from penetration of seawater via the resulting cracks, widened cavernous gaps in the earth's crust, gradually spreading laterally until the whole surface was undermined, setting up the great catastrophe. This further explained the overlay of volcanic and sedimentary strata, since there were many such dislocations and resettlements over the vast span of "pre-Adamitic" time. Two outcomes were particularly salient for Deluc in this analysis. First, the age of the *current* continents, measured by observations of "nature's chronometers," like the accumulation of silts in river deltas like that of the Rhine, turned out to be quite recent. And, second, this all converged gratifyingly, but ostensibly without deliberate refitting, with the biblical narrative of Moses (Blumenbach 1780b, 775). With that, Blumenbach accomplished a remarkably concise distillation of Deluc's prolix presentation.

There can be little doubt from this account that Blumenbach drew heavily on Deluc for his theory of the early history of the planet in 1780. He made this explicit in his 1797 edition of the *Handbuch*, in the aftermath of the publications of Deluc's letters to him, which he translated and published in *Magazin für das Neueste aus der Physik und Naturgeschichte*, between 1793 and 1796 (Deluc 1793–1796). In a footnote to §227 of the *Handbuch* of 1797, Blumenbach wrote,

> There is no geognosic system known to me (and one could count already in the year 1764 no fewer than forty-nine) that satisfies this demand [of

> grounding the fundamental knowledge-claims of geognosy in a careful test against physics and chemistry] than that in Mr. DeLuc's geological letters which are translated from the French manuscript in Hr. Professor Voigt's *Magazin*.
>
> (Blumenbach 1797, 511 fn)

From start to finish, then, there are strong reasons to place Deluc's "geohistory" at the fountainhead of Blumenbach's paleontology.[29]

The question that needs exploring is how Blumenbach developed his ideas in the direction of paleontology over the 1780s – that is, *after* the 1782 edition of his *Handbuch*. Deluc was not in close communication with Blumenbach in those years, nor did he publish any new material. We need to cast a wider net. To be sure, Blumenbach always assimilated Deluc's specific theory into the ongoing German discourse of geognosy, as was clear in his review of 1780. But it is also the case that this German discourse was still quite diffuse. The great theoretician of this school, Werner, had indeed published his pioneer text of 1774, and begun teaching at the Freiberg School the following year, but a systematic formulation of the Wernerian theory would not be at hand until much later (Werner 1774; 1971). A great deal of other work was appearing, both in French and in German, and an attentive reader, as Blumenbach was, would have had many sources from which to draw for his understanding. A cardinal illustration of this textual richness and its theoretical possibilities has been provided for the case of Johann Wolfgang von Goethe in these same years, by Wolf von Engelhardt (Engelhardt 1982).[30]

At the same time, of course, that textual richness has been utilized to query Blumenbach's originality, on the one hand, and his importance, on the other (Langer 1985, esp. 91). For the development of paleontology, and more specifically the notion that fossilized life forms could be used to date geological strata (the idea of "lead-fossil" [*Leitfossil-Prinzip*] or "biostratigraphy"), there were several powerful rivals for the crown of originality – most persuasively, in my view, Jean-Louis Giraud-Soulavie in France (Soulavie 1780–1784). Indeed, by 1782 Goethe recognized the potential of that line of thought on the basis of his wide reading, including Soulavie's text (though perhaps only in its censored, less explicit version).[31] Similarly, Rudwick has highlighted Burtin's works of 1784 and 1789. But the case to be made here is less for Blumenbach's originality than for his consolidation and instigation of a research program for *history* of nature, and here, I am confident, he deserves a good measure of recognition.

The next edition of Blumenbach's *Handbuch* appeared in 1788, and it showed primarily an expansion of Blumenbach's documentation of fossil finds (Blumenbach 1788). He had become quite interested, as museum director, in accumulating a collection of such fossils for Göttingen, and he would build, indeed, one of the foremost such collections in Europe by 1800. Theoretically, however, he made no revision in the 1788 edition. The distinction between "known" and "unknown" specimens, which Rudwick dates to 1788, was already tacit in the 1780 and 1782 texts. This was the basis, after all, of Blumenbach's affirmation of extinction already in 1779. By the 1791 edition of the *Handbuch*, Blumenbach did make important changes in the mineralogy section, including a clearer formulation of its taxonomy (Blumenbach 1791). But in the interval, he had published the first volume of his *Beyträge zur Naturgeschichte* (Blumenbach 1790a), which proved his most important publication in the field.

The first edition of the first volume is dated April 1790 (Blumenbach 1790a, VIII). It was accompanied, later that year, by an article in *Magazin für das Neueste aus der Physik und Naturgeschichte*, with the title "Beyträge zur Naturgeschichte der Vorwelt" (Blumenbach 1790b). Only the first five sections of the book dealt directly with questions of earth history and paleontology; the other nine sections dealt with other issues. The article of late 1790 did take up questions of earth history, since it was motivated by Blumenbach's critical resumé of James Hutton's "eternalist" theory of the earth, immediately following in the same volume of *Magazin für das Neueste aus der Physik und Naturgeschichte*, where Blumenbach makes explicit reference to Deluc's critique of Hutton (Blumenbach 1790c).[32] Clearly, Blumenbach preferred the historical-developmental account worked out by Deluc (and others, including the German Wernerians) to the cyclical theory developed by Hutton, especially since the latter made light of species extinction, in stark contrast to Blumenbach's own view. This may have been a central topic in his conversations with Deluc a year later at Windsor in 1791, and their common opposition to Hutton may have been an important motive for Blumenbach in inviting Deluc to draw up a concise summary of his rival theory for a German audience. Notably, that would appear in the same journal in which he had himself critically appraised Hutton's ideas.

A decade later, in 1801, Blumenbach gave the anniversary address to the Göttingen Royal Society, *Specimen archaeologiae telluris*, of which Rudwick has made note. Blumenbach summarized his presentation in a report for the *GAgS* in that same year, and then published the paper in Latin in 1803 (Blumenbach 1801; 1803). A summary of his summary was then appended to the third section of the second edition of *Beyträge zur Naturgeschichte*, Volume 1 (Blumenbach 1806). The burden of these three versions, and often the very phrasing, is roughly identical, with the exception that the Latin paper published in 1803 had a more extended treatment of particular fossilized organisms. This body of material is the most direct evidence we have of the development of Blumenbach's ideas on paleontology and history of the earth.

As he put it in a footnote to Section 2 of the 1790 edition of *Beyträge*, "nearly the only, but therefore all the more important, use of the knowledge of lapidifications [*Versteinerungen*] is the solution which the history of the changes of the earth's surface derives from it" (Blumenbach 1790a; translation [based on the identical text in the 1806 ed.], 283 fn). Blumenbach articulated clear methodological constraints already in 1790 and then more extensively in 1806:

> If petrifactions [*Petrefacten*] can be made of regular use for the archaeology and the physical geography of the earth, as the surest documents of the archives of nature for the fruitful history of the catastrophes which have been connected with our planet since its creation, the study of them [. . .] demands

> as well a thorough critical comparison with the organized bodies of the present creation, as also [. . .] of their different locations, and their geognostical relations.
>
> (Blumenbach 1806; translation, 317–318)

If one observed this methodological propriety,

> a wider examination of these differently made fossils, and of their equally various sorts of condition, brings us to a closer conclusion as to the oldest history of the body of this earth, and [. . .] the numerous catastrophes [. . .] through which its crust has acquired its current appearance [. . .] built out of such great convulsions.
>
> (Ibid., 318)

The most important feature in these texts is that Blumenbach insisted upon a plurality of "revolutions" – "the sequence of the totally different catastrophes [earth] has gone through, by which the numerous fossil remains of former organic creations have come into their present positions" (Blumenbach 1806; translation, 285 fn). In place of a "binary" model, he clearly articulated *three* conceptually distinct phases; moreover, each of these phases in its turn involved *multiple* incidents (Blumenbach 1806; translation, 318). Working backward in time, Blumenbach first collocated those fossilized (and semi-preserved) remains of creatures which could be identified with currently existing species. Second, he identified remains which were *similar* to current species, but not the same, and in addition, whose closest living analogs inhabited distinctly different climatic zones. Here he placed the widely discussed elephant, rhinoceros, and other remains which had been the preoccupation of zoologists over the second half of the eighteenth century, and whose correct reconstruction would become the basis of Georges Cuvier's rise to preeminence in the early nineteenth century. Blumenbach clearly discerned a mass extinction between the first and the second phase:

> A total alteration of the climate took place, which occasioned the destruction of the then living generation of these tropical creatures, as of many other genera and species of organized bodies which existed along with [. . .] as the unknown of Ohio [*Ohio-Incognitum*] among great land-animals [. . .]. This revolution, which seems to have been merely climatic, must be distinguished from those earlier and much more forcible ones [note plural!], from which we must date the petrifactions of the *third* division, the oldest of all.
>
> (Blumenbach 1806; translation, 320–321)

In his third and most ancestral phase, Blumenbach placed fossilized remains utterly different from any known living forms, and which accordingly gave token of a former world [*Vorwelt*] and a discontinuity far more radical than he ascribed to the border between his first and second phases. It was this more ancient and drastic divide that Blumenbach accorded the title of a "total" revolution: "in my opinion it becomes more than merely probable that not only one or more species, but a whole original organized preadamite creation has disappeared from the face of our planet." That is, "as I have said [already in 1780!], [. . .] our earth has already suffered a complete revolution, and experienced one last day" (Blumenbach 1790a; translation, 285). He elaborated: "This general revolution [. . .] is again quite different from the subsequent later one" dividing the first and second phases (ibid., 286). Thus, already in 1790 Blumenbach was thinking in terms of the three phases he discriminated in 1801. But what is important to discern for *all* these phases is his consistent employment of the plural, indicating that there were *many* disruptions, though not all global in scope, that took place both before and after the great catastrophe. Significantly, he wrote of that as "the *last* general catastrophe" (my italics), hence one of many that beset the *Vorwelt* (Blumenbach 1806; translation, 319). He insisted that the fossil remains could not all be "explain[ed] [. . .] by one and the same catastrophe" (Blumenbach 1790a; translation, 285). Clearly, Blumenbach conceived of multiple disruptions in the *Vorwelt*: "these destructive catastrophes themselves were again of more than one sort, and were very far from happening all at the same time." To be sure, Blumenbach was cautious about how much order could be brought to the periodization of the planet's surface forms: "it is scarcely possible at present to determine with any certainty the chronological arrangement of the successive periods [. . .] to say nothing of the causes of them" (Blumenbach 1806; translation, 321).

For Blumenbach the single most important conundrum presented by this periodization of earth history and its living inhabitants was how the extinct organisms came to be *replaced* after the great catastrophes, and especially after the most massive one separating the *Vorwelt* from the recognizable configurations of what Blumenbach took to be the current world (*Schöpfung*, literally: "creation").[33] Mass extinction made the idea of a "total revolution" very problematic, as Rudwick notes, since the question of a "new" creation seemed to require literally a deus ex machina (Rudwick 2005, 299). Blumenbach clung to the fact of the matter, even if he could not offer a compelling causal account. "[A] whole creation of organized bodies has already become extinct, and has been succeeded by a new one," he wrote (Blumenbach 1790a; translation, 288).

> *How* indeed this subsequent creation took place, that I can no more say than how in early times the first spermatic animalcule came into being; that however they *were* subsequently created seems to me undeniable, and I lay that to the account of the great mutability in nature.
>
> (Ibid., 290)

To get some perspective on this, Blumenbach insisted that extinctions and new emergences were *continuing* to take place:

> Creatures enough die away in a locality, and fresh ones again become naturalized and spread themselves. [. . .] So there is nothing contradictory in the idea

> that [. . .] a species may have become extinct; and on the other hand a fresh one may likewise be sometimes very easily created [!] subsequently.
> (Ibid., 289)

This was the vein in which to consider the aftermath of the "last great catastrophe."

> If the former world [*Vorwelt*] suffered a total revolution, as seems unmistakable, and *if* this revolution was probably caused by a general conflagration of the earth [*durch einen allgemeinen Erdbrand bewirkt*], afterwards there must have been a very long span of time before the newly changed crust of our planet had cooled down and its surface once again became at all ready to be enlivened with a fresh vegetation, and vivified with a new animal creation.
> (Blumenbach 1790a, sect. 4 *Umschaffung der Vorwelt*, 24 [my translation]; see Bendyshe's translation, 287)

Blumenbach surmised that "the Creator" would then have "permitted the same natural forces in general to achieve the production [*Hervorbringung*] of a new organic creation" to fulfill the same purposes that the older order of life had done in the prior world (Blumenbach 1790a, 24 (my translation); compare Bendyshe's translation, 287). Noteworthy here is not only the *transcendent intervention* but also the *teleological construction* of the order of organized life in both worlds. That teleological structure was, moreover, ascribed to *natural forces* which, even more importantly, remained constant across the catastrophe.

This allowed Blumenbach to have recourse to his own master concept, the *Bildungstrieb*, to grasp the phenomenon. "Just that, after such a total revolution had so changed the materials at hand, the *formative drive* clearly was compelled to take a direction differing more or less from the old one in the production of new species" (Blumenbach 1790a, 25, modifying Bendyshe's translation, 287). That is, "the formative drive in these two creations to be sure functioned in a similar but not in the identical manner" (Blumenbach 1790a, 26 [my translation]; compare Bendyshe's translation, 287). The language of an "alteration of direction of the formative drive" became a centerpiece of Blumenbach's teaching in the 1790s, as evidenced in its prominence in his student Christoph Girtanner's representation in his key monograph (Girtanner 1796). In his text of 1790, Blumenbach was at pains to distinguish that mode of explanation from a rival view proposing "*degeneration* [*Degeneration*] acting for a long series of thousands of years" (Blumenbach 1790a, 26; translation, 287).[34] Evoking the contrasting direction of the spiral in fossil seashells relative to living ones, he firmly asserted, "Such a thing is not a consequence of degeneration, but a remodelling [*Umschaffung*] through an altered direction of the formative drive" (Blumenbach 1790a, 27; translation, 288). He considered the possibility, based on the similarity of ancient fossils to currently existing life forms, that *some* organisms might have persisted across *all* the catastrophes into the present, but he asserted a preference for the hypothesis that these were *only* similarities induced by the formative drive of nature, which was indeed perennial, but which adapted to the new circumstances of the post-catastrophe earth (Blumenbach 1801, 1983–1984). Thus, Blumenbach proposed that "an altered direction of the formative force [*eine veränderte Richtung des Bildungstriebes*]" could explain the phenomena (Blumenbach 1790a, 26; translation, 288). In a footnote added to the second edition, he elaborated:

> The formative power of nature in these remodellings [*Umschaffungen*] partly reproduces again creatures of a similar type to those of the old world, which however in by far the greatest number of instances have put on forms more applicable to others in the new order of things, so that in the new creatures the laws of the formative drive have been somewhat modified.
>
> (Blumenbach 1806, 20 fn; translation, 287 fn)

Rather than abandon his commitment to the immutability of species (though not of *varieties* within a stem-line), Blumenbach insisted on the formation of entirely new life forms, with only structural parallelism to earlier forms due to their common formation by the ubiquitous *Bildungstrieb*. The notion of perennial forces generating different particular forms in differing ecological circumstances has some resonance with Buffon's notion of the persistence of indestructible "organic molecules" and the action of various "interior molds" which allowed for similar emergences of new life forms across geological change.[35] That assumption of perennial forces, which took into account altered circumstances, seemed to Blumenbach more consistent with "physiology" than the notion of immanent developmental change leading to new species (Blumenbach 1801, 1983–1984). The functionally constrained (teleological) "animal economy" of each species (or its stem-line) was so finely integrated in terms of the relations of its various organic parts that any significant change would induce drastic dysfunction, in his view.[36] The force of his "comparative anatomy" thus induced conservatism about biological transformism, just as it would in the parallel postures of Cuvier.[37]

It was this that led Dougherty to suggest that Blumenbach could not be assimilated to *our notion* of a history of nature. That is, Blumenbach could not subscribe to an immanent transformation of species, which Dougherty identifies with our "developmental history" and its sense of "historicity" (*Geschichtlichkeit*) (Dougherty 1986, 104). All Blumenbach could affirm was a "historical model" whereby a history of nature undertook "to present the natural causes of change in order to recognize the general laws to which the relation between milieu and form is subject" (ibid., 103). As empirical science, natural history could only approximate the *rational idea* of a unity of nature. While natural *description* could deal only with individual life forms and thus could never achieve the concept of a species (*allgemeine Gattung*), natural *history* could, via the combination of the empirical evidence of common descent and a teleological conception of the functional integration of the organism, work toward a *theory* of the "*history* of the [life] forms." Thus, "generation [*Zeugung*] and final purpose together are the exclusive criteria for a determination of species" (ibid., 103). That could be only a *model*, according to Dougherty, grounded in "reflective judgment" (ibid., 105).[38] This model was constrained by the *historicity of natural science* in Blumenbach's era, which could only with violence be assimilated to our own contemporary framework.[39] Thus Blumenbach had a thoroughly different conception of history from ours. Indeed, Dougherty finds him far closer to ancient views of organismic development, from Aristotle to the Scholastics.

Dougherty stresses that for Blumenbach the coherence of the organism was nothing short of miraculous – hence the best evidence of intelligent design, not just in the internal coordination of parts but also in the *functional* integration of the entire organism in its environment (1986, 101). If comparative anatomy highlighted the first integral, natural history targeted the second.

> The concept of natural history in Blumenbach can only be grasped if we presuppose an absolute specificity of each created form of life, a final purpose in the organization of that form, and a harmony between the parts of the individual form as well with the various forms collectively.
>
> (Ibid., 101)

Only living things could demonstrate such "a specific, goal-directed form of life"; hence only they belonged to natural history.

Dougherty situates natural history for Blumenbach at the confluence of three conceptual orders: the organic, life, and creation. The organic concretely betokened "a harmonious agreement of organization and function" which had to be unique to each entity. This is, for Dougherty, a "model" relation which establishes "first of all the possibility of any given organized form" (1986, 101). Life, he continues, "is a formative drive, a process that drives [*treibt*] an organic form into existence." Natural history cannot explain this *Bildungstrieb*, but only describe the ecological constraints in which it works. The only real explanation is *Schöpfung*, "creation" in the strong sense of Divine Providence (ibid., 102). Hence, for Dougherty, "the task of natural history" according to Blumenbach was "ultimately the recognition of an eternal lawfulness and a transcendent order [. . .] the guidance of the world by an almighty and all-wise Creator" (ibid., 97). It was this shared ultimate view, Dougherty contends, that made Blumenbach sympathetic to Deluc's more explicit physico-theology, even if the *theoretical convergence* of their views was "not easy to grasp" (ibid., 98). For Dougherty, it is difficult to reconcile the theoretical edifices of Blumenbach and Deluc, but they shared a commitment to traditional Divine Providence as the ultimate point of inquiry into natural history.

Where they parted substantially was in their rhetoric of scientific presentation. Blumenbach was not comfortable with the *directness* with which Deluc wished to link his geological research and his biblical allegiances. That came to a head in the project of translating Deluc's letters for the German audience. Dougherty offers us the chronological frame of their correspondence.[40] We can supplement that with Hübner's bibliographical chronology of the publications of Deluc's letters in English and German versions.[41] The upshot is that the letters were published in parallel up through Letter IV, composed in March of 1794 and published that year in both languages in their respective journals. But then the English translation followed quite closely the composition of the next two letters, in 1794 and 1795, while the German translation stopped for a two-year interval. The interruption, we discover, arose from a hesitation on Blumenbach's part, which is important to elaborate.

During the interval in question, Blumenbach sent Deluc a copy of his substantially revised study on human variety, with a cover letter that began to articulate the difficulty that put a pause to the translation project of Deluc's letters into German. He wrote,

> I have treated my subject [. . .] simply as an anatomist and naturalist, and yet the result of my research is absolutely consistent with what Revelation teaches us about the identity of our species. But I have carefully kept myself from touching on this conformity explicitly; I did not once cite Moses, and this is because I know all too well the unfortunate prejudices of the party of our public for whom I hoped to be most useful through my research.
>
> Those persons would have taken me to be preoccupied [*préoccupé* – a delicate word for *doctrinaire*] if they had seen passages from revelation cited, and they would probably have rejected my book without having read it or learned anything from it.[42]

One of the interesting historical questions to pose, here, concerns the precise identity of those in the "party of our public" caught up in "unfortunate prejudices," those "cultured despisers of religion," as Rudwick has termed them (Rudwick 2005, 332). In the context of Blumenbach's actual engagements in the decade 1785–1795, there is strong reason to believe that at the center of that circle would be Georg Forster and his bosom friend, Thomas Soemmerring. The correspondence between these two in the context of Soemmerring's book on the black race and Forster's controversy with Kant suggests the salient tension between especially Forster and Blumenbach on questions of the relation of religion to science.

The key point for the Deluc translations is that the latter did not find Blumenbach at all persuasive in the stance described in the letter of June 1795, and Blumenbach had to try to make himself better understood in his letter of November of that year. In that letter, Blumenbach stressed his hope of influencing the wayward audience, and he argued that they would not listen if he wrote explicitly of Moses. It was the miraculous character of the organism, demonstrated by comparative anatomy, which should of its own force drive these skeptics to realize that only a providential hand could bring this about.[43]

Dougherty suspects that Blumenbach was not being altogether candid in his professions to Deluc about the motivations actually driving his work, and instead seemed to be adapting to the dispositions of his correspondent. In any event, Blumenbach explained to Deluc that he wished to close the translation of the latter's work at §10 of Letter V.

> I dared to [make this proposal] because I thought that this would sufficiently furnish German naturalists with a concentrated sketch of your properly

> geological system, and that by contrast the rest, notwithstanding that it is extremely interesting, would not be appropriate in a journal of physics and natural history.[44]

This suggests strongly that Blumenbach's motivation for translating Deluc had to do not with physico-theology but with the history of nature, and in all likelihood with rebutting Hutton's rival system.[45] To be sure, the religious implications of both systems were clear to Blumenbach, and even important. But they were tacit, not foremost, in his view of the project of natural history and of the translation project more specifically. It is all the more interesting, then, that he relented, and went forward with the translation of Letter VI, where Deluc explicitly endeavored a naturalization of the scriptural account of Moses.[46]

Conclusion: Blumenbach and the history of nature

I would like to conclude with a few remarks on Blumenbach and the history of nature, returning to the point of departure of my essay. Baron's essay of 1963 stands as one of the earliest and strongest assertions of the importance of Blumenbach in the development of a historical approach to nature, and one that, moreover, recognizes "the rise of a developmental-historical or historicized way of thinking already in the second half of the eighteenth century" (Baron 1963, 19). These are both claims that I have essayed to buttress in my discussion. A core recognition of the changing character of nature instigated a dynamic view which displaced the timeless, static model of physical laws and also began to shake up the long-standing commitment to "species essentialism" – as well as the traditional religious notion of a single, comprehensive moment of "creation" (as in preformation) – offering instead the notion of a "never ending process of becoming and new emergence" (ibid., 20). Blumenbach charged even his revered model, Albrecht von Haller, with a misguided dread that proposing any change in the order of nature had to be both heretical and unscientific.[47] The essential move of the epoch – and of Blumenbach in particular – was, according to Baron, to realize that the historicization of *biology* depended upon its systematic alignment with the emergent *geological* sense of time (and its differentiation into "epochs," to invoke the term made famous by Buffon in 1778) (ibid., 21).[48]

What triggered Blumenbach's thinking was the empirical evidence for extinction, and not simply rare instances of it, but massive hordes of no-longer-to-be-found organic forms, illustrated, in his own works, by some 200 species of Ammonites for which no correlative life forms could be found in the living world (Baron 1963, 23). One can hardly argue that Blumenbach was the first to take up the issue of extinction. Already in 1779 he wrote of "a few famous men" who advocated the idea, and with whom he chose to align himself (ibid., 22).[49] But the fact is that Blumenbach made the question of extinction the leading principle not only for the study of fossils but for the theory of life forms in general, the foundation, in conjunction with comparative anatomy, for a true life science.[50] In Baron's words, "Blumenbach's periodization of fossils represents one of the earliest efforts to draw a parallel between the history of the earth and the history of organisms" (ibid., 25). Thus, Blumenbach was among the first to see that the reconstruction of the "individual phases" of the historical process of the development of organisms could be achieved through the study of fossils (ibid., 26). The *direction* of this relation is crucial, here. In the history of geology, the significance of "biostratigraphy" has always centered round dating strata sequences – that is, *geological* insight. But strata sequences *also* created effective empirical evidence for periodization in the sequence of *life* forms – *evolution* in our most general sense of shifts in the basic form and function of living things across the span of earth history. To be sure, Blumenbach knew he was far from a good causal account.[51] Yet, by that very token, such a realization obviously *called* for one (ibid., 21).

The great impediment in history of science to recognizing the late eighteenth-century historicization of nature has been the strong resistance to the anachronistic pursuit of "transformism" or, worse still, of "forerunners" of Darwin, in these developments (Wenzel 1982). This was the motive behind Dougherty's effort to discriminate the "historical model" Blumenbach could affirm from our "historicity," which he takes to be altogether informed by such a "transformist" model. In the German historiography, the most important theoretical exponent of a hard and fast distinction between the impulses toward historicization in the eighteenth century and such "real transformism" has been Dietrich von Engelhardt.[52] He developed his position by contrasting the emergence of a historical sense of *science* – that is, a sense of disciplinary change and progress in the various fields of natural-scientific inquiry, as a key feature of the eighteenth-century Enlightenment – with a related historical sense of *nature* itself (Engelhardt 1979). Engelhardt summed up his view in an essay entitled "Temporalization of Nature and Historicization of Natural Science [*Naturwissens*] in the Eighteenth Century" (Engelhardt 1995). The sense of progress in scientific inquiry proved far stronger (as well as earlier) than the sense of nature's own developmental change. But the latter did emerge in the course of the eighteenth century, as well: scientists recognized that different forms appeared in different times across the history of the earth. What that did not mean, Engelhardt insisted, was that there was a *real* (i.e., phylogenetic) mutation of earlier life forms into later ones. "Real transformation was overwhelmingly deemed speculative" in the eighteenth century, he argued, and its explicit proponents – Benoît de Maillet, Jean-Baptiste-René Robinet, Antoine Nicolas Duchesne – were largely dismissed (ibid., 237–238). Buffon did change the project of natural history from a static to a dynamic reconstruction of life forms. Still, Engelhardt notes, even Buffon could not affirm "a real-genetic connection [*realgenetischen Zusammenhang*]" (ibid., 239).

Blumenbach followed Buffon both in discriminating different life forms in a sequence of different epochs and in rejecting "real transformation" of species. "He never thought of a real emergence of one out of the other [*ein reales Auseinanderhervorgehen*]." But Blumenbach's students, Engelhardt admits, "did not exclude that possibility" (1995, 240). Kielmeyer and Friedrich Sigmund Voigt were examples of German life scientists around the turn of the nineteenth century who clearly made room for this idea. In France, Engelhardt notes, Soulavie had also made that transition (ibid., 244). Lamarck would become its great advocate after 1800. Ironically, in Engelhardt's view, Romantic *Naturphilosophie*, despite its equation of nature with history, turned away from "real descent," preferring the "ideal" and the conceptual over the actual and genealogical (ibid., 244ff.). Engelhardt offers a powerful and persistent reconstruction of the conceptual possibilities of the epoch from 1750 to 1830. Moreover, this captures Blumenbach's ambivalence well. Yet my suggestion is that we should retrieve the substantial impetus toward history of nature in Blumenbach (and Buffon and Kant before him), even as we recognize that it is not yet Darwin's or our own.

Rudwick claims that Blumenbach's preoccupations – in particular with physical anthropology – made it impossible for him to be a major contributor in the field of paleontology (Rudwick 2005, 428). That is a misunderstanding of the relation between these two domains for Blumenbach. The intimacy and mutual reinforcement of these research programs are illustrated most clearly in the structure of the two volumes of Blumenbach's *Beyträge zur Naturgeschichte*, where there is a natural and substantial sequence from his paleontological ideas to his ideas on physical anthropology, race, and human cultures (Blumenbach 1790a; 1806; 1811). The whole discourse of anthropology in the later eighteenth century set out from the idea of a "natural history" of man or of human mind, terms that feature prominently in the writings of John Locke and the Comte de Buffon (Zammito 2002). The thrust of this, already clear in Linnaeus, was to treat humankind as a natural species (Linnaeus 1735).[53] Thus the development of life forms was essential to the understanding of the development of man. From Buffon through Peter Camper to Blumenbach, this was the crucial path for "natural history" as a paradigm of inquiry. It shaped the debates about vertebrate fossils over the course of the eighteenth century (Rupke 1983; O'Malley and Magoun 1962).[54] More specifically, it found expression particularly in the debates about the anthropoid apes, a discourse that, as Robert Wokler has established, led to the debate about human races (Wokler 1976; 1980; 1988; 1995; see also Niekerk 2004; and Smith 2007).

The recent annotated edition of Blumenbach's *Beyträge* by Mario Marino documents clearly the coherence of these concerns within the volume itself and across not only Blumenbach's career but also the reception of his work in subsequent eras (Blumenbach 2014; see Zammito, forthcoming b). Clearly, preoccupation with Blumenbach's work on race has overshadowed his other contributions. In my view, that has not only distorted his ideas on race, which have been seriously misrepresented, but also more significantly failed to register properly his role in the development of life sciences more widely. The most recent work on Blumenbach is now undertaking to rectify these disproportions and distortions (Michael 2017). I have tried to support this endeavor in my forthcoming monograph, *The Gestation of Biology in Germany* (Zammito, forthcoming c). My effort in this essay is to establish firmly that paleontology was a domain in which Blumenbach played a seminal role, not only in his own work, which I highlight here, but also in the inspiration he gave to a number of his greatest students, whose work in paleontology is unquestionable.

Notes

1 Lamarck has become the landmark figure in the history of "transformism"; see Corsi (1988) and Conry (1980).

2 Humboldt and Schlotheim were actually classmates at the Freiberg school. Humboldt studied there in 1791–1792, and Schlotheim 1791–1793. Schlotheim is not frequently discussed among Blumenbach's students, but his role particularly here seems quite important. Rudwick writes of him immediately after discussing Blumenbach, but mainly in terms of his publications in the 1820s. What is important is the *formation* of his interest in the field, which was established under Blumenbach and led him to study with Werner at the Freiberg School of Mines, already in the early 1790s. On Schlotheim, see Langer (1982) and Oschmann (1964).

3 Deluc, long languishing in obscurity, has suddenly become a major focus of research interest. Perhaps the triggering study was the collaborative essay by François Ellenberger and Gabriel Gohau (1981). The first international conference on his work came only in 2008, the results of which have now been published (Heilbron and Sigrist 2011). The first thorough monograph on Deluc is Hübner (2010).

4 This is a major part of the early career of Georges Cuvier, as Martin Rudwick develops it (Rudwick 1997).

5 The discovery of the wooly mammoth in Siberia by Peter-Simon Pallas and the discovery of the "Ohio creature" celebrated by Thomas Jefferson (which Cuvier eventually labeled a "mastodon") were prominent topics of discussion among naturalists of the later eighteenth century, and led to the direct exchange of ideas between Blumenbach and Cuvier.

6 While Blumenbach considered him prolix, this did not diminish his esteem for Zimmermann's contribution. He certainly welcomed the affirmation of his own work by Zimmermann, especially in contrast to the criticisms the latter made of Kant, on the question of human "races."

7 Blumenbach was also remotely related to Walch, whose father was Blumenbach's maternal granduncle (Dougherty and Klatt 2006–2015, I, 27).

8 As René Sigrist aptly puts it, "Fossils at first meant anything coming from the earth. Whatever their origin or their connection to living organisms, 'petrifications' or 'figured stones' were classified as stones" (Sigrist 2011a, 115 fn). A classic instance of this blur is Bertrand (1766). On the meaning of "fossil" see Rudwick (1972).

9 Thus the famous line from his *Beyträge zur Naturgeschichte* in 1790: "Every paving stone in Göttingen is a proof that species, or rather whole genera, of creation have disappeared" (Blumenbach 1790a, 6; transl. Bendyshe 1865, 283). I will cite the English translation, amending it when appropriate.

10 For another construction of Deluc's intentions, see Hübner (2011, esp. 42 fn): "In the early 1790s Deluc dedicated his geological letters to Blumenbach to draw the attention of the learned public to his theory of the cause of the flood."

11 The key thrust of this essay was to *pluralize* the single great revolution that was central to Deluc's geohistory. Burtin criticized Deluc throughout his essay, according to Rudwick (2005, 200–201).

12 Rudwick is taken with the term "archaeology," using it as his section heading for the period 1801–1804: "'Archaeology' of the Earth" (ibid., 417–431). He sees this as a clear consequence of Blumenbach's Göttingen connections with colleagues in ancient history (ibid., 425), but a far more likely source, by 1801, was Kant's coining of the phrase "archaeology of nature," in his *Kritik der Urtheilskraft* (1790), the importance of which for Blumenbach is indisputable.

13 For a thorough assessment of Rudwick's masterwork, see Oldroyd (2006, esp. 493): "Cuvier is the centrepiece of the whole argument."

14 The liberalization of Calvinism in Geneva is associated with the efforts of Jean Alphonse Turrettini, who persuaded the Genevan clergy in 1706 to give up the

formula consensus that bound them to rigid orthodoxy (Hübner 2010, 41). Just how "liberal" that clergy became over the eighteenth century was the crux of the dispute over the *Encyclopédie* article on Geneva (1756) by Jean-Baptiste le Rond d'Alembert. D'Alembert certainly postulated a degree of liberalism that they found offensive, and Jean-Jacques Rousseau, out of characteristically mixed motives, rose to the defense of their more modest liberalism in his vehement *Letter to M. D'Alembert on Theater and the Arts* (Rousseau 1758). All that and more is the subject of the lengthy Gargett (1994).

15 Their relation was close enough that Deluc sought his guidance in 1774 in developing a course of readings for the queen of England (letter from Le Sage to Deluc, June 22, 1774, as reported in Hübner 2010, 94).

16 See the very instructive discussion by Oldroyd (2009).

17 See esp. Hübner (2010) and Heilbron and Sigrist (2011).

18 René Sigrist observes, "Moro's was the first system to explain the differences between primary mountains, without fossils, and secondary ones containing the remains of marine bodies" (Sigrist 2011a, 119).

19 On Saussure, see Sigrist (2001) and esp. Marcil (2001).

20 On the politics of the Deluc family and Rousseau, see Whatmore (2006) and Creighton (1982).

21 For a discussion of Guillaume's contributions, see Sigrist (2011a). A useful bibliography of Guillaume's work is given by Sigrist in Heilbron and Sigrist (2011, 325–328). Sigrist notes that significant contributions from Guillaume appeared under Jean-André's name, including several in a German context that would be relevant for Blumenbach.

22 Lichtenberg, an ardent Anglophile, frequently visited England. In that context, he introduced Deluc to Blumenbach in 1776, when he brought Deluc to Göttingen. On Lichtenberg, see Clark (2002).

23 Hübner notes the important assistance provided Deluc by the mining engineer Claus Friedrich von Reden, concerning rock forms of the Harz region (Hübner 2010, 124).

24 "Fiction" was often a code word to disparage Buffon's key work, *Époques de la nature* (1778). Buffon was, indeed, a prominent target of Deluc, sufficiently so as to provoke a rejoinder: *Défense de Mr. De Buffon contre les attaques injustes et indécentes de M. Deluc & Sage*. La Haye: de Groot, 1793.

25 See also Johann Albert Heinrich Reimarus (1802), who uses Kant (and Blumenbach) to rebut what he clearly sees as an excess of religious enthusiasm in Deluc's natural science. Even Le Sage had his reservations about the interfusion of scripture and science in Deluc (Hübner 2010, 118).

26 Blumenbach endorses Kant's boundary-keeping: "as Mr. Kant so rightly observes, 'it is not the advancement but the disarray [*nicht Vermehrung sondern Verunstaltung*] of the sciences, when one allows their borders to run over into one another'" (Blumenbach 1791, 516, §224n, citing Kant 1787, VIII).

27 See Blumenbach (1782, 543–561, §§245–251). For Blumenbach's discrimination of the various states of decomposition and metamorphosis of organic remains, see esp. §248. Thus we can see here one of the clarifying endeavors which made Blumenbach's work more than a mere "textbook," but rather a guide to the establishment of a more rigorous field of inquiry.

28 One sentence in the original has been divided, for English sensibilities, into two in my translation. This passage is replicated verbatim in Blumenbach (1782, 476–477). It remains the basis for all subsequent versions up through 1803.

29 But this has not gone uncontested. In the great anniversary celebration of Blumenbach's fiftieth year in academic life, the prominent geologist Karl Ernst Adolf von Hoff suggested in 1826 that Blumenbach's theory was in fact a rejection of Deluc, offering a causal theory at sharp variance from the latter's (Hoff 1826). Hübner summarizes von

Hoff's claim as follows: "Blumenbach argued in his *Handbuch der Naturgeschichte* (Göttingen: Dietrich, 1799), 474f., that not collapsing continents, as in Deluc's theory, but rather subterranean fires and water had caused the extinction of species by raising the ocean floors and flooding the continents. Some marine life became extinct in the process and terrestrial life perished in this 'judgment'" (Hübner 2011, 42 fn). While it may be that von Hoff chose in 1826 to see this as a *departure* by Blumenbach from Deluc's theory, I suggest that Blumenbach rather constructed a sympathetic *assimilation* of that theory; we have no evidence that he ever criticized Deluc's history of the earth. Similarly, I disagree with Dougherty's claim that Blumenbach's notion of catastrophe was significantly different from Deluc's. Dougherty writes, "Not fire and formative drive [*Bildungstrieb*], as in Blumenbach, but water and chemical factors of the environment, which cause the modifications of organisms, are for DeLuc the causes of a self-changing nature" (Dougherty 1986, 98). "Fire" versus "water" seems too crude a contrast to capture *either* theory of catastrophe, slipping into an argot that evokes the neptunism/vulcanism dispute of the 1790s that really displaces the important issues. To be sure, there is a real contrast between *Bildungstrieb* and "chemical factors of the environment," but that would need a substantial articulation that Dougherty does not provide. More persuasive, for me, is his contrast between the explicit "transformism" of Deluc (albeit for the purpose of evading problems of spontaneous generation) and the species essentialism to which Blumenbach was indeed committed.

30 For another very illuminating account of the ambient discourse of geology in that moment, see Rupke (1983).

31 "The time will soon come when one will no longer just lump fossils together but arrange them relative to the epochs of the earth." Goethe to Johann Heinrich Merck, October 27, 1782, cited in Engelhardt (1982, 189–190).

32 Notable in the essay is Blumenbach's sharp criticism of fossil collectors who proved oblivious of the location from which the specimens were extracted and who did not undertake careful enough comparisons with living forms. This permitted many writers to indulge in serious errors concerning extinction, among whom, tacitly, he placed Hutton (Blumenbach 1790c).

33 In his report in *GAgS* concerning his lecture of 1801, Blumenbach stressed this as the final point to which his exposition had led: the "most striking [*merkwürdiges*] problem" of species persistence versus extinction (Blumenbach 1801; 1803).

34 This topic, degenerations or *Ausartungen*, would carry Blumenbach for the rest of the exposition of his *Beyträge* into matters more remote to paleontology, especially his notions of racial variety.

35 On Buffon and the indestructibility of "organic molecules" and the perennial possibilities of "internal molds" see Eddy (1994).

36 Dougherty develops this aspect of Blumenbach's thought extensively (1986, 101–103).

37 This would result in the conflict between Cuvier and Lamarck, and then the famous controversy with Étienne Geoffroy Saint-Hilaire; see Appel (1987).

38 Here, as with the notion of "rational idea," there is an overlay of Kantian terminology which I believe to be misleading in making sense of Blumenbach.

39 Hence the curious subtitle of Dougherty's essay: "Ein Beitrag zur Wissenschaftsgeschichte bei der Entdeckung der Geschichtlichkeit *ihres* Gegenstandes" (A contribution to the history of science as she discovers the historicity of her object)."

40 Four letters from Blumenbach to Deluc from this period survive: June 3, 1795; November 2, 1795; January 2, 1796; and May 4, 1796. Dougherty surmises that Deluc's lost replies would have been composed immediately upon receipt of the first letter, and then on December 2 and 21, 1795, for the second; February 18, 1796, for the third; and May 19, 1796, for the last.

41 According to Hübner (2010, 253), Deluc's first letters appeared in both English and German translations in 1793; the second and third letters appeared more swiftly in

German: 1793 versus 1794. The fourth letter appeared in both languages in 1794. Thereafter, there is a sharp divergence. The English version of Letter V appeared in 1794, but the German version only in 1796. Similarly, Letter VI appeared in English in 1795, but in German only in 1796, some two years after its composition, dated by Deluc in Windsor to September of 1794. The composition dates of Deluc's letters are provided in Dougherty (1986, 98 fn).

42 Blumenbach to Deluc, June 3, 1795, cited in Dougherty (1986, 100).

43 Blumenbach to Deluc, November 2, 1795, cited in Dougherty (1986, 100).

44 Blumenbach to Deluc, November 2, 1795, cited in Dougherty (1986, 98).

45 Blumenbach to Deluc, January 2, 1796: Blumenbach assured Deluc that his letters would be a good countermeasure for those "numerous readers of a certain class" who could be brought away from "errors ultimately equally harmful for morals as for the true science of nature"; cited in Dougherty (1986, 99).

46 That was the last letter published in English. But Deluc then composed a final letter for the German series, and one that dealt far more directly with Blumenbach's own paleontological concerns. The meaning of this seventh letter deserves more exploration. Was it Deluc's reward to Blumenbach for having gone forward with the translation, or was it, rather, a retaliation, in which Deluc disputed some of Blumenbach's most important contentions in the field?

47 Baron points to the ironic opening of Blumenbach's *Beyträge*, where Haller is explicitly called out for unwarranted anxieties on this score (Blumenbach 1790a, 3). Haller's resistance to any extension of the temporal span of earth history beyond the biblical was already explicit in his condemnation of Buffon's history of the earth in his reviews of the latter's first volume of *Histoire naturelle*.

48 The clearest formulation of this linkage as a historical development is Hooykaas (1957).

49 We do not know exactly who Blumenbach had in mind, but Buffon must surely have been one key figure, here.

50 Blumenbach was "very swiftly and everyday more convinced of the truth of Rousseau's [corrected to *Leibniz's* in later editions] explanation that such *anatome comparata* is the living soul of the entire natural history of animals" (Blumenbach 1805, preface, V–VI).

51 Baron notes not only Blumenbach's hesitancies but also his factual errors in the course of his discussion.

52 For other accounts, see Seifert (1983) and Geus (1994).

53 In 1763, a dissertation associated with his student Christian Emanuel Hoppius, entitled "Anthropomorpha," was published by Linnaeus in *Amoenitates accademicae* (vol. 6, 63–76), elaborating the typology. See Broberg (1994, 69).

54 Miriam Meijer notes, "Camper himself became active in paleontology after the fossil remains of the giant land vertebrates were confronted, i.e. recognized for what they were, around 1770" (Meijer 1999, 63).

Bibliography

Anonymous. 1793. *Défense de M. De Buffon contre les attaques injustes et indécentes de M. Deluc & Sage*. La Haye: de Groot.

Appel, Toby. 1987. *The Cuvier-Geoffroy Debate: French Biology in the Decades Before Darwin*. Oxford: Oxford University Press.

Baron, Walter. 1963. "Die Anschauung Johann Friedrich Blumenbachs über die Geschichtlichkeit der Natur." (= Part I of Baron, Walter, and Bernhard Sticker: "Ansätze zur historischen Denkweise in der Naturforschung an der Wende vom 18. zum 19. Jahrhundert") *Sudhofs Archiv* 47: 19–26.

Bendyshe, Thomas, trans. 1865. *The Anthropological Treatises of Johann Friedrich Blumenbach*. London: Longman, Green, Longman, Roberts, and Green. (Reprint: Elibron Classics, 2005).

Bernasconi, Robert. 2014. "Heredity and Hybridity in the Natural History of Kant, Girtanner, and Schelling During the 1790s." In *Reproduction, Race, and Gender in Philosophy and the Early Life Sciences*, edited by Susanne Lettow, 237–258. Albany: SUNY University Press.

Bertrand, Elie. 1766. *Recueil de divers traités sur l'Histoire naturelle de la terre et des fossiles*. Avignon: Chambeau.

Blumenbach, Johann Friedrich. 1779. *Handbuch der Naturgeschichte*. [Erster Teil]. Göttingen: Johann Christian Dieterich.

———. 1780a. *Handbuch der Naturgeschichte. Zweyter Theil*. Göttingen: Johann Christian Dieterich.

———. 1780b. [Review of Jean-André Deluc, Lettres physiques et morales]. *Göttingische Anzeigen von gelehrten Sachen. Zugabe* 49 (December 2): 769–776.

———. 1782. *Handbuch der Naturgeschichte*, 2nd edn. Göttingen: Johann Christian Dieterich.

———. 1788. *Handbuch der Naturgeschichte*, 3rd edn. Göttingen: Johann Christian Dieterich.

———. 1790a. *Beyträge zur Naturgeschichte: Erster Theil*. Göttingen: Johann Christian Dietrich.

———. 1790b. "Beyträge zur Naturgeschichte der Vorwelt." *Magazin für das Neueste aus der Physik und Naturgeschichte* 6 (4): 1–17.

———. 1790c. "Vorerinnerung des Einsenders" to "Dr Hutton's Theorie der Erde." *Magazin für das Neueste aus der Physik und Naturgeschichte* 6 (4): 17–27.

———. 1791. *Handbuch der Naturgeschichte*, 4th edn. Göttingen: Johann Christian Dieterich.

———. 1797. *Handbuch der Naturgeschichte*, 5th edn. Göttingen: Johann Christian Dieterich.

———. 1801. "Report on Specimen archaeologiae telluris." *Göttingische Anzeigen von gelehrten Sachen* 199 (December 12): 1977–1984.

———. 1803. *Specimen archaeologiae telluris terrarumque inprimis Hannoverarum*. Göttingen: Heinrich Dieterich.

———. 1805. *Handbuch der vergleichenden Anatomie*. Göttingen: Heinrich Dieterich.

———. 1806. *Beyträge zur Naturgeschichte. Erster Theil*, 2nd edn. Göttingen: Heinrich Dieterich. Translation in Bendyshe 1865, 277–324.

———. 1811. *Beyträge zur Naturgeschichte. Zweyter Theil*. Göttingen: Heinrich Dieterich. Translation (extracts) in Bendyshe 1865, 325–340.

———. 2014. *Beyträge zur Naturgeschichte*. Mit einer Einleitung herausgegeben von Mario Marino. New York: Olms-Weidmann.

Broberg, Gunnar. 1994. "*Homo sapiens*: Linnaeus's Classification of Man." In *Linnaeus: The Man and His Work*, edited by Tore Frängsmyr, 156–194. Canton, MA: Science History Publications.

Buffon, Georges-Louis Leclerc de. 1749a. "De la manière d'étudier et de traiter l'histoire naturelle." In *Histoire Naturelle, générale et particulière, avec la description due cabinet du roi*. 44 vols. Vol. I, 3–64. Paris: Imprimerie royale.

———. 1749b. "Histoire et théorie de la terre." In *Histoire Naturelle, générale et particulière, avec la description due cabinet du roi*. 44 vols. Vol. I, 65–126. Paris: Imprimerie royale.

———. 1749c. “Histoire générale des animaux.” In *Histoire Naturelle, générale et particulière, avec la description due cabinet du roi.* 44 vols. Vol. II, 1–426. Paris: Imprimerie royale.
———. 1749d. “Histoire naturelle de l‘homme.” In *Histoire Naturelle, générale et particulière, avec la description due cabinet du roi.* 44 vols. Vol. II, 429–603. Paris: Imprimerie royale.
———. 1778. *Histoire Naturelle, générale et particulière, avec la description due cabinet du roi.* 44 vols. Supplement.Vol. V: Des époques de la nature. Paris: Imprimerie royale. (Critical edition: Les époques de la nature, edited by Jacques Roger. Paris: Éditions du Muséum, 1962).
Burtin, François-Xavier. 1789. “Réponse à la question physique, proposée par la Société de Teyler, sur les révolutions générales qu’a subies la surface de la terre, et sur l’ancienneté de notre globe.” In *Over de algemeene omkeeringen aan der oppervlakte der aarde, en over den oudheid van onze aardkloot* [. . .] uitgegeven door Teylers Tweede Genootschap, 1–242. Haarlem: Joh. Enschede en zoonen, en J. van Walre.
Camper, Petrus. 1779. “Account of the Organs of Speech of the Orang Outang.” *Philosophical Transactions* 69: 139–159.
———. 1782. *Natuurkundige verhandelingen over den orang outang: en eenige andere aap-soorten; over den rhinoceros met den dubbelen horen; en over het rendier.* Amsterdam: de Erven P. Meijer en G. Warnars. (German: 1791. Naturgeschichte des Orang-utang und einiger anderer Affenarten, des africanischen Nashorns, und des Rennthiers. Trans. J. F. M. Herbell. Düsseldorf: Johann Christian Dänzer; French: 1803. “De l’orang-outang et de quelques autres especes de singes [. . .].” In *Oeuvres de Pierre Camper, qui ont pour objet l’histoire naturelle* [. . .]. Vol. 1, 1–204. Paris, H-J. Jansen).
Carozzi, Albert, and Marguerite Carozzi. 1991. *Pallas’ Theory of the Earth in German (1778): Translation and Reevaluation: Reaction by a Contemporary: H.-B. de Saussure.* Geneva: Société de Physique et d’Histoire naturelle de Genève.
Clark, William. 2002. “From Enlightenment to Romanticism: Lichtenberg and Göttingen Physics.” In *Göttingen and the Development of the Natural Sciences*, edited by Nicolaas Rupke, 72–85. Göttingen: Wallstein.
Conry, Yvette. 1980. “L’idée d’une ‘marche de la nature’ dans la biologie prédrawinienne au XIXe siècle.” *Revue d’histoire des sciences* 33: 97–149.
Cornaz, Henri. 1992. “Fortunato Bartolomeo De Felice and the *Encyclopédie d’Yverdon.*” In *The Encyclopédie and the Age of Revolution*, edited by Clorinda Donato and Robert Maniquis, 39–48. Los Angeles: University of California; Yverdon: Société du Musée et Vieil Yverdon.
Corsi, Pietro. 1988. *The Age of Lamarck: Evolutionary Theories in France 1790–1830.* Berkeley: University of California Press.
Creighton, Douglas. 1982. *Jacques-François Deluc of Geneva and His Friendship with Jean-Jacques Rousseau.* Oxford, MS: University of Mississippi Press.
Deluc, Jean-André. 1772. *Recherches sur les modifications de l’atmosphère, contenant l’histoire critique du baromètre et du thermomètre, un traité sur la construction de ces instrumens, et expériences relatives à leurs usages, et principalement à la mesure des hauteurs et à la correction des réfractions moyennes.* 2 vols. Geneva: n.p.
——. 1779/1780. *Lettres physiques et morales sur l’histoire de la terre et de l’homme: addressées à la Reine de la Grande-Bretagne.* Paris: V. Duchesne, The Hague: De Tune.
———. 1790–1791. “Letters to Dr. James Hutton on the *Theory of the Earth.*” *The Monthly Review* 2 (1790): 206–227, 582–601; 3 (1790): 573–585; 5 (1791): 564–585.

———. 1793–1796. "Herrn de Luc's geologische Briefe an Hrn. Prof. Blumenbach: Aus der französischen Handschrift." *Magazin für das Neueste aus der Physik und Naturgeschichte* 8 (4) (1793): 1–41; 9 (1) (1794): 1–123; 9 (4) (1794): 1–49; 10 (3) (1796): 1–20; 10 (4) (1796): 1–104; 11 (1) (1796): 1–71. Engl. version: "Geological Letters, Addressed to Professor Blumenbach." *The British Critic and Quarterly Theological Review* 2 (1793): 231–238, 351–358; 3 (1793): 110–118, 226–237; 3 (1793): 476–478, 589–598; 4 (1794): 212–218, 328–336, 447–459, 569–578; 5 (1795): 197–207, 316–326; French version: *Lettres sur l'histoire physicque de la terre addressées à M. Le Professeur Blumenbach refermant de nouvelles preuves géologiques et historiques de la mission divine de Moyse*. Paris: Nyon, 1798.

Doig, Kathleen Hardesty. 1994. "The Yverdon Encyclopédie." In *Notable Encyclopedias of the Late Eighteenth Century: Eleven Successors of the Encyclopédie*, edited by Frank Kafker, 85–116. Oxford: The Voltaire Foundation.

Donato, Clorinda. 1993. "L'Encyclopédie d'Yverdon et l'Encyclopédie de Diderot et de D'Alembert: éléments pour une comparaison." *Annales Benjamin Constant* 14: 75–83.

Dougherty, Frank William Peter. 1986. "Der Begriff der Naturgeschichte nach J.F. Blumenbach anhand seiner Korrespondenz mit Jean-André DeLuc – Ein Beitrag zur Wissenschaftsgeschichte bei der Entdeckung der Geschichtlichkeit *ihres* Gegenstandes." *Berichte zur Wissenschaftsgeschichte* 9: 95–107.

Dougherty, Frank William Peter, and Norbert Klatt, eds. 2006–2015. *The Correspondence of Johann Friedrich Blumenbach*. 6 vols. Göttingen: Klatt.

Ebach, Malte Christian. 2015. *Origins of Biogeography: The Role of Biological Classification in Early Plant and Animal Geography*. London: Springer.

Eddy, John. 1994. "Buffon's Histoire Naturelle: History? A Critique of Recent Interpretations." *Isis* 85: 644–661.

Ellenberger, François, and Gabriel Gohau. 1981. "A l'aurore de la stratigraphie paléontologique: Jean-André de Luc, son influence sur Cuvier." *Revue d'histoire des sciences* 34: 217–257.

Engelhardt, Dietrich von. 1979. *Historisches Bewußtsein in der Naturwissenschaft von der Aufklärung bis zum Positivismus*. Munich: Alber.

———. 1995. "Verzeitlichung der Natur und Historisierung des Naturwissens im 18. Jahrhundert." In *Naturauffassungen in Philosophie, Wissenschaft und Technik*, edited by Lothar Schäfer. Vol. III: *Aufklärung und späte Neuzeit*, 225–253. Munich: Alber.

Engelhardt, Wolf von. 1982. "Goethes Beschäftigung mit Gesteinen und Erdgeschichte im ersten Weimarer Jahrzehnt." In *Genio huius loci: Dank an Leiva Petersen*, edited by Dorothea Kuhn and Bernhard Zeller, 169–204. Vienna: Böhlau.

Félice, Christian de. 1999. *L'Encyclopédie d'Yverdon: Une encyclopédie suisse au siècle des lumières*. Yverdon: Fondation de Felice.

Feuerstein-Herz, Petra. 2004. "Eberhard August Wilhelm von Zimmermann (1743–1815) und die Tiergeographie." Diss., Tech. Universität Braunschweig.

Forster, Georg. 1777. *A Voyage Round the World*. 2 vols. London: B. White.

Forster, Johann R. 1778. *Observations Made on a Voyage Round the World*. London: G. Robinson. (Reprint with commentaries by Nicholas Thomas, Harriet Guest, and Michael. Dettelbach. Honolulu: University of Hawai'i Press, 1996).

François, Alexis. 1924. "Jean-Jacques Rousseau et la science genevoise au XVIIIe Siècle: Ses rapports avec les naturalistes de Luc." *Revue d'histoire littéraire de la France* 31: 206–224.

Fritscher, Bernhard. 1992. "Kant und Werner: Zum Problem einer Geschichte der Natur und zum Verhältnis von Philosophie und Geologie um 1800." *Kant-Studien* 83: 417–435.

Gargett, Graham. 1994. *Jacob Vernet, Geneva, and the Philosophes*. Oxford: Voltaire Foundation.

Gascoigne, John. 2002. "Blumenbach, Banks, and the Beginnings of Anthropology." In *Göttingen and the Development of the Natural Sciences*, edited by Nicolaas Rupke, 76–98. Göttingen: Wallstein.

Geus, Armin. 1994. "Von der Naturgeschichte zur Geschichtlichkeit der Natur." In *Macrocosmos in Microcosmo: Die Welt in der Stube: Zur Geschichte des Sammelns 1450 bis 1800*, edited by Andreas Grote, 733–746. Opladen: Lask and Budrich.

Gillispie, Charles C. 1996. *Genesis and Geology*, 2nd edn. Cambridge, MA: Harvard University Press (1st edn., 1951).

Girtanner, Christoph. 1796. *Über das Kantische Prinzip für die Naturgeschichte*. Göttingen: Vandenhoeck and Ruprecht.

Häner, Flavio. 2017. *Dinge sammeln, Wissen schaffen: Die Geschichte der naturhistorischen Sammlungen in Basel, 1735–1850*. Bielefeld: transcript.

Heilbron, John L. 2011. "Savior of Science and Society." In *Jean-André Deluc: Historian of Earth and Man*, edited by John L. Heilbron and René Sigrist, 186–240. Geneva: Slatkine.

Heilbron, John L., and René Sigrist, eds. 2011. *Jean-André Deluc: Historian of Earth and Man*. Geneva: Slatkine.

Hoff, Karl Ernst Adolf von. 1826. *Erinnerung an Blumenbach's Verdienste um die Geologie, bey der fünfzigjährigen Jubelfeyer seines Lehramtes am 24*. Februar 1826. Gotha: Engelhard-Reyer.

Hoffmann, Christoph. 2003. "The Ruin of a Book: Jean André de Luc's Recherches sur les modifications de l'atmosphère (1772)." *Modern Language Notes (MLN)* 118 (3): 586–602.

Hooykaas, Reijer. 1957. "The Parallel Between History of the Earth and the History of the Animal World." *Archives internationales d'histoire des Sciences* 10: 3–18.

Hoppius, Christian Emanuel. 1763. "Anthropomorpha." In *Amoenitates Accademicae*, edited by Carolus Linnaeus. Vol. 6, 63–76. Holmiae: Salvius.

Hoquet, Thierry. 2005. *Buffon: histoire naturelle et philosophie*. Paris: Champion.

Hübner, Marita. 2010. *Jean-André Deluc (1727–1817): Protestantische Kultur und moderne Naturforschung*. Göttingen: Vandenhoeck und Ruprecht.

———. 2011. "Between the Republic of Letters and the Kingdom of God." In *Jean-André Deluc: Historian of Earth and Man*, edited by John L. Heilbron and René Sigrist, 19–44. Geneva: Slatkine.

Kant, Immanuel. 1787. *Critik der reinen Vernunft*, 2nd edn. Riga: Hartknoch: 1787. (*Kant's Gesammelte Schriften*, herausgegeben von der Preußischen Akademie der Wissenschaften [Akademie-Ausgabe]. Vol. 3. Berlin and Leipzig: Reimer, 1904).

———. 1790. *Critik der Urtheilskraft*. Berlin: Lagarde und Friederich. (*Kant's Gesammelte Schriften*, herausgegeben von der Preußischen Akademie der Wissenschaften [Akademie-Ausgabe]. Vol. 5, 165–486. Berlin: Reimer, 1913).

———. 1791. "Über das Misslingen aller philosophischen Versuche in der Theodicee." *Berlinische Monatsschrift* 18 (1791): 194–225. (*Kant's Gesammelte Schriften*, herausgegeben von der Preußischen Akademie der Wissenschaften [Akademie-Ausgabe]. Vol. 8, 253–271. Berlin: de Gruyter, 1923).

Langer, Wolfhart. 1982. "Ernst Friedrich von Schlotheim (1764–1832): Zur Erinnerung an seinen 150. Todestag." *Natur und Museum* 112: 77–80.

———. 1985. "Verzeitlichungs- und Historisierungstendenzen in der frühen Geologie und Paläontologie." *Berichte zur Wissenschaftsgeschichte* 8: 87–97.

Lehmann, Johann Gottlob. 1756. *Versuch einer Geschichte von Flötz-Gebürgen*. Berlin: Klütersche Buchhandlung.

Lenoir, Timothy. 1978. "Generational Factors in the Origin of Romantische Naturphilosophie." *Journal of the History of Biology* 11: 57–100.

———. 1980. "Kant, Blumenbach, and Vital Materialism in German Biology." *Isis* 71: 77–108.

———. 1981. "The Göttingen School and the Development of Transcendental Naturphilosophie in the Romantic Era." *Studies on History of Biology* 5: 111–205.

———. 1989. *The Strategy of Life: Teleology and Mechanism in Nineteenth-Century Biology*, 2nd edn. Chicago, IL: University of Chicago Press.

Lepenies, Wolf. 1976. *Das Ende der Naturgeschichte: Wandel kultureller Selbstverständlichkeiten in den Wissenschaften des 18. und 19 Jahrhunderts*. Munich: Hanser.

———. 1978. "Von der Naturgeschichte zur Geschichte der Natur: Erläutert an drei Schriften von Barthez, Buffon und Georg Forster aus dem Jahre 1778." *Schweizer Monatshefte* 58: 787–795.

Le Sage, Georges Louis. 1784. *Lucrèce Newtonien. Berlin: Decker. (Orig. in Mémoires de l'Académie royale des Sciences et Belles-Lettres de Berlin*. Année 1782, 404–427. Berlin: Decker, 1784.

Linnaeus, Carolus. 1735. *Systema naturae, sive regna tria naturae systematice proposita per classes, ordines, genera, et species*. Leiden: Haak, Groot. (10th ed., 1758).

Lyon, John, and Phillip R. Sloan, eds. 1981. *From Natural History to the History of Nature: Readings from Buffon and His Critics*. Notre Dame: University of Notre Dame Press.

Maccabez, Eugène. 1903. *F. B. de Felice, 1723–1789, et son encyclopédie: Yverdon 1770–1780*. Basel: Birkhaeuser.

Marcil, Yasmine. 2001. "Saussure savant ou voyageur? Les Voyages dans les Alpes dans les périodiques des années 1780." In *H.-B. de Saussure (1740–1799): Un regard sur la Terre*, edited by René Sigrist, 351–365. Geneva: Georg.

Marino, Luigi. 1995. *Praeceptores Germaniae: Göttingen 1770–1820*. Göttingen: Vandenhoeck and Ruprecht.

Mazzolini, Renato, and Shirley Roe, eds. 1986. *Science Against the Unbelievers: The Correspondence of Bonnet and Needham, 1760–1780*. Oxford: Voltaire Foundation.

Meijer, Miriam. 1999. *Race and Aesthetics in the Anthropology of Petrus Camper (1722–1789)*. Amsterdam: Rodopi.

Michael, John S. 2017. "Nuance Lost in Translation: Interpretations of J. F. Blumenbach's Anthropology in the English Speaking World." *NTM Zeitschrift für Geschichte der Wissenschaften, Technik und Medizin*. Online only: Springer, doi 10.1007/s00048-017-0173-8.

Montandon, Cléopâtre. 1975. *Le développement de la science à Genève aux XVIIIe et XIXe siècles*. Vevey: Delta.

Moro, Anton-Lazzaro. 1740. *De' crostacei e degli altri marini corpi che si truovano su' Monti*. Venice: Monti.

Neuburger, Max. 1943. "British Medicine and the Göttingen Medical School in the Eighteenth Century." *Bulletin of the History of Medicine* 14: 449–466.

Niekerk, Carl. 2004. "Man and Orangutan in Eighteenth-Century Thinking: Retracing the Early History of Dutch and German Anthropology." *Monatshefte* 96: 477–502.

Oldroyd, David. 2006. " 'The Geohistorical Revolution': The Emergence of Geology as an Historical Science." *Annals of Science* 63: 493–501.

———. 2009. "Jean André Deluc (1727–1817): An Atheist's Comparative View of Historiography." In *Geology and Religion: A History of Harmony and Hostility*, edited by Martina Kölbl-Ebert, 7–15. London: The Geological Society (Special Publications 310.

O'Malley, Charles Donald, and Horace Winchell Magoun. 1962. "Early Concepts of the Anthropomorpha." *Physis: revista di storia della scienza* 4: 39–64.

Oschmann, M. 1964. "Ernst Friedrich von Schlotheim – das Lebensbild eines grossen Paläontologen." *Bergakademie* 16: 444–448.

Popkin, Richard. 1987. *Isaac La Peyrère (1596–1676): His Life, Work, and Influence.* Dordrecht: Brill.

Ratcliff, Marc. 2011. "*Gens du haut et gens du bas*: Deluc and the politics of accuracy." In *Jean-André Deluc: Historian of Earth and Man*, edited by John L. Heilbron and René Sigrist, 45–62. Geneva: Slatkine.

Reichler, Claude. 1994. "Science et sublime dans la découverte des Alpes." *Revue de géographie alpine* 82 (3): 11–29.

———. 2002. *La découverte des Alpes et la question du paysage*. Geneva: Georg.

———. 2011. "Theorist of the Science and Sensibility of Landscape." In *Jean-André Deluc: Historian of Earth and Man*, edited by John L. Heilbron and René Sigrist, 63–80. Geneva: Slatkine.

Reimarus, Johann Albert Heinrich. 1802. *Über die Bildung des Erdballes und ins Besondere über das Lehrgebäude des Herrn de Luc*. Hamburg: Bohn.

Rousseau, Jean-Jacques. 1758. "A M. D'Alembert, [. . .] sur son article Genève dans le VIIme volume de l'Encyclopédie, et particulièrement sur le projet d'établir un théâtre de comédie en cette ville" (so-called "Lettre a M. D'Alembert sur les spectacles"). Amsterdam: Marc-Michel Rey. (Modern English edition: Bloom, Allan, ed. 1960. *Politics and the Arts, Letter to M. d'Alembert on the Theatre*. Glencoe, IL: The Free Press).

———. 1761. *Julie ou la Nouvelle Héloïse: lettres de deux amants habitants d'une petite ville au pied des Alpes*. Amsterdam: Marc-Michel Rey. (Modern French edition: Paris, Garnier: 1981.

Rudwick, Martin. 1972. *The Meaning of Fossils: Episodes in the History of Palaeontology*. London: MacDonald.

———. 1997. *Georges Cuvier, Fossil Bones, and Geological Catastrophes*. Chicago, IL: University of Chicago Press.

———. 2005. *Bursting the Limits of Time: The Reconstruction of Geohistory in the Age of Revolution*. Chicago, IL: University of Chicago Press.

Rupke, Nicolaas A. 1983. "The Study of Fossils in the Romantic Philosophy of History and Nature." *History of Science* 21: 389–413.

———. 1994. "A Second Look: C. G. Gillispie's Genesis and Geology." *Isis* 85: 261–270.

Saussure, Horace Bénédict de. 1779. *Voyages dans les Alpes, précédés d'un essai sur l'histoire naturelle des environs de Genève*. Neuchâtel: Fauche. (Modern edition: Geneva: Georg, 2002).

Schneider, Sabine M. 2001. "Komplexere Ordnung. Die Dynamisierung der Natur im naturkundlichen Diskurs bei Buffon." In *Der imaginierte Garten*, edited by Günter Oesterle and Harald Tausch, 115–134. Göttingen: Vandenhoeck and Ruprecht.

Seifert, Arno. 1983. "'Verzeitlichung': Zur Kritik einer neueren Frühneuzeitkategorie." *Zeitschrift für historische Forschung* 10: 447–477.

Selle, Götz von. 1937. *Die Georg-August-Universität zu Göttingen, 1737–1937*. Göttingen: Vandenhoeck and Ruprecht.

Sigrist, René, ed. 2001. *H.-B. de Saussure (1740–1799): Un regard sur la Terre*. Geneva: Georg.

———. 2011a. "Collecting Nature's Medals." In *Jean-André Deluc: Historian of Earth and Man*, edited by John L. Heilbron and René Sigrist, 105–146. Geneva: Slatkine.

———. 2011b. "Scientific Standards in the 1780s: A Controversy Over Hygrometers." In *Jean-André Deluc: Historian of Earth and Man*, edited by John L. Heilbron and René Sigrist, 147–184. Geneva: Slatkine.

Smith, Justin E.H. 2007. "Language, Bipedalism, and the Mind-Body Problem in Edward Tyson's *Orang-Outang* (1699)." *Intellectual History Review* 17 (3): 291–304.

Soemmerring, Samuel Thomas. 1781. "Etwas vernünftiges vom Orang Utang." *Göttinger Taschen-Kalender auf das Jahr 1781*: 40–64.

Soulavie, Jean Louis. 1780–1784. *Histoire naturelle de la France méridionale, ou, Recherches sur la minéralogie du Vivarais, du Viennois, du Valentinois, du Forez, de l'Auvergne, du Velai, de l'Uségeois, du Comtat Venaissin, de la Provence, des diocèses de Nismes, Montpellier, Agde, &c.: sur la physique de la Mer méditerranée, sur les météores, les arbres, les animaux, l'homme & la femme de ces contrées : avec cinq planches doubles par volume, & une carte géographique des trois règnes*. Nîmes: C. Belle.

Trembley, Jacques, ed. 1987. *Les savants genevois dans l'Europe intellectuelle du XVIIe au milieu du XIXe siècle*. Geneva: Editions du Journal de Genève.

Treviranus, Gottfried Reinhold. 1802–1822. *Biologie oder Philosophie der lebenden Natur für Naturforscher und Aerzte*. 6 vols. Göttingen: Röwer.

Walch, Johann Ernst Immanuel, ed. 1768–1774. *Die Naturgeschichte der Versteinerungen: zur Erläuterung der Knorrischen Sammlung von Merkwürdigkeiten der Natur*. 4 vols. Nuremberg: Felßecker.

Wenzel, Manfred. 1982. "Verzeitlichungstendenzen im Vorfeld des Evolutionismus." *Natur und Museum* 112: 15–25.

Werner, Abraham Gottlob. 1774. *Von den äusserlichen Kennzeichen der Fossilien*. Leipzig: Crusius.

———. 1971. *Short Classification and Description of the Various Rocks*. Translated with an introduction and notes by Alexander M. Ospovat. New York: Hafner. (Original: Kurze Klassifikation und Beschreibung der verschiedenen Gebirgsarten. Dresden: Waltherische Hofbuchhandlung, 1787).

Whatmore, Richard. 2006. "Rousseau and the représentants: the politics of the Lettres écrites de la montagne." *Modern Intellectual History* 3: 385–413.

Wingler, Gundolf. 1958. "Die Tiergeographie des Gottfried Reinhold Treviranus und ihre Vorgänger." Ph.D. Diss., University of Frankfurt am Main.

Wokler, Robert. 1976. "Tyson, Buffon, and the Orangutan." *Studies on Voltaire and the Eighteenth Century* 155: 2301–2319.

———. 1980. "The Ape Debates in Enlightenment Anthropology." *Studies on Voltaire and the Eighteenth Century* 192: 1164–1175.

———. 1988. "Apes and Races in the Scottish Enlightenment: Monboddo and Kames on the Nature of Man." In *Philosophy and Science in the Scottish Enlightenment*, edited by Peter Jones, 145–168. Edinburgh: Donald.

———. 1995. "Anthropology and Conjectural History in the Enlightenment." In *Inventing Human Science*, edited by Christopher Fox et. al., 31–52. Berkeley: University of California Press.

Zammito, John H. 2002. *Kant, Herder, and the Birth of Anthropology*. Chicago, IL: University of Chicago Press.

———. 2012. "The Lenoir Thesis Revisited: Blumenbach and Kant." *Studies in History and Philosophy of Science, Part C: Studies in History and Philosophy of Biological and Biomedical Sciences* 43: 120–132.

———. forthcoming a. "From Natural History to History of Nature: Kant Between Buffon and Herder." In *Rethinking Kant*. Vol. 5, edited by Oliver Thorndike and Pablo Muschnik. Newcastle upon Tyne: Cambridge Scholars.

———. forthcoming b. "Review of Johann Friedrich Blumenbach." In *Beyträge zur Naturgeschichte*. Mit einer Einleitung herausgegeben von Mario Marino (Hildesheim, Zurich, New York: Olms-Weidmann, 2014). *Scientia Poetica: Jahrbuch für Geschichte der Literatur und der Wissenschaften*.

———. forthcoming c. *The Gestation of Biology in Germany: Philosophy and Physiology from Stahl to Schelling*. Chicago, IL: University of Chicago Press.

Zimmermann, Eberhard August Wilhelm. 1778–1783. *Geographische Geschichte des Menschen und der allgemein verbreiteten vierfüßigen Thiere*. 3 vols. Leipzig: Weygand.

12 The origins of scientific racism and Huxley's rule

Nicolaas Rupke

Introduction

In rounding off our collection, I would like to address two interconnected sets of questions related to the assessment of Blumenbach's physical anthropology in terms of racist *vs.* non-racist.[1] The first set stems from his classic depiction of the five representative skulls in a horizontal row, with the Caucasian specimen placed in the center of that row (this volume, Figure 5.1) or, to use Blumenach's words, in "first place":

> I have allotted the first place [in a list of five human varieties] to the Caucasian, for the reasons given below, which make me esteem it the primeval one. This diverges in both directions into two, most remote and very different from each other; on the one side, namely, into the Ethiopian, and on the other into the Mongolian. The remaining two occupy the intermediate positions between that primeval one and these two extreme varieties; that is, the American between the Caucasian and Mongolian; the Malay between the same Caucasian and Ethiopian.
>
> (Bendyshe 1865, 264–265; transl. of Blumenbach 1795, 286–287)

Thus the reasons for the central position of the Caucasian variety were its beauty and autochthony, both related to the belief that the Caucasus region was the cradle of humanity:

> I have taken the name of this variety from Mount Caucasus, both because its neighbourhood, and especially its southern slope, produces the most beautiful race of men, I mean the Georgian; and because all physiological reasons converge to this, that in that region, if anywhere, it seems we ought with the greatest probability to place the autochthones of mankind.
>
> (Bendyshe 1865, 269; transl. of Blumenbach 1795, 303)

How are we to make sense of Blumenbach's visual representation of the five skulls and of the particular place assigned to each of them in the horizontal row?

Did the image of a central, primordial skull, with two strands emanating from it, imply a vertical, tapered, and racist hierarchy with the white race at the top, as Gould has argued? Or are we dealing with a non-racist family tree that shows branches growing up and away from the original form at the bottom, as Junker (this volume) has conjectured?

In further development of the importance of geography (Spary, this volume) I would like to highlight a third possibility: Blumenbach's arrangement should be seen less in the light of taxonomy and more of geographical distribution and migration, and of the scientific construction of Eurocentricity that gained popularity during his lifetime. Göttingen took a leading part in these developments. Although in many instances patronizing or pejorative, the explicit and implicit contentions with respect to non-European populations of the world were not racist, especially not in the context of monogenism. Instead they expressed cultural conceit. In other words, the central and first place given by Blumenbach to the Caucasian skull reflected a Eurocentric perception that saw in certain features of the non-Caucasian varieties the wear-and-tear of migrations from the Caucasus to other continental regions, and had little if anything to do with taxonomic ranking.

The second set of questions relates to the following: if indeed it is wrong to see in Blumenbach's doctoral dissertation and his collection of human skulls a fountainhead of scientific racism, where then should we locate its beginnings? If not in Blumenbach's Göttingen, where else? What was the scientific context within which the notion of "race" became "racist"? Was it polygenism, with such representatives as the American physician and skull collector Samuel George Morton and his *Crania Americana* (1839), and his follower, the physician and surgeon Josiah Clark Nott, and his *Types of Mankind* (1854), and *Indigenous Races of the Earth* (1857), both co-authored with the English-born Egyptologist George Robbins Gliddon?

In a majority of instances, polygenism did go hand in hand with racist views, even though the theory as such did not require this. Yet my contention is that racism became scientific when the reverse took place – namely, when the scientific theory of human origins itself became an inherently racist thesis, even though the theory's advocates may have been on the side of the abolition of slavery and for "Negro emancipation." This happened – that is, a scientific theory of origins acquired racism as an intrinsic component – when Charles Darwin published his *On the Origin of Species* (1859) and some of his followers based the gradual evolution of *Homo sapiens* from ape-like ancestors on the interpretation of human varieties as lower-to-higher levels of humanization. More specifically, the turn to scientific racism took place in the wake of the so-called hippocampus controversy about the relationship between humans and apes. The protagonists were Thomas Henry Huxley, who formulated the Darwinian, racist "law" that came to be known as "Huxley's rule," and his bête noire, Richard Owen, who held on to a Blumenbachian/Prichardian belief that the differences between human varieties are insignificant compared to human-ape ones. Huxley was judged the winner, and scientific racism unfurled its banner.

Eurocentrism

In his *Handbuch der Naturgeschichte* as well as the *Beyträge zur Naturgeschichte*, Blumenbach described and illustrated his human varieties in terms of their cultures and geographical locations – ethnographically – more than he had done in his doctoral dissertation. This geographical approach was part of a wider attempt to describe and interpret nature as well as human variety not only in terms of the taxonomy of their structural properties but more so in relation to the spaces of their occurrence. One of the scientific breakthroughs of the late eighteenth and early nineteenth centuries was the discovery of a lawlike distribution of life's diversity across the globe. This discovery went hand in hand with the development of cartography and the use of isolines (Rupke 2001; see also Withers 2017, and on spatiality in the historiography of science, Livingstone 2003, and Livingstone and Withers 2011).

Late eighteenth-century Göttingen was a hub of this preoccupation with the spatial distribution of natural and cultural phenomena. One of Blumenbach's senior colleagues, Johann Friedrich Gatterer, a leading figure in the school of universal history, used global physiographical maps to illustrate his lectures and help elucidate historical events (Plewe 1940). The encouragement by several professors of learned travel to foreign parts considerably contributed to the development of the geographical paradigm, not least in the case of Alexander von Humboldt and his journey of exploration of the equatorial Americas (1799–1804). Humboldt's scientific style represented a major component of the nineteenth-century study of nature, and was characterized by a preoccupation with precision measurements of environmental parameters and their lawlike interrelations on a global scale. Land masses, rock formations, ocean currents, rainfall, temperature, geomagnetism, plants, and animals were all studied in terms of their spatial interrelations. Perhaps the most famous mapping venture of the period was the international Geomagnetic Project, organized and led by Humboldt and the Göttingen astronomer and mathematician Carl Friedrich Gauß, in cooperation with the Royal Society and the British Association for the Advancement of Science. Measurements of environmental parameters in distant parts of the globe, which in the past had remained mostly isolated observations, were now plotted on isoline maps and made an integral part of the international effort, connected to Göttingen, London, Paris, and other European centers (Rupke, in press).

An exceptionally successful collection of Humboldtian charts and maps was the *Physikalischer Atlas* (Berghaus 2nd edn. 1849–1852; the first maps were circulated as early as 1837) by the cartographer Heinrich Berghaus.[2] He separately published selections from the main atlas, such as the *Allgemeiner anthropographischer Atlas* (Berghaus 1852) (General anthropographical atlas), which contained four maps of the global distribution of *Homo sapiens*, in terms of both physical anthropology and ethnography. In this context the Humboldtians turned geographical spaces into ideologically colored, sociopolitical spaces, constructing the maps in such a way as to make their own location appear at once central, superior, and the natural site of world domination (Rupke 1999; see Kiernan 1969).

In a variety of ways the interest in global distribution served to place the people who made the maps in the geographical center. For example, Berghaus's atlas and its several imitations showed Europe (or just northwestern Europe) as the world's pivotal region where race, salubrious climate, means of subsistence, clothing, mental development, religion, form of government, and more all reached their global optimum. Crudely Eurocentric subtexts underscored the message. One inset graphically documented the onset of puberty in different regions of the northern hemisphere, from equator to north pole. It was believed that in "primitive" society adolescent boys and girls reach sexual maturity at an earlier age than in "advanced" society, and accordingly the diagram showed that in Europe's temperate zone, puberty is reached relatively late, indicative of the height of its civilization. The very latest onset in fact was shown for Göttingen (Rupke 1996).

To rephrase: the Georgia Augusta was a major site in the construction of scientific Eurocentrism. Blumenbach's row of five horizontally arranged skulls, with the Caucasian variety in the center, is best understood as a geographical representation of human variety. His forerunners already had used geography as a criterion of classification, in particular Carl Linnaeus. Blumenbach refined this approach, adding the element of migration. His five varieties loosely corresponded to "the five parts of the world" (Bendyshe 1865, 267–268; transl. of Blumenbach 1795, 296–302). Eurocentric prejudice made him place his own variety in the center, being well aware of the cultural relativity of doing so. White people, he wrote, are "*to European perceptions of beauty* the best formed humans" (italics added; Blumenbach 1790, 82).[3] Deviations from the original had happened in the course of migrations and were due to environmental influences. These explained why the Mongolian and Ethiopian varieties were the extremes and the American and Malay the intermediary forms; moreover, resettlement also explained why the American variety is a link between the Caucasian and Mongolian, and not between the Caucasian and Ethiopian.

> It was justly observed by the first Europeans who visited the new continent, that the Americans came very near to the Mongolians, which adds fresh weight to the very probable opinion that the Americans came from northern Asia, and derived their origin from the Mongolian nation. It is probable that migrations of that kind took place at different times, after considerable intervals, according as various physical, geological, or political catastrophes gave occasion to them; and hence, if any place is allowed for conjecture in these investigations, the reason may probably be derived, why the Esquimaux have still much more of the Mongolian appearance about them than the rest of the Americans: partly, because the catastrophe which drove them from northern Asia must be much more recent, and so they are a much later arrival; and partly because the climate of the new country, which they now inhabit, is much more homogenous with that of their original country.
>
> (Bendyshe 1865, 274; transl. of Blumenbach 1795, 317–318)

Perhaps tongue-in-cheek, Blumenbach compared humans with pigs, in the sense that both are omnivores with a cosmopolitan distribution and have developed into geographical varieties under the influence of environmental factors:

> No other of our commonly called domestic animals has experienced such a manifold influence of climate as the hog; for no other has been so widely scattered as this over the five parts of the world. None has been subjected so much to the operation of variety of aliment; for no animal is so omnivorous as the hog, etc. There is only one domestic animal besides (domestic in the true sense, if not in the ordinary acceptation of this word) that also surpasses all others in these respects, and that is man. The difference between him and other domestic animals is only this, that they are not so completely born to domestication as he is, having been created by nature immediately a domestic animal. The exact original wild condition of most of the domestic animals is known. But no one knows the original wild condition of man. There is none, for nature has limited him in no wise, but has created him for every mode of life, for every climate, and every sort of aliment, and has set before him the whole world as his own and given him both organic kingdoms for his aliment. But the consequence of this is that there is no second animal besides him in the creation upon whose *solidum vivum* [totality of life or life force] so endless a quantity of various *stimuli*, and therefore so endless a quantity of concurring causes of degeneration, must needs operate.
>
> (Bendyshe 1865, 293–294; transl. of Blumenbach 1790, 47–49)

Thus the Caucasian variety is located in its original, temperate region, and least affected by the ravages of migration and severe climate. The extreme varieties of humankind are the ones that suffered the full brunt of the tropical climate (Ethiopian) or of both the tropics and the Arctic, combined with high altitude conditions in Central Asia (Mongolian, which included a geographical range of peoples from the Arctic to the Indian subcontinent). The Americans, migrating from the Asian North to the milder, mostly temperate regions of the New Continent, became intermediary; while the Malay, spread over more than just the severe tropics, were not as much impacted as the African variety, only in certain regions, such as Australia.

The climate- and food-induced modifications did not change the taxonomic status of the non-Caucasian varieties, even though their beauty and level of civilization were considered to have become inferior to Europe's. Blumenbach, like-minded colleagues, and pupils such as Humboldt were not racists, but followed the then common trend of Eurocentrism – a fashion of cultural conceit that accompanied power – a conceit that today is present, for example, in the notion of American exceptionalism. Such self-regard may be reprehensible, but need not be racist, however ramshackle the intellectual scaffolding of centristic and exceptionalist constructions was and continues to be – ramshackle, because the criteria for comparison and contrast are selected and weighted by those who see themselves at the hub of power or in "a city that is set on a hill." The horizontal arrangement of the five skulls and the centrality of the Caucasian one reflected Eurocentric perceptions rather than a racist taxonomy. A Caucasian was not an intrinsically higher human being, but *primus inter pares* – first among equals.

Blumenbach's physical anthropology was conceived within a geographical paradigm, not a taxonomic one that involved Gould's "fateful geometric shift [. . .] from cartography to hierarchy" (Gould 1996, 406). In his study of feral children, in particular of "Wild Peter" or "Peter the Wild Boy" of Hameln, Blumenbach asserted that a truly wild human being could never be generated through "Abartung" or degeneration (Blumenbach 1811, 13–44). All humans, he maintained, are domesticated by nature and therefore human beings do not merge with animals such as apes. This view was authoritatively supported by Friedrich Tiedemann, whose anti-racist monograph "On the Brain of the Negro, Compared With That of the European and the Orang-outan" (Tiedemann 1836) in recent decades has begun to receive the attention it deserves (Tiedemann 1984).

Huxley's rule

Still, also the geographical approach to human variety could go hand in hand with racism, mainly depending on one's view of human origins – monogenist or polygenist (for various complexities see Haller 1971, 70–88). Through the second half of the eighteenth century, supernatural causes – for example, God-the-Creator – were gradually removed from the study of the physical world, not least under the influence of Immanuel Kant and his pre-critical *Allgemeine Naturgeschichte und Theorie des Himmels* ([Kant] 1755) (*General History of Nature and Theory of the Heavens*). A Supreme Designer was no longer appealed to in explaining the solar system, the earth, and life or its diversity; instead, nature itself, natural processes, were believed to hold the keys to the origin of the world. Not uncommonly, Darwin and his *On the Origin of Species* have been credited with the introduction of biological naturalism; this may be true for the restricted sphere of England (Scotland requires separate attention) but does not apply to the European continent, where, well before Darwin, Blumenbach was the academic teacher to cross the Rubicon that separated creationist biology from evolutionary life sciences, followed by several of his many students, among whom Gottfried Reinhold Treviranus (Treviranus 1802). A central figure in the development of these speculations, Blumenbach was one of the early naturalists cautiously to express the likelihood of the naturalistic generation of species, tactfully transitioning from special creation to natural origin, retaining the word "Schöpfung" while investing it with the meaning of "Natur" (Rupke 2010, 146–148).

With the removal of "divine fiat" from the epistemological tool kit of natural history, physicochemical agents were increasingly called upon. The primary process by which species had come into existence – it was widely held – had been *mutterlose Urzeugung*, a form of primordial self-organization that had produced specific germs (see also Seehan and Wahrmann 2015). Thus each species had originated by the autogenesis of one, two, or many germs of its kind. Several lines of evidence were brought to bear on the issue. Probably more than any other field, biogeography adopted autogenesis as a central, organizing concept. Some species might have migrated from one region to another, but the majority of them were believed to be true natives or autochthons. Most dramatic in its consequences was the adoption of autogenesis as an organizing concept in Romantic anthropology. Nearly to a man, its star representatives in the German-speaking world believed in the *mutterlose Urzeugung* of humans (Rupke 2005, passim; 2008, 79–81).

As we have seen, Blumenbach, although a monogenist, believed that only the Caucasian variety was a true autochthon. The polygenists, by contrast, considered every one of the five racial categories as autochthonous, each having originated in their own province of distribution. Karl Asmund Rudolphi, for example, argued that all human races are true "aborigines" (Rudolphi 1821, 50–57). A similar view was expressed by Carl Gustav Carus; humans had neither originated as told in the Bible, as a single individual or a single pair of adults, like Adam and Eve in the Garden of Eden, nor evolved from apes by the transformation of monkeys into men.[4] The human species – he asserted – had its origins in primordial vesicles that developed in enormous numbers in water under mild and stable climatic conditions (Carus 1838, 112–113); the different climatic zones had given rise to tribes with different, unequal levels of mental talent (Carus 1849, passim). Thus polygenism made it easy – although not scientifically necessary – to separate the races into distinct lower and higher taxonomic categories. Morton, Nott, and others were polygenetic racists, working within the geographical model of independent, multiple, and autochthonous origins. By geographically clustering "indigenous races" with wildlife, especially apes and monkeys that were typical of each province of distribution, they contributed to the animalization of races other than the Caucasian and bolstered white supremacist science and sociopolitics (Nott and Glidden 1857, see especially p. 641, "Illustrative of the Geographical Distribution of Monkeys, in Their Relation to That of Some Inferior Types of Men").

Not infrequently, the term "scientific racism" has been equated with the use by racially prejudiced scientists of miscellaneous evidence from physical and social anthropology.[5] I would like to propose a more precise definition by focusing less on the prejudices of the scientists and more on the racism implicit in the scientific theories of human origins to which they adhered. Agreeing with Gould that the crucial turn toward scientific racism took place when the horizontal, geographical perception of human variety was changed to a vertical, taxonomic one, I add to this definition the moment when the theory of human evolution itself required, as part of its proof, the vertical ranking of Blumenbach's varieties, and when the scientific theory no longer allowed for a non-racist or anti-racist view. As stated earlier, neither the monogenist nor the polygenist theory made racism necessary. Scientific racism received its mature, late-modern expression in the form of what became known as "Huxley's law" or, more accurately, "Huxley's rule," and was the outcome of an (in)famous clash between Huxley and Owen that took place during the early 1860s, known as the "hippocampus controversy." This clash represented a frontline battle over the question of the origin of *Homo sapiens*, pitching Huxley's Darwinian, racist model against Owen's Blumenbachian, non-racist one.

Surreptitiously introducing into English science a Germanic, naturalistic theory of organic origins, Owen was concerned to preserve the Blumenbachian/Prichardian notion of humankind's unity, and of its separateness from anthropoid apes.[6] He demonstrated the "men-animals" gap in the form of cross sections of the skulls of humans and apes (Figure 12.1), also using for the same purpose other anatomical features, such as the muscles and tendons in the human foot compared to those of the gorilla. Additionally, Owen compared and contrasted the brains of humans to those of anthropoid apes, based largely on his own unique chimpanzee and gorilla material, provided by the explorer Paul du Chaillu and others. He also drew on the work of continentals, such as Louis Pierre Gratiolet in Paris, the leading expert on cerebral convolutions, and Rudolf Wagner, Blumenbach's successor at Göttingen University, who in his collection had the "ultimate" human brain – namely, of Gauss. As with the skulls, Owen argued for a monogenist similarity among human varieties and a contrasting gap with apes, stressing various features, in particular the hippocampus minor (Rupke 2009, 182–208).

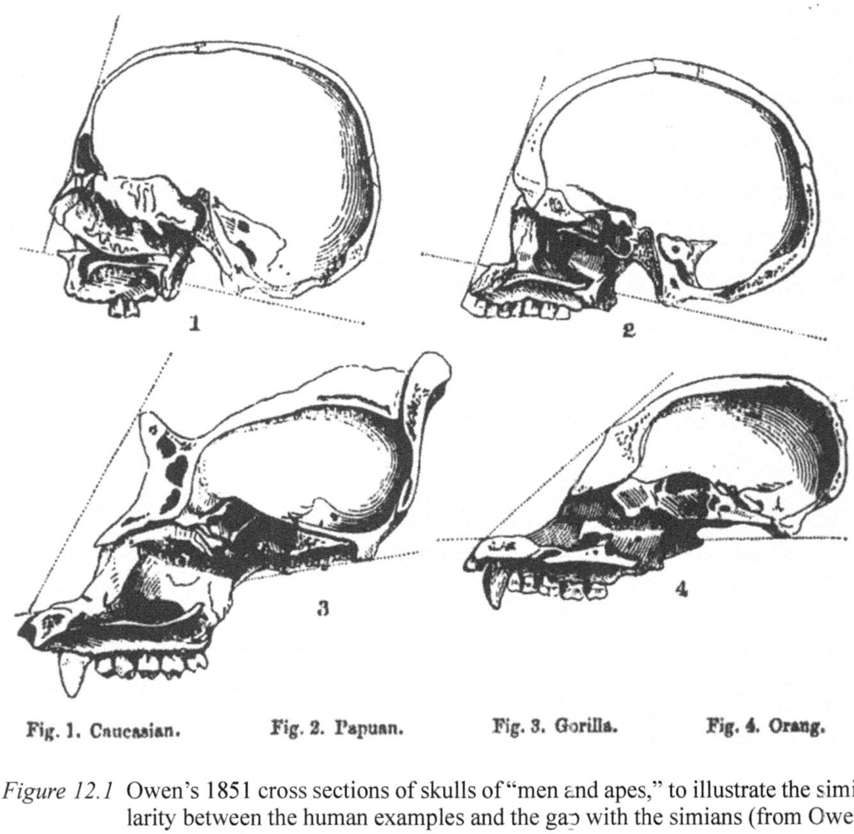

Figure 12.1 Owen's 1851 cross sections of skulls of "men and apes," to illustrate the similarity between the human examples and the gap with the simians (from Owen 1851, 777, figs. 1–4).

When in 1859 Darwin published *On the Origin of Species*, Huxley, who through the late 1840s and the 1850s had been a staunch anti-evolutionist, threw in his lot with the Darwinians. The same was true for the German zoologist and materialist philosopher Carl Vogt. Both Huxley and Vogt changed their party allegiance and took on one of the two challenges that Darwin did not meet in his magnum opus – namely, human evolution (the other was the origin of life). In the Darwinian-Lyellian model of evolution by small and gradual changes, cumulatively preserved in the struggle for life, the evolution of *Homo sapiens* had occurred in a series of small steps from animal to human; proof of this would exist in the form of intermediary stages and, for the purpose of providing this proof, Blumenbach's non-Caucasian varieties were used, either to represent successive stages of humanization or different levels of parallel lines of evolution from one or more ape ancestors. In this way, racism became baked into the scientific cake.

Huxley took on Owen at a number of consecutive meetings of the British Association for the Advancement of Science, in Oxford, Manchester, and Cambridge – the "hippocampus controversy" (Rupke 2009, 182–243). The results appeared in papers and in Huxley's *Evidence as to Man's Place in Nature* (Huxley 1863). Vogt followed suit with *Vorlesungen über den Menschen, seine Stellung in der Schöpfung und in der Geschichte der Erde* (Vogt 1863) (Lectures about man, his place in creation and in the history of the earth). The differences between Blumenbach's human varieties were changed from geographical ones, monogenously clustered around a Eurocentric Caucasian archetype, to hierarchical stages in the Darwinian evolution of ape-like ancestors to humans, and ultimately to Europeans. The horizontal line was irrevocably changed to the vertical, showing indigenous races as closest to anthropoids – in Huxley's case the indigenous people were Australian aborigines (Figure 12.2). In Darwinian evolution theory, racist ranking constituted a pillar of proof of human descent from animal ancestors.

While attacking Owen's classic picture of the contrast between the skulls of humans and of anthropoid apes, Huxley developed a "law" or "rule" that canonically encapsulated modern scientific racism. It states that, anatomically speaking, the difference between the purportedly highest human race and the supposedly lowest is larger than the difference between the lowest human race and the highest ape. The reverse perspective, starting with the apes, expressed the exact same racist thing: the difference between the lowest and highest anthropoids is larger than the difference between the highest anthropoid and the lowest human (which is a false comparison, implying that human races are equivalent to species; a proper comparison would have been of humans with races of gorilla or, instead, races of chimp) (Rupke 2009, 209–213). Huxley concluded,

> Thus, even in the important matter of cranial capacity, Men differ more widely from one another than they do from the Apes; while the lowest Apes differ as much, in proportion, from the highest, as the latter does from Man.
>
> (Huxley 1863, 95)

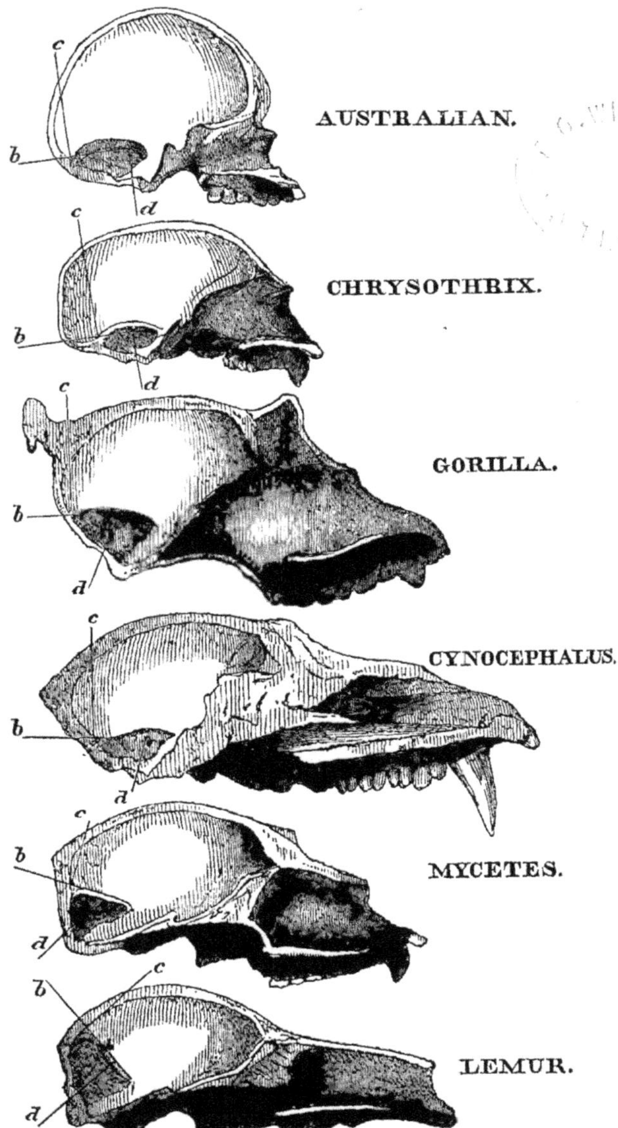

FIG. 17.—Sections of the skulls of Man and various Apes, drawn so as to give the cerebral cavity the same length in each case, thereby displaying the varying

Figure 12.2 Huxley's depiction of a monkeys-to-humans skull sequence, highlighting cranial capacity (from Huxley 1863, 79, fig. 17). Note that the top skull is identified not as human but as "Australian."

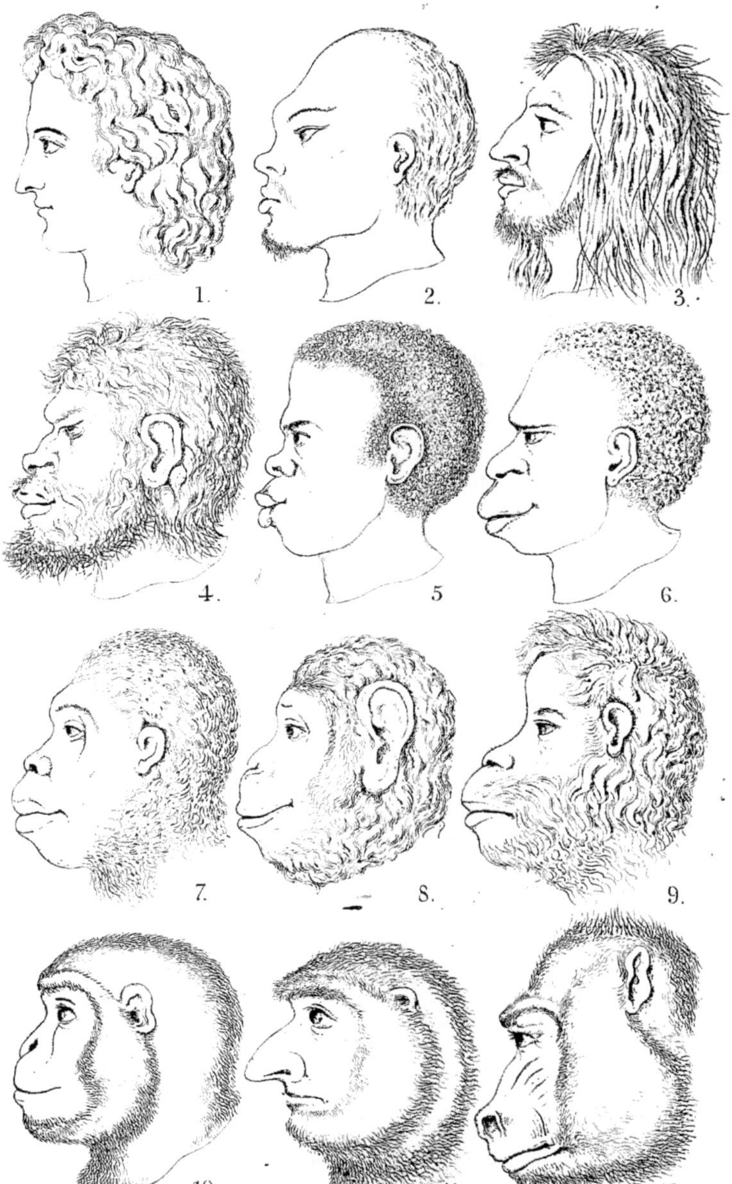

Die Familiengruppe der Katarrhinen (siehe Seite 555).

Figure 12.3 Huxley's rule as depicted by Haeckel, animalizing "lower" humans by making them seem closer to higher apes than to "higher" humans (from Haeckel 1868, frontispiece).

> Whatever part of the animal fabric – whatever series of muscles, whatever viscera might be selected for comparison – the result would be the same – the lower Apes and the Gorilla would differ more than the Gorilla and the Man.
>
> (Ibid., 101)

> Thus, whatever system of organs be studied, the comparison of their modifications in the ape series leads to one and the same result – that the structural differences which separate Man from the Gorilla and the Chimpanzee are not so great as those which separate the Gorilla from the lower apes.
>
> (Ibid., 123)

In Victorian London, Huxley was crowned the winner of the hippocampus debate. Human races had to be defined by their place in the hierarchy of morphological sequences. The lowest were considered half animals, the highest demigods – a view that provided scientific legitimacy for treating native peoples in the overseas colonies as part of anthropoid wildlife.[7] The so-called Hottentot Venus, Saartjie Baartman, supposedly was a closer relative of gorillas than of William Shakespeare or Carl Friedrich Gauß (Crais and Scully 2009). Darwinian opinion concurred that among human races a gradual animalization downwards and humanization upwards existed. Ernst Haeckel, Jena's leading evolutionary biologist and talented illustrator, formulated Huxley's rule with the following translation: "that in every morphological respect the differences between the highest and the lowest apes are greater than the respective differences between the highest apes and man"[8] (Haeckel 1902, 706; see also 1868, 491, 496; 1874, 489); the frontispiece to his *Natürliche Schöpfungsgeschichte* (1868) visually made the same point (Figure 12.3).

In conclusion, the fountainhead of scientific racism was not Blumenbach's doctoral dissertation but Huxley's contributions to the hippocampus controversy, which resulted in Huxley's rule. Bronowski's "particular cloud" (Rupke and Lauer, this volume) – that is, a connection between scientific racism and the Jewish holocaust – did not hang over Göttingen. The extent to which it hung over Huxley's London or Haeckel's Jena is a matter of ongoing controversy, exceeding the scope of this volume.

Notes

1 This chapter picks up on points made in preceding chapters and in the process repeats a few quotations for the sake of clarity of argumentation.

2 World distribution maps were of course no Humboldtian invention (Robinson 1982; Browne 1983). An early example of a global zoogeographical map was the "Tabula mundi geographico zoologica sistens quadrupedes hucusque notos sedibusque suis adscriptos," prepared by the Brunswick professor of mathematics and physics and patron to young Gauß E.A.W. Zimmermann, to accompany his three-volume *Geographische Geschichte des Menschen und der allgemein verbreiteten vierfüssigen Thiere* (1778–1783).

3 "Nach den Europäischen Begriffen von Schönheit die bestgebildetsten Menschen" (Blumenbach 1790, 82).

4 As widely known, Jean Baptiste de Lamarck, in his *Philosophie zoologique* (1809), adopted the theory of autogenetic, independent origins of species, but added the notion of transformation of existing forms of life, an idea that was carried to its extreme by Darwin, who avoided the problem of the origin of life and put forward a theory of descent-with-modification from "some one primordial form, into which life was first breathed" (Darwin 1859, 484; see also 490).
5 In recent decades, the very notion of race has been repudiated as pseudoscientific and racist (see Wikipedia entry).
6 The development of Owen's evolutionary thinking and that of many like-minded colleagues from the autogenesis of species towards evolution by descent is poorly known and needs more work.
7 The impact of evolutionary biology on racist policies in the colonies – New Zealand, Australia, South Africa, Canada – towards the aborigines was limited. "Each colony cobbled together a particular racial policy without reference to an overarching imperial message. The intellectual and policy vacuum permitted the colonies to indulge in any conduct they wished towards their native peoples during the late nineteenth century. The racialist policies which ensued were not British imperial ones, but Australian, New Zealand, Canadian and South African" (Francis 1994, 212).
8 "Daß in jeder morphologischen Beziehung die Unterschiede zwischen den höchsten und niedersten Affen größer sind als die betreffenden Unterschiede zwischen den höchsten Affen und dem Menschen" (Haeckel 1902, 706).

Bibliography

Bendyshe, Thomas, trans. 1865. *The Anthropological Treatises of Johann Friedrich Blumenbach*. London: Longman, Green, Longman, Roberts, and Green.

Berghaus, Heinrich. 1849–1852. *Physikalischer Atlas. Eine, unter den fördernden Anregung Alexander's von Humboldt verfasste, Sammlung von 93 Karten, auf denen die hauptsächlichsten Erscheinungen der anorganischen und organischen Natur nach ihrer geographischen Verbreitung und Vertheilung bildlich dargestellt sind*. 2 vols. Gotha: Justus Perthes Verlag.

———. 1852. *Allgemeiner anthropographischer Atlas. Eine Sammlung von vier Karten, welche die Grundlinien der, auf das Menschen-Leben bezüglichen Erscheinungen nach geographischer Verbreitung und Vertheilung abbilden und versinnlichen*. Gotha: Justus Perthes Verlag.

Blumenbach, Johann Friedrich. 1790. *Beyträge zur Naturgeschichte*. Part 1. Göttingen: Dieterich.

———. 1795. *De generis humani varietate nativa: Editio tertia*. Göttingen: Vandenhoek et Ruprecht.

———. 1811. *Beyträge zur Naturgeschichte*. Part 2. Göttingen: Dieterich.

Browne, Janet. 1983. *The Secular Ark. Studies in the History of Biogeography*. New Haven: Yale University Press.

Carus, Carl Gustav. 1838. *System der Physiologie für Naturforscher und Ärzte*. Part 1. Dresden: Fleischer.

———. 1849. *Denkschrift zum hundertjährigen Geburtsfeste Goethe's: Über ungleiche Befähigung der verschiedenen Menschheitsstämme für höhere geistige Entwickelung*. Leipzig: Brockhaus.

Crais, Clifton, and Pamela Scully. 2009. *Sara Baartman and the Hottentot Venus: A Ghost Story and a Biography*. Princeton, NJ: Princeton University Press.

Darwin, Charles. 1859. *On the Origin of Species by Means of Natural Selection: Or, the Preservation of Favoured Races in the Struggle for Life*. London: Murray.

Francis, Mark. 1994. "Anthropology and Social Darwinism in the British Empire: 1870–1900." *The Australian Journal of Politics and History* 40: 203–215.

Gould, Stephen J. 1996. *The Mismeasure of Man*, 2nd edn. London: Norton.

Haeckel, Ernst. 1868. *Natürliche Schöpfungsgeschichte*. Berlin: Reimer.

———. 1874. *Anthropogenie, oder Entwickelungsgeschichte des Menschen*. 2 vols. Leipzig: Engelmann.

———. 1902. *Natürliche Schöpfungsgeschichte*. 2 vols. Berlin: Reimer.

Haller, John S. 1971. *Outcasts from Evolution: Scientific Attitudes of Racial Inferiority, 1859–1900*. London: University of Illinois Press.

Huxley, Thomas Henry. 1863. *Evidence as to Man's Place in Nature*. London: Williams and Norgate.

———. 1906. *Man's Place in Nature and Other Essays*. London: Dent, New York: Dutton.

Kant, Immanuel. 1755. *Allgemeine Naturgeschichte und Theorie des Himmels*. Königsberg, Leipzig: Petersen.

Kiernan, V. G. 1969. *The Lords of Human Kind: European Attitudes Towards the Outside World in the Imperial Age*. London: Weidenfeld and Nicolson.

Lamarck, Jean-Baptiste Pierre Antoine de. 1809. *Philosophie zoologique*. Paris: Germer Baillère.

Livingstone, David N. 2003. *Putting Science in Its Place: Geographies of Scientific Knowledge*. Chicago, IL: University of Chicago Press.

———, and Charles W.J. Withers, eds. 2011. *Geographies of Nineteenth-Century Science*. Chicago, IL: University of Chicago Press.

Morton, Samuel George. 1839. *Crania Americana: Or a Comparative View of the Skulls of the Various Aboriginal Nations of North and South America*. Philadelphia: Dobson.

Nott, Josiah Clark, and George Robins Gliddon. 1854. *Types of Mankind: Or, Ethnological Researches Based Upon the Ancient Monuments, Paintings, Sculptures, and Crania of Races*. Philadelphia: Lippincott, Grambo.

———. 1857. *Indigenous Races of the Earth*. Philadelphia: Lippincott.

Owen, Richard. 1851. "Proceedings of Societies: Zoological Society." *Literary Gazette, and Journal of Science and Art* 1817: 777–778.

Plewe, Ernst. 1940. "Ein verschollener Atlas von Johann Christoph Gatterer." *Petermann's geographische Mitteilungen* 12: 393–399.

Robinson, Arthur H. 1982. *Early Thematic Mapping in the History of Cartography*. Chicago, IL: University of Chicago Press.

Rudolphi, Karl Asmund. 1821. *Grundriss der Physiologie*. Vol. 1. Berlin: Dümmler.

Rupke, Nicolaas. 1996. "Humboldtian medicine." *Medical History* 40: 293–310.

———. 1999. "L'Homme est-il le parent du singe?" *Les Cahiers de Science et Vie* 49 (February 1999): 74–83.

———. 2001. "Humboldtian Distribution Maps: The Spatial Ordering of Scientific Knowledge." In *The Structure of Knowledge: Classifications of Science and Learning since the Renaissance*, edited by Tore Frängsmyr, 93–116. Berkeley: University of California, Office for History of Science and Technology.

———. 2005. "Neither Creation Nor Evolution: The Third Way in Mid-Nineteenth Century Thinking About the Origin of Species." *Annals of the History and Theory of Biology* 10: 143–172.

———. 2008. "The Origin of Species from Linnaeus to Darwin." In *Aurora Torealis: Studies in the History of Science and Ideas in Honor of Tore Frängsmyr*, edited by Marco Beretta, Karl Grandin, and Svante Lindqvist, 73–87. Sagamore Beach, MA: Science History.

———. 2009. *Richard Owen: Biology Without Darwin*. Chicago, IL: Chicago University Press.
———. 2010. "Darwin's Choice: The Contingency of the Theory of Evolution by Natural Selection." In *Biology and Ideology from Descartes to Dawkins*, edited by Denis Alexander and Ronald L. Numbers, 139–164. Chicago, IL: University of Chicago Press.
———. Forthcoming. "The Geomagnetic Project: Internationalism in Science Between the French Revolution and the Franco-Prussian War." In *The Cambridge History of Science. Vol. 8: Modern Science in National, Transnational, and Global Context*, edited by Hugh R. Slotten, Ronald L. Numbers, and David N. Livingstone. Cambridge: Cambridge University Press (submitted for publication).
Sheehan, Jonathan, and Dror Wahrman. 2015. *Invisible Hands: Self-Organization and the Eighteenth Century*. Chicago, IL: Chicago University Press.
Stepan, Nancy. 1982. *The Idea of Race in Science: Great Britain 1800–1960*. London: Palgrave Macmillan.
Tiedemann, Friedrich. 1836. "On the Brain of the Negro, Compared with That of the European and the Orang-Outang." *Philosophical Transactions of the Royal Society of London* 126: 497–527.
———. 1984. *Das Hirn des Negers mit dem des Europäers und Orang-Outangs verglichen: With an Introduction by Hans-Konrad Schmutz*. Marburg: Basilisken-Presse.
Treviranus, Gottfried Reinhold. 1802. *Biologie, oder Philosophie der lebenden Natur für Naturforscher und Ärzte*. Vol. 1. Göttingen: Röwer.
Vogt, Carl. 1863. *Vorlesungen über den Menschen, seine Stellung in der Schöpfung und in der Geschichte der Erde*. 2 vols. Giessen: Ricker.
Withers, Charles W.J. 2017. *Zero Degrees: Geographies of the Prime Meridian*. Cambridge, MA: Harvard University Press.
Zimmermann, Eberhard August Wilhelm. 1778–1783. *Geographische Geschichte des Menschen, und der allgemein verbreiteten vierfüssigen Thiere*. Leipzig: Weygand.

Appendix: Biographical sketch

Family background

Johann Friedrich Blumenbach (May 11, 1752–January 22, 1840; Figure A1) came from a well-to-do family of academics and civil servants in Gotha, the capital of the small German duchy of Saxe-Gotha-Altenburg. His father, Johann Heinrich Blumenbach (1709–1787), was a professor at the Gotha grammar school, the *Gymnasium illustre*, and its prorector; his mother, Charlotte Eleonore Hedwig Blumenbach (1727–1794), was the daughter of Karl Franz Buddeus (1695–1753), vice-chancellor of the government of Saxe-Gotha, and the granddaughter of the Jena professor of theology Johann Franz Buddeus (1667–1729). Blumenbach had a sister (Charlotte Sophie Henriette, died 1802) and a brother (Friedrich Wilhelm Carl Ernst, died 1806). There were also family ties to another dynasty of scholars at Jena, the Walch family: Johann Georg Walch (1693–1775), a professor of theology at Jena, was the brother-in-law of Blumenbach's maternal grandfather and therefore Blumenbach's great-uncle. One of his sons, Johann Ernst Immanuel Walch (1725–1778), was Blumenbach's teacher in natural history at Jena University, and another son, Christian Wilhelm Franz Walch (1726–1784), became professor of theology at Göttingen University in 1754.

Blumenbach retained his close connection with Gotha in later life. He stayed there for fairly long visits once or twice a year and on these occasions was invited to the duke's court. He corresponded with several members of the ducal family and with Gotha scientists, especially Franz Xaver von Zach, and contributed to scientific journals published there. Toward the end of 1814, a breakup with the new duke of Gotha, August, occurred. Two years later, Blumenbach bought the house in Göttingen where he had been living since 1784 – an investment that usually involved the acquisition of the town's citizenship (Figure A2).

In 1779 Blumenbach married Louise Amalie Brandes (1752–1837), thus establishing family relations with influential circles at Göttingen University and in the administration of the Electorate of Hanover, the territory to which Göttingen belonged. Her father was Georg Friedrich Brandes (1709–1791), the Hanoverian state secretary for university affairs. Louise Amalie was also the sister-in-law of the classical scholar Christian Gottlob Heyne (1729–1812), the central figure of the academic scene at Göttingen, given his influence as director of the University Library and secretary of the Academy of Sciences.

Figure A1 *Johann Friedrich Blumenbach.* Mezzotint by Johann Elias Haid. Earliest known portrait of Blumenbach.

Blumenbach and his wife had four children: Georg Heinrich Wilhelm (1780–1855), who later was a senior civil servant in Hanover and in 1818 married Helene Ludovike Friederike Henriette, née Cleve (1797–1875); Emma Marie Hedwig (1783–1819), who in 1807 married Carl Wilhelm Friedrich Theodor von Jasmund (1782–1847); Charlotte Friederike Adelheid (Adele) (1787–1837); and Carl Ludwig Edmund (1788–1814), lieutenant in the "King's German Legion" and killed in the Battle of Toulouse.

Education and career

From 1759 to 1769 Blumenbach attended the *Gymnasium illustre* at Gotha, afterwards studying medicine at Jena, the state university of the three Saxe-Ernestine duchies, one of which was Saxe-Gotha-Altenburg. In the autumn of 1772 he moved to Göttingen, where the lectures by Christian Wilhelm Büttner inspired his dissertation *De generis humani varietate nativa.* When Büttner's vast natural history collections were purchased by the Hanoverian state in order to create an

Figure A2 Blumenbach's home in Göttingen. Historical photograph from D. Koch: *Das Göttinger Honoratiorentum vom 17. bis zur Mitte des 19. Jahrhunderts*. Göttingen: Vandenhoeck and Ruprecht, 1958, table VIII. Blumenbach moved into this house in 1784 and bought it in 1816. The present-day name of the street is "Neustadt." The historical buildings of this part of the town, including Blumenbach's house, were demolished around 1970.

Academic Museum at Göttingen, Blumenbach was hired to catalogue them. In September 1775 he received his doctoral degree with Ernst Gottfried Baldinger as his supervisor, and in February 1776 Blumenbach became associate professor of medicine, prosector, and sub-curator of the Academic Museum, the official director of which was Heyne.

Having been appointed full professor in 1778 and member of the Göttingen Academy of Sciences in 1784, when still in his early thirties, Blumenbach's position in academic Göttingen was secure and settled, although in later years he was able to expand his influence (1792 member of the Faculty of Medicine; 1812 head of the Academic Museum; 1812 secretary of the Göttingen Academy of Sciences; 1816 "professor primarius" of the Faculty of Medicine).

Blumenbach and Göttingen

Like Christian Gottlob Heyne and Carl Friedrich Gauß, Blumenbach devoted his entire academic life to Göttingen University. For several decades his lectures on natural history attracted students from all over Europe and from a range of social strata. In return, the Georgia Augusta provided him with excellent working conditions, generous funding, and exceptional intellectual freedom to carry out research and make his ideas public by means of lecture courses and published writings.

Although the scope of Blumenbach's research was global, he never left Göttingen for long. Except for summer sojourns in nearby spas (Pyrmont; Driburg), excursions with students to the Harz Mountains, and visits to Gotha and Weimar (where he saw Johann Wolfgang von Goethe), he undertook only two major scientific journeys: to Switzerland (April–October 1783) and to London (November 1791–April 1792). The main purpose was to establish contact with scientific institutions and their leading representatives.

Blumenbach enjoyed similar opportunities when in September 1807 he took part in a diplomatic mission to Paris on behalf of Göttingen, which at the time belonged to the Napoleonic satellite kingdom Westphalia (1807–1813). Because of his international reputation and diplomatic skill, Blumenbach, together with the Göttingen jurist and diplomat Georg Friedrich von Martens, was entrusted with pleading the case of his town and university when during the Napoleonic Wars the French army was closing in on Göttingen. Also, in October 1813, after Napoleon's defeat in the Battle of Leipzig, Blumenbach's ambassadorial expertise was called upon when he was sent to the temporary headquarters of the anti-Napoleonic coalition in Heiligenstadt in an effort to prevent military damage to his town and university. In recognition of Blumenbach's services, the city in 1824 granted him a twenty-year exemption from all communal property taxes.

Blumenbach is buried in Göttingen's former Albani Cemetery (today Cheltenham Park). The grave still exists.

Blumenbach and the Göttingen Academy of Sciences

Blumenbach's career is closely connected with the Göttingen Academy of Sciences (at the time called "Society of Sciences"). His first publications were based on papers read at its 1773 and 1774 sessions when he was twenty-one and twenty-two years of age respectively, and before he had obtained his doctoral degree (see Figure A3). In 1784, Blumenbach was elected a member of the Academy and from 1786 onward regularly contributed to its *Commentationes*. This outlet also served him as a publication platform for a long-term project in the field of physical anthropology, the *Decades collectionis suae craniorum diversarum gentium illustratae*, a series of seven papers with images and descriptions of human skulls from all over the world, published between 1790 and 1828. Additionally, Blumenbach wrote more than 400 articles for the Academy's reviewing journal, the *Göttingische gelehrte Anzeigen*.

The key position in the Academy during the eighteenth and early nineteenth centuries was the office of secretary. It was a permanent appointment, whereas the position of director rotated on an annual basis. When Heyne, who had been secretary for more than forty years, died in 1812, Blumenbach succeeded him and retained the post for the rest of his life.

Blumenbach maintained innumerable professional and private contacts with scholars and scientists, as well as with people in politics and society at large. By around 1820, he was a member of more than forty, and toward the end of his life some seventy academies and scientific societies in Europe and in the United

Figure A3 Bronze medal celebrating J. F. Blumenbach's doctoral jubilee in 1825, with his portrait by Heinrich Gube.

States. More than 2,200 of the letters he wrote or received are known, documenting a network of correspondence that included the centers of the academic world of his time, like Paris and London, but also its outposts – for example, in Petropavlovsk on the Kamchatka Peninsula, in South Africa, and in Indonesia.

Scientific collections

Blumenbach explored the natural world primarily by collecting, not so much by traveling or experimenting in a laboratory. He gained access to nature on a global scale by bringing together a large collection of natural history specimens, requesting correspondents, colleagues, friends, and students to send him minerals, fossils, meteorites, anatomical specimens, cultural artifacts, and even living animals. Especially fruitful were Blumenbach's contacts with London, the center of the British Empire, and with Saint Petersburg, from where the tsarist government organized the exploration of its territories in Central and Northern Asia, Siberia, and Alaska. Former students who set off on journeys of scientific exploration, among whom Alexander von Humboldt, Georg Heinrich von Langsdorff, and Maximilian zu Wied-Neuwied, contributed to his collections.

Blumenbach not only enlarged the collections of the Academic Museum but also accumulated substantial private collections. Upon his death in 1840, these were bought for, and added to, the museum. Prominent among Blumenbach's private objects of natural history was his collection of human skulls. When he died, it contained some 240 specimens, among these the skulls of pre-Columbian inhabitants of Peru, ancient Romans and Germans, artificially deformed skulls from Anatolia, and skulls from Polynesia, South Asia, and Australia. Blumenbach used them to validate his doctrine that all human "varieties" belong to a single biological species. Many of the objects studied by Blumenbach can be found in today's research and teaching collections at Göttingen University.

Wolfgang Böker

Biographical literature about Blumenbach

Autobiographical texts

"Einige biographische Notizen zu Hrn: Hofraths Blumenbach in Göttingen Portrait (Auszug eines Schreibens an d. H.)." *Allgemeinen geographischen Ephemeriden* 32 (2) (June 1810): 214–217.

Marx, Karl Friedrich Heinrich: Zum Andenken an Johann Friedrich Blumenbach: eine Gedächtniss-Rede gehalten in der Sitzung der Königlichen Societät der Wissenschaften den 8. Februar 1840. Göttingen: Druck und Verlag der Dieterichschen Buchhandlung, 1840, 4–6 (Blumenbach' short report about his education and studies until 1778).

"IV. Ioannes Fridericvs Blvmenbach, Gothanus." In *Tertio indicit octo candidatis doctissimis gradum doctoris medicinae contulisse ordinem medicorum*, edited by Baldinger, Ernst Gottfried, 29–33. Göttingen: Dieterich, 1775 (CV in Latin, composed on the occasion of Blumenbach's doctoral exam, with information about his family background and his studies at Jena and Göttingen; printed from the original manuscript and annotated in

Dougherty, Frank William Peter, *The Correspondence of Johann Friedrich Blumenbach*. Rev., augm. and edited by Norbert Klatt. Volume 1 (1773–1782). Göttingen: Klatt, 2006 (Brosamen zur Blumenbach-Forschung, 2), 27–33.

Contemporary biographical accounts

Callisen, Adolph Carl Peter. 1830. *Medicinisches Schriftsteller-Lexicon der jetzt lebenden Aerzte, Wundärzte, Geburtshelfer, Apotheker und Naturforscher aller gebildeten Völker*. Vol. 2, 345–356. Copenhagen: n. p. Supplements: Vol. 26 (1838), 334–339.

Versuch einer academischen Gelehrten-Geschichte von der Georg-Augustus-Universität zu Göttingen. Vol. 2. Göttingen: Vandenhoeck und Ruprecht, 1788, 148–149; Vol. 3. Hannover: Helwingsche Hofbuchhandlung, 1820, 303–307; Vol. 4. Göttingen: Vandenhoeck und Ruprecht, 1838, 421–424.

Obituary by Karl Friedrich Heinrich Marx (1840) (see above)

Modern biographical studies

Dougherty, Frank William Peter. 1984. *Commercium epistolicum J. F. Blumenbachii: Aus einem Briefwechsel des klassischen Zeitalters der Naturgeschichte* [exhibition catalogue]. Göttingen: Arbeitsstelle zur Edition des Blumenbach-Briefwechsels in der Niedersächsischen Staats- und Universitätsbibliothek Göttingen, 1984.

Ehlers, Ernst. 1901. "Göttinger Zoologen." In *Festschrift zur Feier des hundertfünfzigsten Bestehens der Königlichen Gesellschaft der Wissenschaften zu Göttingen*. [Vol. 3] Beiträge zur Gelehrtengeschichte Göttingens, 391–494. Berlin: Weidmann

Klatt, Norbert. 2008–2013. *Kleine Beiträge zur Blumenbach-Forschung*. Vol. 1–5. Göttingen: Klatt; containing "Zum frühen Verhältnis von Johann Friedrich Blumenbach und Alexander von Humboldt" (Vol. 1); "Blumenbachs Aufenthalt in England – Versuch einer Rekonstruktion" (Vol. 4); "Johann Friedrich Blumenbach und Franz-Joseph Gall – Eine Begegnung" (Vol. 5); "August Emil Leopold von Sachsen-Gotha und Altenburg und Johann Friedrich Blumenbach – eine Beziehung sui generis" (Vol. 6).

Index

Page numbers in *italics* indicate figures and in **bold** indicate tables on the corresponding pages.

For Product Safety Concerns and Information please contact our EU
representative GPSR@taylorandfrancis.com
Taylor & Francis Verlag GmbH, Kaufingerstraße 24, 80331 München, Germany

www.ingramcontent.com/pod-product-compliance
Ingram Content Group UK Ltd.
Pitfield, Milton Keynes, MK11 3LW, UK
UKHW020013200926
475702UK00014B/1858
9780367588823